APPLIED MYCOLOGY

APPLIED MYCOLOGY

Edited by

Mahendra Rai
SGB Amravati University, India

and

Paul Dennis Bridge
British Antarctic Survey, UK

www.cabi.org

CABI is a trading name of CAB International

CABI Head Office
Nosworthy Way
Wallingford
Oxon OX10 8DE
UK

Tel: +44 (0)1491 832111
Fax: +44 (0)1491 833508
E-mail: cabi@cabi.org
Website: www.cabi.org

CABI North American Office
875 Massachusetts Avenue
7th Floor
Cambridge, MA 02139
USA

Tel: +1 617 395 4056
Fax: +1 617 354 6875
E-mail: cabi-nao@cabi.org

A catalogue record for this book is available from the British Library, London, UK.

Library of Congress Cataloging-in-Publication Data

Applied mycology / edited by Mahendra Rai and Paul Dennis Bridge.
p. cm.
ISBN 978-1-84593-534-4 (alk. paper)
1. Mycology. I. Rai, Mahendra. II. Bridge, P. D. III. Title.

QK603.A647 2009
579.5--dc22

2009001597

ISBN-13: 978 1 84593 534 4

Typeset by Columns Design, Reading, UK.
Printed and bound in the UK by the MPG Books Group.

The paper used for the text pages in this book is FSC certified. The FSC (Forest Stewardship Council) is an international network to promote responsible management of the world's forests.

Contents

Contributors

Zsuzsanna Antal, *Department of Microbiology, University of Szeged, Faculty of Science and Informatics, Közép fasor 52., H-6726 Szeged, Hungary*

T. Benítez, *Departamento de Genética, Facultad de Biología, Universidad de Sevilla, Avda. Reina Mercedes s/n, 41012, Sevilla, España; Tel: 0034 954557109 Fax: 0034 954557103, E-mail: tahia@us.es*

Paul Bridge, *British Antarctic Survey, NERC, High Cross, Madingley Rd, Cambridge, CB3 0ET, UK; E-mail: pdbr@bas.ac.uk*

A.C. Codón, *Departamento de Genética, Facultad de Biología, Universidad de Sevilla, Avda. Reina Mercedes s/n, 41012, Seville, Spain; Tel: 0034 954557112, Fax: 0034 954557103, E-mail: accodon@us.es*

Freddy R. Delvaux, *Centre for Malting and Brewing Science, Faculty of Bioscience Engineering, Katholieke Universiteit Leuven, Kasteelpark Arenberg 22, 3001 Heverlee, Belgium*

D. Diouf, *Université Cheikh Anta Diop (UCAD), Faculté de Sciences et Techniques. Département de biologie végétale, BP 5005, Dakar, Sénégal*

Robin Duponnois, *IRD, UMR 113 CIRAD/INRA/IRD/SUPAGRO/UM2, Laboratoire des Symbioses Tropicales et Méditerranéennes (LSTM), Montpellier, France; E-mail: robin.duponnois@ird.sn*

Aniket Gade, *Department of Biotechnology, SGB Amravati University, Amravati, PIN-444 602, Maharashtra, India*

A. Galiana, *IRD. Laboratoire Commun de Microbiologie IRD / ISRA / UCAD, Centre de Recherche de Bel Air, BP 1386, Dakar, Sénégal*

Ileana V. García, *Museo Argentino de Ciencias Naturales "Bernardino Rivadavia" (MACN-CONICET), Argentina, Av. Ángel Gallardo 470 (C1405DJR); Tel: 4982-6595, int 160, Fax: int 172, E-mail: igarcia@macn.gov.ar*

Lóránt Hatvani, *Department of Microbiology, University of Szeged, Faculty of Science and Informatics, Közép fasor 52., H-6726 Szeged, Hungary*

David L. Hawksworth, *Departamento de Biología Vegetal II, Facultad de Farmacia, Universidad Complutense de Madrid, Plaza Ramón y Cajal, Ciudad*

Universitaria, Madrid 28040, Spain; Natural & Social Sciences, University of Gloucestershire Campus, Swindon Road, Cheltenham GL50 4AZ, UK; University of Gloucestershire, Cheltenham, UK; and Department of Botany, Natural History Museum, Cromwell Road, London SW7 5BD, UK; E-mail: d.hawksworth@nhm.ac.uk

Dai Hirose, *Faculty of Pharmacy, Nihon University, Chiba 274-8555, Japan*

Sándor Kocsubé, *Department of Microbiology, University of Szeged, Faculty of Science and Informatics, Kozep fasor 52, H-6726 Szeged, Hungary; E-mail: shigsanyi@gmail.com*

László Kredics, *Department of Microbiology, University of Szeged, Faculty of Science and Informatics, Közép fasor 52., H-6726 Szeged, Hungary; Tel: 3662544516; Fax: 3662544823; E-mail: kredics@bio.u-szeged.hu*

Anna Li, *Department of Plant Pathology, Shandong Agricultural University, Taian, Shandong 271018, China*

Duochuan Li, *Department of Plant Pathology, Shandong Agricultural University, Taian, Shandong 271018, China; Tel: 86-538-8249071, Fax: 86-538-8226399, E-mail: lidc20@sdau.edu.cn*

Zonglin Lewis Liu, *U.S. Department of Agriculture, Agricultural Research Service, National Center for Agricultural Utilization Research, 1815 N University Street, Peoria, IL 61604 USA; Tel: 1 309 681 6294; Fax: 1 309 681 6693; E-mail: ZLewis.Liu@ars.usda.gov*

V. Loureiro, *Departamento de Botânica e Engenharia Biológica Instituto Superior de Agronomia, Technical University of Lisbon, Tapada da Ajuda 1349-017, Lisboa, Portugal*

Manuel Malfeito-Ferreira, *Departamento de Botânica e Engenharia Biológica Instituto Superior de Agronomia, Technical University of Lisbon, Tapada da Ajuda 1349-017 Lisboa, Portugal; Tel: +351 21 3653448, Fax: 21 3653238; E-mail: mmalfeito@isa.utl.pt*

László Manczinger, *Department of Microbiology, University of Szeged, Faculty of Science and Informatics, Közép fasor 52., H-6726 Szeged, Hungary*

Carol M. McClelland, *Department of Biology, McMurry University, Abilene, TX, USA*

Rodolfo E. Mendoza, *Museo Argentino de Ciencias Naturales "Bernardino Rivadavia" (MACN-CONICET), Argentina; E-mail: rmendoza@macn.gov.ar*

M.A. Moreno-Mateos, *Departamento de Señalización celular, Centro Andaluz de Medicina Regenerativa y Biología Molecular. Edif. CABIMER - Avda. Américo Vespucio s/n. Parque Científico y Tecnológico Cartuja 93. 41092 Sevilla, España; Tel: 0034 954467838, Fax: 0034954461664, E-mail: miguelangel.moreno@cabimer.es*

Takashi Osono, *Center for Ecological Research, Kyoto University, Shiga 520-2113, Japan, Tel: 81 77 549 8252, Fax: 81 77 549 8201; E-mail: tosono@ecology.kyoto-u.ac.jp*

Zsanett Péteri, *Department of Microbiology, University of Szeged, Faculty of Science and Informatics, Kozep fasor 52, H-6726 Szeged, Hungary; E-mail: zspeteri@gmail.com*

Y. Prin, *CIRAD, UMR 113 CIRAD/INRA/IRD/SUPAGRO/UM2, Laboratoire des Symbioses Tropicales et Méditerranéennes (LSTM), Montpellier, France.*

Mahendra Rai, *Department of Biotechnology, SGB Amravati University, Amravati, PIN-444 602, Maharashtra, India; Tel: 91-721-2667380, Fax: 91-721-2660949, E-mail: mkrai123@rediffmail.com*

Sanjay G. Revankar, *Harper University Hospital, 3990 John R. St., 5 Hudson Detroit, MI 48201, USA; Tel: (313) 745-8599, Fax: (313) 993-0302, E-mail: srevankar@med.wayne.edu*

A.M. Rincón, *Departamento de Genética, Facultad de Biología, Universidad de Sevilla, Avda. Reina Mercedes s/n, 41012, Sevilla, España; Tel: 0034 954556230 Fax: 0034 954557103, E-mail: amrincon@us.es*

Robert A. Samson, *CBS Fungal Biodiversity Centre, Uppsalalaan 8, 3584 CT Utrecht, The Netherlands; E-mail: r.samson@cbs.knaw.nl*

Daniel Sliva, *Cancer Research Laboratory, Methodist Research Institute, 1800 N Capitol Ave, E504, Indianapolis, IN 46202, USA; Tel: (317) 962-5731; Fax: (317) 962-9369; E-mail: dsliva@clarian.org*

Mingzhou Song, *Department of Computer Science, New Mexico State University, P.O. Box 30001, MSC CS, Las Cruces, NM 88003 USA*

K.R. Sridhar, *Microbiology and Biotechnology, Department of Biosciences, Mangalore University, Mangalagangotri, Mangalore 574 199, Karnataka, India; E-mail: sirikr@yahoo.com*

Csaba Vágvölgyi, *Department of Microbiology, University of Szeged, Faculty of Science and Informatics, Közép fasor 52., H-6726 Szeged, Hungary*

János Varga, *Department of Microbiology, University of Szeged, Faculty of Science and Informatics, Kozep fasor 52, H-6726 Szeged, Hungary; Tel: 3662544515; Fax: 3662544823; E-mail: jvarga@bio.u-szeged.hu*

Pieter J. Verbelen, *Centre for Malting and Brewing Science, Faculty of Bioscience Engineering, Katholieke Universiteit Leuven, Kasteelpark Arenberg 22, 3001 Heverlee, Belgium; Fax: 32-16-321576; E-mail: Pieter.Verbelen@biw.kuleuven.be*

Brian L. Wickes, *Department of Microbiology and Immunology, The University of Texas Health Science Center at San Antonio, 7703 Floyd Curl Drive, San Antonio, TX 78229-3900; Tel: (210) 567-3938, Fax: (210) 567-6612; E-mail: wickes@uthsca.edu*

Alka Yadav, *Department of Biotechnology, SGB Amravati University, Amravati, PIN-444 602, Maharashtra, India*

Preface

The fungal kingdom consists of a wide variety of organisms with a diverse range of forms and functions. They are ubiquitous in nature, having been recovered from almost every ecological niche – from deep-sea sediments to the jet stream. The utilization and application of fungi by mankind has a long and varied history that probably predates any written records. This utilization includes the very early use of edible fruiting bodies as foods, and the 2000–3000-year-old histories of fungi in brewing, baking and as medicinal treatments. Mankind's use of fungi has broadened and grown considerably in the last 100 years. Well-known examples of this include the discovery and production of antibiotics and the wider utilization of fungi in the food industry, where they are used commercially, both as food products and in the production of compounds and enzymes for food processing (e.g. citric acid and pectinases).

The association of many fungi with plants and invertebrates has resulted in their use in agriculture as both biological control agents against plant pests and diseases, and as plant growth stimulants such as mycorrhizal inoculants. Although all of these applications could be considered as relatively long-established processes, new applications continue to be developed. Recent examples include the development of the commercial production of a mycoprotein food product, continuing advances in the environmental remediation of soils and industrial wastes, and the identification and production of a wide range of enzymes and compounds for biotechnology. Within the biotechnology area there has been a particular growth in the application of mycology, and current research includes the use of fungi in the production of biofuels, and fungally mediated chemical transformations.

This book is intended to provide both students and researchers with a broad background to some of the fastest developing areas in current applied mycology. It was clearly not going to be practical to incorporate all of the different aspects and directions of such a broad area into a single volume, and so we have brought together a range of contributions to highlight the diverse nature of current applied mycology research. Environmentally, fungi are of vital importance and their activities are closely linked with those of bacteria and lower plants in undertaking the nutrient and chemical cycles needed to maintain life on Earth. Despite such fundamental importance, mycology is frequently overlooked in many scientific courses and treatments. The opening chapter of this volume provides some examples of how mycology is often neglected, and presents a case for considering mycology as a megascience.

The subsequent chapters have been loosely grouped into four sections in order to reflect the wider 'customers' or context of the particular mycological areas or activities. As with many biological groupings, this categorization is largely for the convenience of the user, in this case the reader, and there are aspects of many of the chapters that are of direct relevance in additional areas or processes. In each section we have attempted to include contributions that show either new applications or developments of well-established technology, or novel research into new technology or environments.

The environment, agriculture and forestry are represented by contributions that illustrate novel fungal associations or new aspects of well-known interactions. The role of mycorrhizal fungi in agricultural and other environmental systems has been increasingly recognized in recent years. In this section we have selected two examples to demonstrate how natural mycorrhizal communities may be affected by environmental stress, and how the application of mycorrhizal treatments may benefit forestry productivity. Ochratoxin A is a relatively well-known contaminant of various natural products, and in this section we include an example of the relevance of this contamination in the human food chain. Novel associations are represented by the presence and distribution of aquatic (Ingoldian) fungi in tree canopies, and the potential role of endophytes in litter decomposition.

The section on foods and medicine reflects the long history of applied mycology in the manufacture of alcoholic beverages, with two chapters devoted to beer production and winery spoilage issues. Although the production of alcohol by fermentation for human consumption is a long-established mycological application, the fermentation of agricultural wastes for industrial ethanol and biofuels is a much younger technology, and the chapter on lignocellulosic fermentation also demonstrates how modern genomic technologies can be applied in practice. Medicine is represented by two chapters. The first combines old and new by reviewing the results and indications from recent pharmaceutical and cell-line-based screens for *Ganoderma lucidum*, a fungus that has been associated with medicinal use for some 2000 years. A chapter on dematiaceous fungal infections has been included, as fungi causing infections in humans are increasingly being recognized as significant pathogens. Looking to the future, the increase in cases of human immunocompromisation as a result of medical treatments and conditions has led to such mycotic infections being considered as 'emerging' diseases.

Chapters in the section on biotechnology and emerging science reflect some of the current interests in fungal enzymes and their importance in broader environmental processes and applications. At a very broad level, multiple activities of a single fungus may be utilized in a variety of processes, and this is illustrated by considering some of the biotechnological applications of *Trichoderma harzianum*. In contrast, a wide range of fungi may all show a similar property or activity that has biotechnological applications, and particular aspects of these are reviewed in the chapters on chitinases and proteases. A common topic in many of the chapters in this book is the need for, or the potential benefits of, genetically manipulated strains in research and development. In this section the chapter on the use of the *Agrobacterium* prokaryotic model system to transfer fungal genes provides a number of insights into both DNA transfer and transformation methods. Collectively, the fungal kingdom contains a very diverse range of biochemical properties and pathways. The biotechnological relevance of some of these, such as proteases, has long been recognized, and has been extensively studied. It is likely that fungi have many other pathways and properties with as yet unrecognized biotechnological significance, and one aspect of this is demonstrated by the chapter reviewing the very recent work on the formation of potentially industrially useful metal nanoparticles in fungal mycelia.

This volume demonstrates that there is a long history in the identification and utilization of a wide variety of 'mycotechnology'. However, it is also clear that the existing processes are likely to represent only a small fraction of those technologies that are potentially available. It would seem likely that the overall application and use of fungi will increase in both the environmental and biotechnology fields. Environmental changes and increasing pressure on agriculture are both areas where mycological processes for soil remediation, fertility and biocontrol are likely to be further developed. In biotechnology, the increased interest in biofuel production, waste composting and the wider applications of fungal enzymes and metabolites all provide many opportunities for mycology.

Since the original discovery of penicillin there has been a long history of identifying and isolating bioactive and pharmaceutically useful compounds from fungi. Now, and in the future, the increased knowledge of fungal ecology and biochemistry can provide markers to help identify candidate strains for targeted screening programmes. Developments in technology are also likely to involve applied mycology, and it is worth noting that two of the commonly used enzymes in the recently developed molecular biology and biotechnology fields, Proteinase K and Novozym 435, are largely obtained commercially from fungal material. There are many positive indicators for the future development of applied mycology, but there are also concerns, such as those raised in Chapter 1, for the future of the underlying systematic and mycological skills necessary to allow further developments.

We are grateful to Stefanie Gehrig and Sarah Mellor of CABI for their patience, encouragement and suggestions. MR also acknowledges the help and support provided by Alka Karwa, Aniket Gade, Ravindra Ade, Avinash Ingle, Dnayeshwar Rathod, Vaibhav Tiwari, Alka Yadav and Jayendra Kesharwani.

Mahendra Rai and Paul Bridge

1 Mycology: a Neglected Megascience

DAVID L. HAWKSWORTH

Departamento de Biología Vegetal II, Facultad de Farmacia, Universidad Complutense de Madrid, Madrid, Spain; Department of Natural and Social Sciences, University of Gloucestershire, Cheltenham, UK; and Department of Botany, Natural History Museum, London, UK

Introduction

Objectives

This contribution aims to demonstrate why mycology should be regarded as a megascience: a subject requiring international collaboration to overcome barriers that need to be confronted in the interests of global security and human well-being. The reasons why this megascience has been neglected are considered, and attention is drawn to the consequences of the continued negligence of mycology for humankind. Finally, MycoActions are tentatively suggested that may contribute to the recognition by international and national funding agencies of mycology as a megascience.

Definitions

'Mycology' is the study of all aspects of fungi, and is used here to embrace all organisms traditionally studied by mycologists, regardless of their current classification. That is, downy mildews, filamentous fungi, lichens, moulds, mushrooms, slime moulds, water moulds and yeasts; these are dispersed through three kingdoms of life. The filamentous fungi, lichens, mushrooms, moulds and yeasts belong to the kingdom *Fungi*; downy mildews and water moulds belong to the kingdom *Straminipila* (syn. *Chromista, Heterokonta*), which also includes many algal groups, and slime moulds are members of the *Protista* (syn. *Protoctista*), along with many other protozoans. The key differences between the kingdoms are summarized in Kirk *et al.* (2008).

'Megascience' is a concept developed by the Organization for Economic Cooperation and Development (OECD), an intergovernmental organization established in 1961, and now with 30 member countries, a staff of 2500 and a budget of €342.9 million in 2008. In 1992, the OECD initiated the Megascience Forum series of meetings, which address topics considered large enough to demand international coordination. These have included biodiversity informatics, large-scale research facilities, neuroinformatics, neutron sources, nuclear physics and radio astronomy. It is charged by the ministers of science in the member countries to identify opportunities, challenges and obstacles in science, and to make recommendations as to how they might be addressed through international collaboration[1].

Mycology as a Megascience

The aspects of mycology that merit recognition as megascience are many and diverse. As the topic of the present book is applied mycology, the focus here is on features relevant to human kind. For an entré into the role of fungi in natural ecosystems, and their interactions with other organisms in nature, see Dighton (2003) and Dighton *et al.* (2005).

Megadiversity

The diversity of fungi can be considered both in terms of the range of organisms and the number of kingdoms to which they belong, and also with respect to the number of species on Earth. While the total of described species is relatively modest, at around 100,000, it is the estimated number that places them firmly in the megadiverse category along with the nematodes and insects. Estimates of the number of described species range up to 9.9 million, but the figure of 1.5 million, derived from several different data sets (Hawksworth, 1991) is generally accepted as a working figure (Hawksworth, 2001). Even taking a particularly lowest-possible number on a group-by-group basis yields a minimum figure of 712,000 species (Schmit and Mueller, 2007). This means that we know at best 13% of the fungus species on Earth, and probably only around 7%. These figures are in marked contrast to the extent of the non-described biodiversity of, for example, plants, where it is estimated that 300,000 occur worldwide, of which 270,000, that is 90%, are already known (Rodríguez, 2000).

Fungi new to science continue to be discovered and described apace, currently at the rate of about 800 species per year. This figure is, however, limited only by the human resources devoted to inventorying, exploring and describing. Somewhat worrying, however, is that 26% of the fungi described since the 1980s have been by a mere 50 authors, several of whom are now dead or retired (Hawksworth, 2006).

1 The term 'megascience', as used by the OECD, is not to be confused with the popular US television series Mega Science, which puts out programmes on specific sensationalist topics from 'What do animals feel?' and 'Mosquito Wars' to 'Time Travel', 'Ultimate Thrill Rides' and 'War 2020'.

The non-described species are not only to be found in the tropics, but continue to be discovered even in the best-studied countries of Europe, and these include macrofungi as well as microscopic species. Additional species even of conspicuous genera such as *Xerocomus* (syn. *Boletus* p. p.) are still being described from south-eastern England (e.g. *X. chrysonemus, X. silwoodensis*) – the best-studied region for fungi in the world. Especially rich in novelties are ecological habitats that are only now starting to be explored in depth. These include extreme environments such as the deep sea, and hitherto little-explored hosts. These hosts include lichens that prove to be hosts to numerous genera and species (Lawrey and Diederich, 2003), often with no close relatives, and arthropods whose hindguts are home for seemingly myriads of yeasts (Suh *et al.*, 2005).

Numerous independent fungal species are proving to be masquerading within recognized 'species', from which they cannot often be discriminated on morphological features alone. Speciation in fungi evidently commonly proceeds ahead of any morphological differentiation. Such so-called 'cryptic species' are often distinct biologically in their mating behaviour, and also fall into different monophyletic clades in molecular analyses. Molecular studies of large numbers of cultures or specimens from diverse regions are now also revealing that the distributions of many fungi are geographically much more restricted than has often been assumed. It is now clear that most fungi are not 'everywhere' waiting for appropriate hosts or ecological conditions. For example, 'species' for which the same name has been used on different continents are proving increasingly not to be conspecific, except where long-distance dispersal by human or other agencies has occurred (Lumbsch *et al.*, 2008).

It is important to stress that the task of describing the fungi on Earth is not just an esoteric academic exercise, but necessary if we are to understand what fungi are 'out there' that may have beneficial attributes or potential threats to, for example, food security. Without names and descriptions, organisms cannot be recognized when encountered or communicated about. If substantial inroads are to be made into describing the huge numbers of unnamed fungi, it is clear that a megascience approach is required, i.e. concerted international action and funding on a massive scale to address the knowledge gap.

Earth processes

The extent to which fungi have a critical role in Earth processes is not always appreciated. Most important is their major contribution to global carbon cycling, as they are involved in returning around 85 Gt of carbon into the atmosphere through decomposition processes (cf. Houghton, 2004). Furthermore, it has been suggested elsewhere that the total fungal biomass on Earth amounts to some 15 Gt, of which about 5 Gt is carbon (Hawksworth, 2006). In order to place these figures in perspective, it should be noted that the total carbon input to the atmosphere attributable to human activities is around 5.5 Gt. Yet we know little of how fungi are likely to respond to changes in global ambient temperatures or increases in carbon dioxide concentrations. This is far from an easy equation to resolve, as the net result will depend on the effect on rates of decomposition and fungal growth. If the fungal biomass increases, as it may do through indirect and/or direct effects on mycorrhizas and other soil fungi (Jones *et al.*, 1998; Egerton-Warburton *et al.*, 2002), so will the annual input of carbon to the atmosphere from decomposition but, at the same time, the role of fungi as a carbon sink will increase. The carbon sink role is a matter not only of the carbon locked in fungal tissues per se, but also the amount of oxalic acid released by fungi into soil and also rock, where silicate minerals are converted to carbon-containing oxalates (Gadd, 2007). When it is recalled that this activity is being undertaken by an estimated 10 Gt of soil fungi and lichens covering about 8% of the land surface of Earth, including exposed rock surfaces, the significance starts to become apparent.

There is little basic research on these issues in either the laboratory or climatically controlled chambers, and even less in field situations. Getting better approximations will require a major international effort involving studies in all major biomes of the planet using the same methodologies. This would clearly be a megascience activity that could only be put in train by concerted actions of national research funding agencies working within a globally established framework with agreed protocols.

Macroeconomics

'Macroeconomics' is the branch of economics that deals with the performance, structure and behaviour of a national or regional economy as a whole (Blaug, 1985). There is no estimate of the overall figure for the contribution of fungi to the global economy, nor to individual nations or continents, but it is certainly vast. Their roles are diverse, and some are beneficial and others deleterious. A balance sheet does not seem to have been attempted, and only a flavour of the effects on a macroeconomic scale can be presented here to justify this claim.

The credit side of the balance sheet most conspicuously relates to human health and food production. Pharmaceutical sales amounted to US$634 Bn in 2006, and this includes many products that are manufactured by fungi or based on fungal compounds (see below). In the chemical sector, fungi are important as organic acid producers. The global production of citric acid, which utilizes *Aspergillus niger,* was 880,000 t in 1998; the main uses are as food and drink additives (E330), cleaning products, soaps, shampoos and recharging of water softeners. Chemical companies generate huge amounts of ethanol, primarily for fuel, amounting to 51 Bnl (billion litres) in 2006; this figure includes that produced by the petrochemical industry from oil as well as fungal (yeast) fermentations of vegetables and other plant products. The UK drinks market stood at US$102.66 (£51.85) Bn in 2004. This figure includes alcoholic beverages from fungal (yeast) fermentations as well as soft drinks using, for example, citric acid to make them effervesce. A global extrapolation of the UK figures would lead to one at least an order of magnitude higher.

Timber and wood products are a major component of international trade. In 2006 this involved a staggering 1.35 Bn m^3 of material (UNECE/FAO, 2006). The value of this trade in 2000 was US$400 Bn. Mycorrhizas are crucial to forest productivity and, without these fungi, it is self-evident that this trade could not continue at the same levels without exploiting substantially more land, as tree growth would be much less. It would not be unreasonable to attribute at least 10% of this production to the increased growth of trees made possible by mycorrhizal fungi, i.e. US$40 Bn.

Edible and medicinal mushrooms are traded, some being collected in the wild by local peoples, and others cultivated in massive farming operations. Collecting sought-after species can be a crucial part of local economies (Boa, 2004). Some mushrooms command exorbitant prices amongst gastronomes – a single 0.5 kg fruit body of the Alba white truffle (*Tuber magnatum*) sold at auction for US$115,920 in 2005, making it the most expensive food on Earth and commanding a price six times that of gold. Commercial mushroom production has been rising exponentially in the last few decades, with 'exotic' species now regularly to be found in upmarket supermarkets in Europe and North America. The world production of cultivated mushrooms is cited as 6.6 Mt globally in 1997 (Chang and Miles, 2004), but has surely increased substantially since then; the major producer is now China. Similarly, there is bread, where the annual production of bakers' yeast alone has been estimated at 1.5 Mt each year. Fungus-ripened cheeses, and oriental fermented foodstuffs such as tempe, are also important. It must also be remembered that fungi have a fermentation role in the production of chocolate, coffee and tea. Novel products can become economic successes: Quorn ('mycoprotein' from continuous fermentation of *Fusarium venenatum*) was first sold in the UK in 1985, and the retail value of the brand was indicated as US$200 M in 2006; 500,000 meals containing Quorn are now consumed daily in the UK where it is purchased by one in five households.

Industrial enzymes obtained from fungi, especially *Aspergillus* and *Trichoderma* spp., are diverse. These include cellulases, laccases, lipases, pectinases, proteases and xylanases. Applications of such enzymes are as far-ranging as biological washing powders, biopolishing and de-pilling products, and the creation of stone-washed effects in denim. More information on many of these aspects can be found in An (2005). More specialized, but nevertheless significant, uses include an oil prepared from the lichen *Evernia prunastri* used as a carrier of volatile compounds in more expensive French perfumes.

The debit side of the balance sheet has, as its most important entry, the biodeterioration of manufactured goods and constructions. These include losses due to mould growth on all kinds of materials. Wood-rotting fungi (e.g. *Serpula lacrymans*) ravage the indoor timber of buildings and destroy wooden fences, while lichens can disfigure and damage stone, brick, tile and other outdoor construction materials. In the warm and humid southern USA, homes where 'toxic mould' (*Stachybotrys chartarum*) flourishes have been the subject of court actions for damages against builders: a staggering US$32 M was initially awarded in one case (Money, 2004). Fungi growing in electrical and other sensitive equipment have to be prevented to avoid potentially disastrous results, and regulations involving challenge-testing by sets of fungi are now required in many countries (Allsopp *et al.*, 2004). The 'kerosene fungus' (*Amorphotheca resinae*), that thrives in aircraft fuel tanks at the fuel/air interface, is the subject of an ongoing battle as its hyphae can clog filters and may cause aeroplanes to crash; antifungal additives are required, and regular inspection of the state of fuel tanks has to be undertaken.

Consequently, considering the importance of fungi in trade and manufacture, it is evident that they have the potential to affect national and regional economies. They are unquestionably significant at the macroeconomic level.

Human health

Fungal products are critical in human health care. In addition to penicillins, cephalosporins and other antibiotics, success stories of recent decades include cyclosporin A, which reduces rejection in organ transplants, and statins for the control of cholesterol levels. Some overall estimates of the monetary value of fungal products are given above, and two examples will provide an impression of the scale in which fungal products feature in health care. The annual sales of amoxicillin (a penicillin) in the USA alone amounts to around US$500 M, and statins are being taken by some 30 M people worldwide and generating US$25 Bn annually for the producing companies. The full range of compounds beneficial to health produced by fungi is staggering and expanding (Peláez, 2005). Just how many people living today owe their lives, or quality of life, to fungal products does not appear to have been estimated. However, many bacterial infections that would have been fatal before the 1950s are no longer so and, consequently, one in ten would be a conservative figure. As the human population stood at 6.6 Bn in 2007, that implies that there are 660 M people alive today because of fungal products.

In comparison with health benefits, the harmful effects of fungi are relatively modest. Allergies due to spores in the air (e.g. hay fever due to *Alternaria* conidia) and superficial mycoses (e.g. thrush, ringworm) are generally a nuisance rather than life threatening, and the symptoms can be alleviated or the condition cured by drugs. Fungi are most harmful in immunocompromised individuals, for example those with AIDS in which pneumocystis pneumonia caused by *Pneumocystis jirovecii* is responsible for about 70% of deaths; the fungus is present in the lungs of most people but causes no problems while their immune system is healthy (Cushion, 1998). However, in the subtropics and tropics some fungi able to affect healthy humans are of concern, especially *Ajellomyces capsulata* causing histoplasmosis, *Coccidioides imitis* coccidiomycosis and *Paracoccidiodes brasiliensis* paracoccidiomycosis. Poisonings from the consumption of wrongly identified basidiomes (fruit bodies) are rare, although probably most frequent in China.

Some mushrooms contain hallucinogenic compounds and have a long history of use in religion and as recreational drugs. The most efficacious of these, which contain psilocybin or psilocin, are now classed as prohibited Class A drugs in some countries (e.g. the UK), where possession carries severe penalties. Of more widespread concern are moulds which contaminate foodstuffs, some of which produce compounds (mycotoxins) that are poisons or carcinogens (Frisvad *et al.*, 2007). Mycotoxins are of particular prevalence and concern in tropical areas with poor food storage facilities, and also in products being exported from those areas. A few fungi also have potential as biological weapons, but fortunately most are unlikely to be usable in situations where a rapid response is desired (Paterson, 2006).

The human health aspects of mycology merit recognition as megascience, not so much in respect of drug discovery or the numbers of people who benefit from them (which are the domains of pharmaceutical companies and public health services), but rather because international standards for monitoring and detecting fungi of human concern need to be developed and implemented.

Food security

World vegetable, cereal, fruit and fibre production was estimated at 4067 Mt in 2004 (UNECE/FAO, 2006). Losses due to fungi occur pre- and postharvest. Those due to plant diseases are estimated at over US$3 Bn annually in North America alone (Barron, 2006). Losses from postharvest infections are estimated at 25% worldwide, but can be as much as 50% in tropical countries (Prusky and Kolattukudy, 2007). Such figures have staggering implications. For example, given world cereal production as 2107 Mt in 2001 (Maene *et al.*, 2008), without pre- and postharvest losses that figure could perhaps have been 25–50% higher, i.e. 2630–3160 Mt that year.

The widespread use of monocultures of particular cultivars or races of staple crops, such as rice and wheat, make supplies of even such basic foodstuffs insecure. Clones are especially vulnerable to disease. There is a continual battle between plant breeders and fungal pathogens. A classic case is that of wheat and its stem rust *Puccinia graminis*, where there remains a time bomb waiting to happen, even in North America (Peterson, 2001). In the case of rice, there was a major problem with a race commended for widespread cultivation in the 1970s. This race proved to be susceptible to a sheath rot fungus, *Sarocladium oryzae*, which produces phytotoxic compounds (Ayyadurai *et al.*, 2005). The fungus was able to infect rice from a reservoir in wild grasses, and the resistance of the race to this fungus had not been tested for in plant breeding trials.

While there are seemingly exhaustive regulations regarding plant quarantine and seed testing for pathogenic fungi, the reality is that there is a major mismatch between the potential risk to food security and the resources and measures devoted to checks on people, their clothes and belongings, and cargoes. Crops in one country may be vulnerable to diseases that are not even known to be a threat in another and so are not searched or tested for at national frontiers. It is salutary to reflect that the race of potato blight,

Phytophthora infestans, responsible for the Irish famine of 1845–1847, came from the Andes in South America (Gómez-Alpizar *et al.*, 2007), and that sudden death of oak due to *P. ramorum* in California is also a result of an introduction (Monahan *et al.*, 2008). Additional risks to plant health may also be expected as a result of climate change. Host plants may come into contact with fungi that are able to attack them when other plants (with their associated fungi) migrate. There is also the possibility of plant pathogens being used in biological warfare to destroy crops in an enemy's territory (Paterson, 2006).

On the positive side, the role of fungi as biocontrol agents of plant pathogens, and especially insect pests, in maintaining food production has to be stressed. This can be within confined areas such as glasshouses (e.g. products from *Lecanicillium lecani* against whitefly) or outside. Of particular importance is the use of a strain of *Metarhizium anisopliae* for the control of desert locusts in arid areas of Africa and Australia (Lomer *et al.*, 2001), although it does affect other grasshoppers as well. Edible mushrooms, in addition to their contribution to economies (see above), also contribute to food security, through material collected from the wild as well as those which are cultivated. Numerous species are collected and eaten from the wild, although they are utilized to different degrees in different cultures (Boa, 2004). The fact that some edible non-mycorrhizal mushrooms can be cultivated on agricultural wastes such as paddy straw, debris of bamboos and wood chips (Chang and Miles, 2004) also improves food security and simultaneously increases the level of sustainability of local peoples.

In order to improve the level of food security against threats from fungal diseases, more attention needs to be accorded to the development of robust plant and plant product inspection protocols and phytosanitary regulations, as well as to listing species of quarantine importance. Consideration needs to be given to increasing the scope and rigour of existing international, regional and national regulatory instruments and practices in order to reduce risks to crops in agricultural areas. As fungi can be moved from country to country, it is evident that such procedures need to continue to be developed by international agreement. Nevertheless, even with such procedures and regulations, successful implementation of new procedures will be dependent on the training of appropriately skilled officers. Further, such officers will also need the back-up of accessible centres of mycological expertise for their training and also advice.

Model organisms

The relative ease of cultivation of some fungi in the laboratory, and that they have many basic biochemical and developmental processes in common with humans, has meant that they have been adopted as experimental models in various types of basic biological research. Examples include: (i) studies on genetic recombination using *Neurospora crassa*; (ii) genome organization and function in *Aspergillus niger*; (iii) cell aggregation and morphogenesis in social organisms in *Dictyostelium discoideum*; (iv) the genetics of ageing in *Podospora anserina*; (v) the genetics of plant host–fungal pathogen interactions in *Magnaporthe grisea*; and (vi) cell division and its control in *Saccharomyces cerevisiae* and *Schizosaccharomyces pombe*. *Saccharomyces cerevisiae* was one of the first organisms to have its genome completely sequenced as a result of a major international collaborative effort, and the sequencing of about 50 fungal genomes is expected to be completed by the end of 2009; these include a selection of plant pathogens and fungi of major medical importance, as well as species with different biologies and systematic positions. Comparisons of whole genomes will increasingly enable the function of different genes to be identified, and are opening up new vistas in experimental mycology (e.g. Gow, 2004; Brown, 2006).

Whole genome studies generally require the collaboration of several laboratories in different countries, and concomitant funding from a consortium or sources. Such initiatives are consequently in the best traditions of megascience.

Large-scale facilities

A common feature of a megascience is that it requires large-scale facilities. This is particularly pertinent to mycology, which requires the underpinning of dried reference collections, genetic resource collections, substantial libraries and public databases.

Dried reference collections underpin the application of scientific names of fungi, constitute comparative material used when making identifications, are the basis of our knowledge as to distribution, ecology and host/substratum ranges, and the place in which voucher material of the organisms actually used in experimental and other studies can be preserved. Numerous museums, universities and research institutes maintain such collections, and the largest are all located in Europe and North America. The total number of institutions with dried plant and/or fungal specimens worldwide, as compiled in Heywood (1995), was 2946; these were collectively indicated to hold over 279 M specimens. Separate figures as to how many of those collections hold fungal as well as plant material show that the number of institutions with fungi is around 500, and that the total number of fungal (including lichen) specimens held worldwide to be in the region of 15 M. Most of the material is, however, held by relatively few centres; details of 70, including all major ones, are given in Hall and Minter (1994). As an estimate, these 70 probably hold around half of the world's total. This massive distributed resource is the key database of our knowledge of fungi on Earth, and the collective cost of the maintenance of these, including buildings, staff and resources, must be enormous and certainly place it on a par with megascience research facilities. Unfortunately, the information included is largely inaccessible as the data on the bulk of the holdings are not available electronically – even for some of the largest collections. While there have been moves to encourage the computerization of specimen data from dried reference collections, significant further progress will be dependent on external funding; many of the collections are poorly resourced and lack the necessary core staff to otherwise complete such tasks in the foreseeable future.

The situation is somewhat different with respect to genetic resource collections, i.e. those which maintain fungi in a living or recoverable state. Material that can be studied in the living state is necessary for comparative work in identification, and also to enable experimental work to be both replicated and built upon. Where voucher material of the actual organisms used in research papers is not preserved and available for use by other mycologists, results cannot be repeated and independently verified in the scientific tradition of reproducibility. Data on the holdings of such living collections are fortunately coordinated through the World Federation for Culture Collections' *World Directory* database. Analysis of the information it holds indicates that there are some 370,251 strains dispersed through 483 collections; these represent about 16,830 species, that is about 16% of the fungi so far described (Hawksworth, 2004). This total, however, excludes the often massive collections built up for screening purposes by some pharmaceutical and bioprospecting companies.

Mycologists involved in systematic work also require access to substantial libraries, as sound descriptive work does not become superfluous or obsolete in the same manner as experimental studies may. Literature dating back to 1753, the internationally agreed starting point date for the scientific naming of fungi, may need to be consulted. This has been a major obstacle to mycologists working outside the largest museums and institutions, but is gradually being alleviated to some extent by electronic access to full texts of books and journals through the World Wide Web on the Internet. Journals are increasingly being made available electronically by their publishers, albeit often at a cost to users. Especially commendable is the scanning of whole mycological books and scarce journals by CYBERLIBER, which is freely available. Information on scientific names of fungi that have been published is now also available free through the *Index Fungorum* web site, which now holds data on almost 434,000 names; this database was initiated by the then International Mycological Institute (now a part of CABI Europe), but is now operated by a consortium of leading mycological centres. *Index Fungorum* is complemented by MycoBank, which was started by the Centraalbureau voor Schimmelcultures (CBS) in The Netherlands, but is now operating under the auspices of the International Mycological Association (IMA). MycoBank works closely with *Index Fungorum*, and differs in being a repository for descriptive and other key information on newly described fungi; leading mycological journals are now making deposition of taxonomic information in MycoBank a requirement for publication, in a parallel manner to the practice of insisting that molecular sequence data are deposited in GenBank.

In the same way as particle physicists need access to the CERN cyclotron large-scale facility in Geneva, mycologists also need access to the holdings of major dried and living collections of fungi, and associated libraries and databases. Indeed, some of the major dried and living reference collections in Europe which contain fungi have already been recognized as 'large-scale facilities' that researchers in other institutions and countries need to consult. The EU-funded SYNTHESYS scheme of the Centre for European Taxonomic Facilities (CTAF), which ran from 2004 to 2008, enabled researchers to spend time in such collections, and merits emulation and extension on a worldwide scale.

Mycology as a Neglected Science

The 'orphan' nature of mycology has been presented previously (Hawksworth, 1997). Fungi lack close relatives, they are misunderstood and often excluded from 'family' events, many are unnamed, they are often overlooked or ignored, have few carers and are inadequately provided for. The neglect is so grave that mycology does not even feature as a separate subject in the categories of science recognized by UNESCO; these only mention fungi as mushrooms under 'botany' and as yeasts under 'microbiology'. This chapter will not reiterate the arguments in support of the 'orphan' status, but consider the reasons why this situation has arisen and reflect on its consequences.

Reasons for neglect

The basis of the neglect is surely the inclusion of fungal groups along with plants in the *Species Plantarum* of Linnaeus (1753). Had fungi been covered in the *Systema Naturae* (Linnaeus, 1758), as is now clear would have been more appropriate in view of their much closer relationship to animals than plants, I believe that the current situation might never have arisen. This led to fungi being embraced in botany departments, and later microbiological[2] ones, in universities, museums and research institutes, and not treated in independent departments. Studies of animals in larger universities and museums in particular were, until recent decades, almost invariably separated into discrete departments such as ones of entomology, herpetology, ichthyology, invertebrate zoology, mammology, ornithology or parasitology; it would have been logical to add ones of 'mycology' for fungi had they not been regarded as the province of botanists. Departments need heads or chiefs, so the recognition of a science in institutional organograms means that there would always be designated positions and staff in those fields. This approach meant that mycology posts often had no security; when staff retired they were often replaced by researchers from other disciplines. Furthermore, mycologists were often lonely individuals amongst teams of botanists or

2 A study of the contents of recent microbiology textbooks (e.g. Schaechter *et al.*, 2006), staffing of microbiology departments or proceedings of microbiological congresses show that fungi (other than yeasts and those of medical importance) are not accepted as full members of the microbiological fold.

microbiologists and viewed with disdain as taking resources from the core activities of their departments or institutions.

In the case of universities, this lack of separate departments meant that mycology teaching rarely extended beyond one term of lectures at most, and that there were no first degrees in mycology on offer. In a few cases universities have run diploma or masters courses in mycology, but these tend to be short-lived and to disappear when particular staff members, and especially heads of department, retire. Now that the fashion is for large biological and life science departments rather than organism-orientated ones, the situation is sadly being exacerbated for many parts of biology.

The result of this poor coverage of mycology in the training of biologists is that their knowledge of fungi, the complexities of their biologies and their importance in earth processes and to human well-being is not commensurate with their significance. The corollary is a disproportionately low profile for fungi in the syllabus of schools and universities.

The slowness with which mycologists have started to organize themselves as separate on the international stage has also been a contributory factor. The International Mycological Association (IMA) was formed as late as 1971, the first time that an International Mycological Congress had been organized[3]. That this initiative took so long is surely in no small measure a consequence of most professional mycologists being isolated and unused to being gregarious. It also needed a champion which, in this case and for other major mycological initiatives, was found in the late Geoffrey C. Ainsworth (1905–1998).

Consequences of neglect

The most important consequence of the relative neglect of mycology in science is that there is insufficient expertise to support the needs for fungal expertise in areas of direct human concern: human health, food security, food safety, plant diseases, forensics, bioterrorism, and biodeterioration. And also more indirect human concern: earth processes, ecological services and processes, vegetation science, and climate change. This means that the quality of investigations in pure and applied sciences that depend on aspects of mycology is suboptimal.

The long history of neglect also means that the accumulated corpus of knowledge of fungi, their properties, ecologies and life cycles is proportionately poor compared with that of other groups of organisms. In terms of described species, for example, it is estimated that even the insects are proportionately about twice better known than the fungi (Heywood, 1995). The inadequate knowledge of fungi that occur in countries, and their status within them, also means that countries will find it impossible to meet

[3] Prior to this date, international gatherings of mycology had always been as a part of International Botanical Congresses or International Congresses of Plant Pathology.

their obligations under the Conservation of Biological Diversity (UNEP, 1994). This is especially so for Article 7 of the Convention, which requires governments to identify components of biodiversity important for its conservation and sustainable use, and further to monitor those components and identify factors that might have adverse impacts on them.

The traditionally isolated position of mycologists in botany and microbiological departments and institutions has also meant that, when departments have been amalgamated into larger biological or life science units, they have tended to be 'lost'. Positions are often not replaced, and at present there is a declining and ageing population of professional whole-organism mycologists in much of the 'developed' Western world. This situation is fortunately not being reflected in, for example, all parts of Asia and Central and South America, which include countries with vigorous populations of mycologists (e.g. Brazil, China, Mexico). However, most living and dried reference collections, reference books, textbooks, monographs and key databases are located in or generated by the ageing and declining mycological community in Europe and North America. In time the centre of mycological expertise may well migrate southwards and eastwards, but at present the subject worldwide is increasingly dependent on a small and reducing number of individuals. The futures of some major collections and key databases are insecure or fragile.

In Europe and North America there is a huge interest in field mycology amongst 'amateur' naturalists. While many of these 'amateurs' are extremely dedicated and talented and produce most valuable monographs, keys, etc., their ability to undertake scientific work is dependent on the back-up of professional mycologists, major reference collections, access to literature and the existence of comprehensive databases. While in the short term one consequence of neglect is the shift of knowledge of fungal organisms from the 'professional' to the 'amateur', in the long term that situation will be unsustainable without the secure basic resources and back-up an 'amateur' needs. There is another consequence as well. In some mycological societies, so few professionals are now 'whole-fungus' mycologists has led to unfortunate friction between perceived 'amateur' and 'professional' factions.

MycoAction Required

The issue of how to enhance the standing of mycology worldwide is one for all mycologists. It is not something that can be left to *somebody*, but an issue for *everybody*, individually and as part of mycological organizations. A series of actions from the individual to the international organization level has been suggested, the MycoAction Plan, to begin redressing the situation (Hawksworth, 2003). These relate to actions for improving collaboration, promotion, education and conservation, have been widely publicized through the International Mycological Association (IMA) web site and are being implemented by some individuals and societies.

Perhaps the most important action is to keep stressing the distinctiveness of mycology as a subject within biology, from the classroom to national and international scientific bodies, and to avoid perpetuating myths such as mycology being a part of botany, or fungi a part of 'flora'. Mycological research has no place in journals with 'botany' in the title unless it concerns plant–fungus interactions, and the fungi in an area or habitat are more appropriately referred to as the 'mycobiota' – where any separate word is necessary at all. At the same time governments and research funding bodies need to be awakened to the importance of fungi in human and environmental affairs. It is to be hoped that this can be achieved without a major crisis such as loss of rice or wheat production, or deaths from fungal toxins, to stimulate action.

Change always takes time to effect and, while working for an improvement in the numbers and standing of professional mycologists, it is also important to ensure that the resources needed to empower 'amateur' mycologists are not only maintained but enhanced.

A major stimulus to the relaunch of mycology would be the recognition of the subject as being in need of concerted actions by the convening of an OECD Megascience Forum on mycology. The subject meets the criteria of requiring both international coordination and large-scale facilities. The publication of this essay is intended to initiate the debate amongst the national government representatives on the OECD as to whether such a forum should be convened.

Acknowledgement

I am indebted to my fiancée Patricia E.J. Wiltshire for providing me with facilities in which to prepare this contribution, and for her improvements to its presentation.

References

Allsopp, D., Seal, K.J. and Gaylarde, C. (2004) *Introduction to Biodeterioration*. 2nd edn., Cambridge University Press, Cambridge, UK.

An Z (ed.) (2004) (2005) *Handbook of Industrial Mycology*. CRC Press, Boca Raton, Florida.

Ayyadurai, N., Kirubakaran, S.I., Srisha, S. and Sakthivel, N. (2005) Biological and molecular variability of *Sarocladium oryzae*, the sheath rot pathogen of rice (*Oryza sativa* L.). *Current Microbiology* 50, 319–323.

Barron, G.L. (2006) *Introductory Mycology: Laboratory Review*. MycoGraphics, Guelph, Canada.

Blaug, M. (1985) *Economic Theory in Retrospect*. Cambridge University Press, Cambridge, UK.

Boa, E. (2004) *Wild Edible Fungi: a Global Account of their Use and Importance to People*. Non-wood Forest Products Report No. 17, Food and Agriculture Organization, Rome.

Brown, A.J.P. (ed.) (2006) *Fungal Genomics* [*The Mycota* Vol. 13]. Springer-Verlag, Berlin.

Chang, S-T. and Miles, P.G. (2004) *Mushrooms: Cultivation, Nutritional Value, Medicinal Effect, and Environmental Impact*. CRC Press, Boca Raton, Florida.

Cushion, M.T. (1998) *Pneumocystis carinii.* In: Ajello, L. and Hay, R.J. (eds) *Topley and Wilson's Microbiology and Microbial Infections.* Vol. 4, *Medical Mycology* 9th edn., Arnold, London, pp. 645–683.

Dighton, J. (2003) *Fungi in Ecosystem Processes.* Marcel Dekker, New York.

Dighton, J., White, J.F. and Oudemans, P. (eds) (2005) *The Fungal Community: its Organization and Role in the Ecosystem.* 3rd edn., Marcel Dekker, New York.

Egerton-Warburton, L.M., Allen, E.B. and Allen, M.F. (2002) Conservation of mycorrhizal fungal communities under elevated atmospheric CO_2 and anthropogenic nitrogen deposition. In: Sivasithamparan, K., Dixon, K.W. and Barrett, R.L. (eds) *Microorganisms in Plant Conservation and Biodiversity.* Kluwer Academic Publishers, Dordrecht, The Netherlands, pp. 19–43.

Frisvad, J.C., Thrane, U. and Samson, R.A. (2007) Mycotoxin producers. In: Dijksterhuis, J. and Samson, R.A. (eds) *Food Mycology: a Multifaceted Approach to Fungi and Food.* CRC Press, Boca Raton, Florida, pp. 135–159.

Gadd, G.M. (2007) Geomycology: biogeochemical transformations of rocks, minerals, metals and radionuclides by fungi, bioweathering and bioremediation. *Mycological Research* 111, 3–49.

Gómez-Alpizar, L., Carbone, I. and Ristaino, J.B. (2007) An Andean origin of *Phytophthora infestans* inferred from mitochondrial and nuclear gene genealogies. *Proceedings of the National Academy of Sciences, USA* 104, 3306–3311.

Gow, N.A.R. (2004) New angles in mycology: studies in directional growth and directional motility. *Mycological Research* 108, 5–13.

Hall, G.S. and Minter, D.W. (1994) *International Mycological Directory.* 3rd edn., CAB International, Wallingford, UK.

Hawksworth, D.L. (1991) The fungal dimension of biodiversity: magnitude, significance, and conservation. *Mycological Research* 105, 1422–1432.

Hawksworth, D.L. (1997) Orphans in 'botanical' diversity. *Muelleria* 10, 111–123.

Hawksworth, D.L. (2001) The magnitude of fungal diversity: the 1.5 million species estimate revisited. *Mycological Research* 95, 641–655.

Hawksworth, D.L. (2003) Monitoring and safeguarding fungal resources worldwide: the need for an international collaborative MycoAction Plan. *Fungal Diversity* 13, 29–45.

Hawksworth, D.L. (2004) Fungal diversity and its implications for genetic resource collections. *Studies in Mycology* 50, 9–17.

Hawksworth, D.L. (2006) Mycology and mycologists. In: Meyer, W. and Pearce, C. (eds) *8th International Mycological Congress, Cairns, Australia, 20–25 August 2006, International Proceedings.* Medimond, Bolonga, Italy, pp. 65–72.

Heywood, V.H. (ed.) (1995) *Global Biodiversity Assessment.* Cambridge University Press, Cambridge, UK.

Houghton, J. (2004) *Global Warming: the Complete Briefing.* 3rd edn., Cambridge University Press, Cambridge, UK.

Jones, T.H., Thompson, L.J., Lawton, J.H., Bezemer, T.M., Bardgett, R.D., Blackburn, T.M., Bruce, K.D., Cannon, P.F., Hall, G.S., Hartley, S.E., Howson, G., Jones, C.G., Kampichler, C., Kandeler, E. and Ritchie, D.A. (1998) Impacts of rising atmospheric carbon dioxide on model terrestrial ecosystems. *Science* 280, 441–443.

Kirk, P.M., Cannon, P.F., Minter, D.W. and Stalpers, J.A. (2008) *Ainsworth and Bisby's Dictionary of the Fungi.* 10th edn., CAB International, Wallingford, UK.

Lawrey, J.D. and Diederich, P. (2003) Lichenicolous fungi: interactions, evolution and biodiversity. *Bryologist* 106, 80–120.

Linnaeus, C. (1753) *Species Plantarum.* 2 vols, L. Salvii, Stockholm.

Linnaeus, C. (1758) *Systema Naturae.* 10th edn., 2 vols, L. Salvii, Stockholm.

Lomer, C.J., Bateman, R.P., Johnson, D.L., Langewald, J. and Thomas, M. (2001) Biological control of locusts and grasshoppers. *Annual Review of Entomology* 46, 667–702.

Lumbsch, H.T., Buchanan, P.K., May, T.W. and Muelle, G.M. (2008) Phylogeography and biogeography of fungi. *Mycological Research* 95, 423–424.

Maene, L.M., Sukalac, K.E. and Heffer, P. (2008) Global food production and plant nutrient demand: present status and future prospects. In: Aulakh, M.S. and Grant, C.A. (eds) *Integrated Nutrient Management for Sustainable Crop Production*. Haworth Press, New York, pp. 1–28.

Monahan, W.B., Tse, J., Koenig, W.D. and Garbelotto, M. (2008) Preserved specimens suggest non-native origins of three species of *Phytophthora* in California. *Mycological Research* 112, 757–758.

Money, N.P. (2004) *Carpet Monsters and Killer Spores: a Natural History of Toxic Mold*. Oxford University Press, New York.

Paterson, R.R.M. (2006) Fungi and fungal toxins as weapons. *Mycological Research* 110, 1003–1010.

Peláez, F. (2005) Biological activities of fungal metabolites. In: An, Z. (ed.) *Handbook of Industrial Mycology*. Marcel-Dekker, New York, pp. 49–92.

Peterson, P.D. (2001) *Stem Rust of Wheat: from Ancient Enemy to Modern Foe*. American Phytopathological Society Press, St Paul, Minnesota.

Prusky, D. and Kolattukudy, P.E. (2007) Cross-talk between host and fungus in postharvest situations and its effect on symptom development. In: Dijksterhuis, J. and Samson, R.A. (eds) *Food Mycology: a Multifaceted Approach to Fungi and Food*. CRC Press, Boca Raton, Florida, pp. 3–25.

Rodríguez, L.O. (2000) *Implementing the GTI: Recommendations from DIVERSITAS Core Programme Element 3, Including an Assessment of Present Knowledge of Key Species Groups*. International Union of Biological Sciences, Paris.

Schaechter, M., Ingraham, J.L. and Neidhardt, F.C. (2006) *Microbe*. American Society for Microbiology, Washington, DC.

Schmit, J.P. and Mueller, G.M. (2007) An estimate of the lower limit of global fungal diversity. *Biodiversity and Conservation* 16, 99–111.

Suh, S-O., McHugh, J.V., Pollock. D.D. and Blackwell. M. (2005) The beetle gut: a hyperdiverse source of novel yeasts. *Mycological Research* 109, 261–265.

UNECE/FAO (2006) *Forest Products Annual Market Review 2005–2006*. Geneva Timber and Forest Strategy Paper No. 21, United Nations, Geneva, Switzerland.

UNEP (1994) *Convention on Biological Diversity*. United Nations Environment Programme, Nairobi, Kenya.

2 Arbuscular Mycorrhizal Fungi and Plant Symbiosis under Stress Conditions: Ecological Implications of Drought, Flooding and Salinity

Ileana V. García and Rodolfo E. Mendoza

Museo Argentino de Ciencias Naturales 'Bernardino Rivadavia' (MACN-CONICET), Argentina

Introduction

Arbuscular mycorrhizal (AM) fungi are obligate biotrophs that live symbiotically in the roots of the majority of angiosperms, gymnosperms, pteridophytes, lycopods and mosses. Only a few members of the Brassicaceae, Caryophyllaceae, Chenopodiaceae and Urticaceae do not have interactions with AM fungi (Smith and Read, 1997). AM fungi are members of the Glomeromycota, a phylum that is as distinct from other fungi as the Ascomycota are from the Basidiomycota (Schüβler *et al.*, 2001).

The Glomeromycota have an ancient origin, and potential fungus/root interactions have been identified in fossil roots from the Devonian period (about 400 Mya; Taylor *et al.*, 1995). Such observations suggest co-evolution of AM fungi together with the first land plants and explain the nearly ubiquitous distribution of the AM symbiosis in the plant kingdom, as well as in globally distributed ecosystems.

Arbuscular mycorrhizal fungi are involved in obtaining and transporting phosphate and other nutrients from the soil to plant roots and, in turn, the plant provides fixed carbon to the fungal partner. In addition, AM fungi are also important in the ecology and physiology of plants, in particular those living under biotic (e.g. pathogen infection) or abiotic stresses (e.g. nutrient or water deficiency). AM symbiosis is characterized by the formation of highly branched intracellular structures known as arbuscules that constitute the primary sites of nutrient and carbon exchange between the symbionts (Smith and Read, 1997). This symbiotic colonization also involves formation of other intra-radical structures such as vesicles, coils and hyphal growth within the root cortex tissue. Vesicles are oval-shaped structures filled with lipid bodies and many nuclei, and are thought to have a storage and/or propagule function. Intra-radical structures are connected to an extra-radical mycelium where any spores are formed. The extra-radical mycelium can also develop runner hyphae that can acquire phosphorus, or nutrients with low mobility such as zinc and copper (Ruiz-Lozano *et al.*, 1996; Al-Karaki and Clark, 1998). In addition to improving host nutrient supply, colonization of roots by AM fungi has been shown to protect plants against pathogens, salt stress in arid and semi-arid areas (Al-Karaki, 2006), moderate drought stress (Augé, 2001), periodic flooding (Miller and Bever, 1999; Mendoza *et al.*, 2005) and to enhance soil aggregation (Rilling, 2004).

Salinity affects some 930 million ha or more than 10% of the world's land area. Global environmental change, incorrect use and poor husbandry of agricultural land can have rapidly deleterious effects on the soil environment, and these can include increased severity and duration of stress conditions such as drought, flooding and salinity. Both low and high soil water contents can reduce plant growth, and the balances of water and salinity in soils are major factors that limit food production worldwide. The ability of a plant to form a symbiotic relationship with AM fungi is a key factor that can improve yields under such environmental stresses. This chapter will describe and explain some of the principal effects and mechanisms involved in AM fungi and plant symbiosis under environmental stresses.

Spatial and Temporal Variations of AM Fungi and Plant Symbiosis

Soils rarely provide ideal conditions for the survival and growth of plants and soil microorganisms. As soil conditions are constantly changing, the soil environment may favour development of arbuscular mycorrhizas at one time and inhibit them at another. AM fungal root colonization and spore density have been shown to have some seasonality, and they have also been associated with host phenology, edaphic properties and/or climate variations (Lugo and Cabello, 2002; Muthukumar and Udaiyan, 2002; Escudero and Mendoza, 2005; García and Mendoza, 2008).

Knowledge of the seasonal dynamics of colonization is necessary to quantify both the function and ecological significance of AM fungi. Periods during which mycorrhizal colonization is high are those when the fungus is most likely to influence the nutrient status of the plant and place a demand upon it for carbon. Mullen and Schmidt (1993) reported that AM root colonization closely followed the nutrient demands generated by the different growth stages of the plant. Similarly, when a rapid increase in the abundance of hyphae and vesicles occurs in the root, this could be a period when the fungus acts as a significant carbon sink.

In west central Ohio, USA, the seasonal dynamics of AM colonization within different wetland habitats was found to be similar among all wetlands and sampling areas within the wetlands (Bohrer *et al.*, 2004). The soil in the sampling areas ranged from mostly dry and nutrient deficient to consistently flooded and nutrient rich. These findings were considered as showing that, contrary to expectation, the AM fungi were not limited and that AM seasonal trends were not diluted by flooding or nutrients in these areas. In this study, all of the sites sampled showed the highest AM colonization levels during periods of higher water tables, and vice versa. This finding contrasts with the previous findings of Miller and Bever (1999), which suggest that the extent of AM colonization was not controlled by seasonal dynamics related to variations in water table, and raised the possibility that AM seasonal variation could be influenced by plant factors rather than by abiotic factors.

Recent advances in molecular techniques now make it possible directly to identify the AM fungi colonizing roots in the field. The diversity of the AM fungal community composition in the roots of two plant species (*Agrostis capillaris* and *Trifolium repens*) that co-occurred in the same grassland ecosystem has been characterized in a molecular study, and a total of 24 different fungal phylotypes were found to have colonized the roots of the two host species (Vandenkoornhuyse *et al.*, 2002). Phylogenetic analyses showed that 19 of these phylotypes belonged to the Glomaceae, three to the Acaulosporaceae and two to the Gigasporaceae. This study identified that the AM fungi colonizing *T. repens* differed from those colonizing *A. capillaris*, providing evidence for fungal host preference and dynamic changes in the fungal community through time.

In grasslands habitats, the seasonality of AM fungal species suggests a temporal partitioning of plant resources and, that by specializing on cool or warm season plants, fungi could minimize interspecific competition for roots. Pringle and Bever (2002) have suggested that fungi derive variable benefits from their associations with plants, and that the divergent seasonalities of AM fungi may allow the coexistence of a diverse group of mycorrhizal species. These authors concluded that the seasonal and spatial heterogeneity of the fungal community might have strong impacts on plant community processes.

AM fungi have been studied in the rhizosphere of three Poaceae with a C_3 metabolic pathway (*Briza subaristata, Deyeuxia hieronymi* and *Poa stuckertii*), two Poaceae with a C_4 metabolic pathway (*Eragrostis lugens* and *Sorghastrum pellitum*) and one Rosaceae (*Alchemilla pinnata*) from a natural mountain grassland in central Argentina. The density of 17 AM fungal taxa was found to be strongly influenced by both seasonality and host metabolic pathway, but the biodiversity and species richness and evenness were not affected (Lugo and Cabello, 2002). Root colonization and the proportion of roots colonized by arbuscules and vesicles reached their maxima in summer (wet season). These values might be due to the associated high plant metabolic activity and soil moisture. The peaks in colonization observed during spring could be due to the sporadic rains that are characteristic of this season, and these may activate fungus colonization. The different host species could be separated into two groups based on the seasonality of radical colonization. These were hosts where there was greater colonization in autumn/winter (*A. pinnata* and *B. subaristata*), and hosts where there was greater colonization in summer/autumn

(*D. hieronymi*, *P. stuckertii*, *E. lugens* and *S. pellitum*). This fungal behaviour and seasonal host patterns would explain the seasonal changes in spore diversity and the little host specificity found in this mountain grassland.

In a 2-year field study, Escudero and Mendoza (2005) analysed seasonal variation in population attributes of AM fungi at four temperate grassland sites in the Argentinean flooding pampas. The sites varied in their soil conditions, vegetation compositions and the period over which the soils were subjected to flooding. Spore density was highest during summer (dry season) and lowest in winter (wet season), with intermediate values in autumn and spring. *Glomus fasciculatum* and *G. intraradices* dominated the spore communities at all four sites, and these species were relatively more abundant in summer than in winter, when flooding is most frequent (Escudero and Mendoza, 2005). There were distinctive seasonal peaks for both fungal species compared with the *Glomus* spp. and *Acaulospora* spp. distribution patterns, suggesting differences among AM fungi species with respect to the seasonality of sporulation (see Fig. 2.1). Spore density and AM root colonization were poorly correlated when measured at any one time. However, spore density was significantly correlated with the root colonization

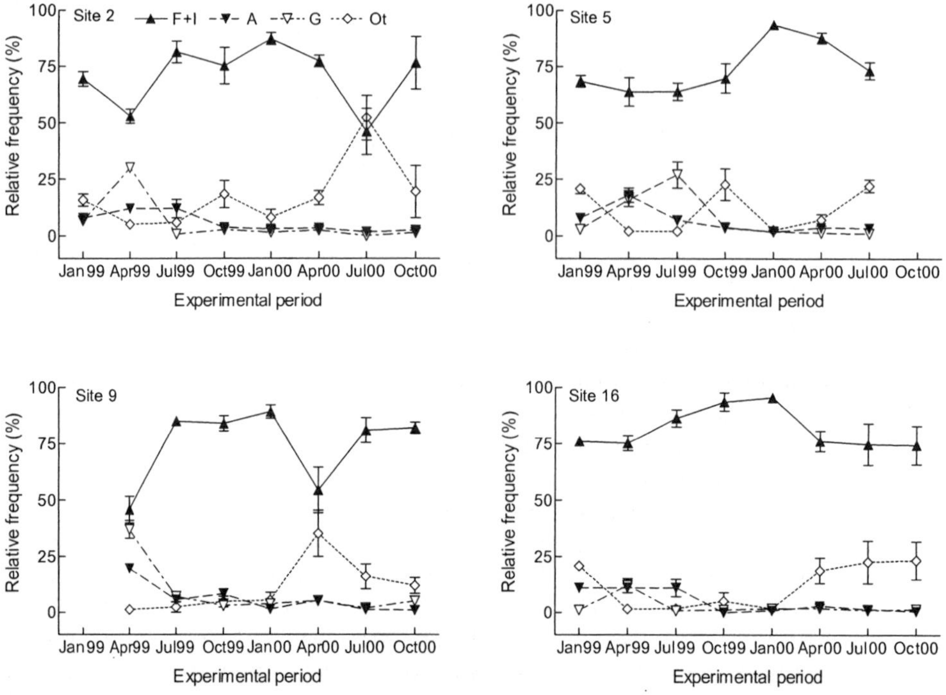

Fig. 2.1. Seasonal changes in the relative spore density of *Glomus fasciculatum* and *G. intraradices* (F+I), *Acaulospora* spp. (A), *Glomus* spp. (G) and other spores (Ot) at the four sites studied. Bars represent the standard error of the mean (from Escudero and Mendoza 2005).

3 months previously, suggesting that high colonization in one season precedes high sporulation in the following.

In these grasslands, some soil factors (pH, Na and P) were closely associated with total spore density, while the relative frequency of *Lotus tenuis* plants and mean air temperature were much more associated with changes in the relative frequency of spore density of AM fungi species (Escudero and Mendoza, 2005). The conclusion made from this study was that the total spore counts for each soil were the result of many soil variables and plant community influences on the AM fungal community, and were not due to the influence of one specific dominant plant species or soil property.

García and Mendoza (2008) recorded temporal changes of AM fungal measures, soil properties and N and P levels in plant tissue along a topographic slope of a temperate grassland in an attempt to explain whether seasonal changes in plant nutrients were associated with AM fungal variables and/or soil characteristics. Soils and plant samples (*L. tenuis, Paspalum vaginatum, Stenotaphrum secundatum*) were collected seasonally across a topographic and saline gradient. The morphology of root colonization by AM showed similar patterns in all of the plants studied. Maximum arbuscular colonization (AC) occurred at the beginning of the growing season in late winter, with a minimum in late summer, whereas maximum vesicular colonization (VC) occurred in summer with a minimum in winter. This could indicate the preferential production of a morphological structure by the fungus in a specific season. The highest AC was associated with the highest N and P levels in plant tissue, suggesting a relationship between increases in the rate of nutrient transfer and symbiosis partners. Soil water content, salinity and sodicity were positively associated with AM root colonization and AC in *L. tenuis,* but negatively associated with AM root colonization and AC in the grasses.

Seasonality appeared to regulate both spore density and AM colonization, and these were independent of combinations of plant species and soil sites. In addition, plant roots of *L. tenuis* and *S. secundatum* were densely colonized by AM fungi in soil with high pH (9.2) and high exchangeable sodium percentages (65%). The authors suggested that AM fungi can survive and colonize plant roots, adapting to extreme saline–sodic soil conditions. The high levels of root colonization suggest that either the plants respond with slow root growth, the fungi colonize roots more completely or the interaction enabled considerable root colonization (García and Mendoza, 2008).

AM colonization and spore numbers in the rhizosphere of *Cyperus iria* and *C. rotundus,* growing in an Indian semi-arid tropical grassland, have been reported as seasonal, with different patterns recorded in subsequent years (Muthukumar and Udaiyan, 2002). This was the first report of AM seasonality in a mycorrhizal-independent host under field conditions. Mycorrhizal colonization of sedges in this study was atypical in that the roots lacked arbuscules, and the mycorrhizal association was characterized by intercellular hyphae and vesicles with occasional intra-radical spores. Changes in the proportion of the roots with these AM structures, the total colonization levels and spore numbers were found to be related to climatic and edaphic factors. However, the strength of climatic and soil factors influence on the AM appeared to vary with the sedge species. The seasonal patterns observed in this study could have been caused by either colonization by different fungal types and phenological (seasonal) variations among them or a seasonal response for a single fungus type.

Moisture is the primary limiting factor in desert ecosystems, and adaptation and function in desert organisms is primarily attributed to their ability to survive long dry periods and to respond rapidly and effectively to moisture inputs. One study of an arid grassland in Namibia has reported that spore abundance and AM root colonization were primarily affected by moisture availability, and that fungal growth and spore production following an isolated rain event were closely associated with moisture availability (Jacobson, 1997). A rapid and opportunistic growth response to moisture, the ability to produce spores in response to declining soil water content and a lack of plant symbiont specificity are characteristics that allow mycorrhizal fungal communities to function under hyper-arid conditions (Jacobson, 1997).

In arid environments, gradients of decreasing soil moisture availability can be correlated with reductions in the population of mycorrhizal plants (Brundrett, 1991). Nevertheless, minimal disturbance or low soil nutrient levels are correlated with a high proportion of plant species with mycorrhizal dependence. AM spore density in soil and the percentage of root length colonized for the most frequent grasses in dry grasslands sites of the Province of Tierra del Fuego (Argentina) have both been shown to be influenced by grazing, soil pH (4.7–5.2) and host plant (Mendoza *et al.*, 2002).

In this study the relationship between spore density and the percentage of root colonized was found to be specific for a particular soil–host combination. *Glomus fasciculatum* was the most abundant AM fungal species present at all the sites, and the total spore counts in soil in the rhizosphere were found to be positively correlated with N, P, Ca and K levels and acidity in soil. *Poa rigidifolia* (neutrophilous), *Festuca gracillima* (acid tolerant) and *Deschampsia flexuosa* (acidophilous) differed in their response to soil acidity. In all the sites higher AM root colonization levels were seen in *Poa*, and differences were larger at the most acidic sites, with *Deschampsia* being the least colonized species. As AM fungi have a role in nutrient uptake by plants in acid soils (Smith and Read, 1997), the results indicate that this may vary with the habitat tolerance of the host.

The soil organic matter content can affect the total soluble N present in soil and is known to be strongly correlated to plant phenology. Soil organic matter can have a positive effect on vesicular and arbuscular colonization, and to a lesser extent on spore density which is, in turn, negatively correlated with total AM and hyphal colonization. When soil conditions are suitable for spore germination, AM colonization increases and spore abundance decreases. Counts of spores and the extent of vesicular and arbuscular colonization are useful indicators for evaluation of changes in desert soil ecosystems. AM fungi have recently been recognized as being an important factor in determining plant species diversity in arid and temperate ecosystems.

Soil Water Deficit and Mycorrhizal Symbiosis

Soil water deficit affects the length and morphology of root colonization, spore production, number of entry points, spread of external mycelium in soil, species distribution and inoculum potential (Augé, 2001). Drought conditions appear to increase root colonization more often than they decrease it, and several studies have reported that spore germination can be increased, decreased or unaffected by soil drying, depending on the fungal species involved (Schellenbaum *et al.*, 1998; Augé, 2001). Under drought conditions the mycorrhizal spore population and the percentage of mycorrhizal colonization differ between different AM fungi.

In a study on the roots of the tropical fruit tree *Ziziphus mauritiana*, *Glomus fasciculatum* showed the most dependency (200%) and the highest values of root colonization (94.2%) and spore population under water stress when compared with *Gigaspora margarita*, *Glomus constrictum*, *G. mosseae*, *Sclerocystis rubiformis* and *Scutellospora calospora*. Therefore, the application of this AM endophyte could be of importance in cultivation of this fruit plant under arid environments. However, mycorrhizal colonization of maize roots has been reported to be not significantly affected by drought, and the percentage of root length colonized and the type of mycorrhizal structures formed (hyphae, arbuscules and vesicles) by *G. mosseae* on maize remained virtually unchanged after 12 days of moderate drought (Schellembaum *et al.*, 1998).

Drought conditions have been seen to enhance arbuscule formation and hyphal development of mixed *Glomus* spp. in *Capsicum annuum* roots, whilst reducing mycorrhizal colonization by *G. fasciculatum* during a 20-day water deficit cycle (Davies Jr. *et al.*, 2002). Arbuscule formation and vesicle development by the mixed *Glomus* spp. was nearly twice that of *G. fasciculatum* during drought, and this was the opposite of that seen under non-drought conditions. In contrast in *Citrus tangerine* roots inoculated with *Glomus versiforme*, water deficit was found significantly to decrease AM colonization, entry points, vesicles and arbuscules after 80 days (Wu and Xia, 2006). This report was in agreement with the findings of García *et al.* (2008), who reported that the symbiosis between an AM fungal community and *L. tenuis* plants was affected by a 35-day drought period, resulting in reduced total root length, root length colonized, arbuscular colonization and number of entry points per unit of colonized root length, although the drought treatment did not affect vesicle colonization (see Table 2.1). Spore density in the dried soil was similar to that in the control soil. It was suggested that, under drought conditions, the strategy of the AM fungal community consisted of reducing external hyphae but moderating the reduction in arbuscules, and investing more in maintaining a similar proportion of vesicles in roots and spores in soil.

Soil hyphal length and biomass are reported to vary with soil moisture, and reductions in the mycelial density during drought periods may be linked to both plant and soil factors (Allen *et al.*, 2005). The negative impact of drought on AM mycelia has two components: (i) the direct effects of soil

Table 2.1. Effect of soil waterlogging and drought on AM fungi variables and nodules in *Lotus tenuis* roots, and spore and hyphal densities in soil compared with field capacity (from García *et al.*, 2008).

		Stressed plants			
Variable	Control plants	Waterlogging	SI[1]	Drought	SI[1]
AM fungal variables					
MC[2] index	0.97 ± 0.01^{a}	0.95 ± 0.02^{a}	ns	0.92 ± 0.02^{a}	ns
AC[2] index	0.77 ± 0.01^{c}	0.35 ± 0.01^{a}	0.55	0.66 ± 0.03^{b}	0.14
VC[2] index	0.16 ± 0.01^{a}	0.57 ± 0.02^{b}	–2.56	0.11 ± 0.01^{a}	ns
HO[2] index	0.14 ± 0.02^{a}	0.14 ± 0.01^{a}	ns	0.21 ± 0.01^{b}	–0.50
mc[3] (m)	163.14 ± 4.34^{b}	116.22 ± 10.91^{a}	0.29	99.61 ± 8.98^{a}	0.39
ac[3] (m)	128.47 ± 2.40^{c}	42.74 ± 4.07^{a}	0.67	63.71 ± 1.28^{b}	0.50
vc[3] (m) *ln*[4]	26.31 ± 1.60^{c}	69.28 ± 6.37^{c}	–1.63	14.63 ± 3.35^{a}	0.44
ho[3] (m) *ln*	23.66 ± 2.94^{a}	16.58 ± 1.80^{a}	ns	28.60 ± 7.17^{a}	ns
Entry points (per mm root)	7.76 ± 0.81^{b}	5.30 ± 0.31^{a}	0.32	5.83 ± 0.46^{ab}	ns
Nodules (per g fresh root)	15.93 ± 1.99^{ab}	31.32 ± 5.56^{c}	–0.97	18.38 ± 1.47^{b}	ns
Spore density (per g soil)	128.68 ± 22.70^{c}	77.56 ± 2.09^{c}	0.40	112.69 ± 7.62^{bc}	ns
Spore viability (%)	1.26 ± 0.38^{a}	7.50 ± 0.57^{b}	–4.95	1.25 ± 0.12^{a}	ns
Hyphal density (m/g dry soil)	11.11 ± 0.99^{b}	7.82 ± 0.62^{a}	0.29	6.07 ± 0.89^{a}	0.45
Relative colonization rates					
RCR[5]mc (per day)	0.0385 ± 0.0007^{b}	0.0283 ± 0.0025^{a}		0.0239 ± 0.0025^{a}	
RCRac (*kw*[4])	0.0382 ± 0.0005^{c}	0.0063 ± 0.0026^{a}		0.0182 ± 0.0006^{b}	
RCRvc (*ln*)	0.0424 ± 0.0017^{b}	0.0698 ± 0.0025^{c}		0.023 ± 0.0063^{a}	
RCRho	0.0406 ± 0.0041^{a}	0.0307 ± 0.0032^{a}		0.0434 ± 0.0070^{a}	
RCRep	0.0059 ± 0.0012^{b}	-0.0101 ± 0.0016^{a}		-0.0075 ± 0.0022^{a}	
RCRsp	0.0072 ± 0.0006^{b}	-0.0073 ± 0.0007^{a}		0.0032 ± 0.0019^{b}	
RCRhy	0.0014 ± 0.0026^{b}	-0.0085 ± 0.0022^{ab}		-0.0167 ± 0.0042^{a}	

[1] Susceptibility index = 1 – (stressed plant/non-stressed plant).
[2] MC, total root length colonized by AM fungi; AC, VC and HO, fraction of root length colonized separately by arbuscules, vesicles and hyphae, respectively.
[3] Root length colonized by AM fungi (mc), arbuscules (ac), vesicles (vc), and hyphae only (ho).
[4] ln, log transformation data; kw, Kruskal Wallis non-parametric test.
[5] Relative colonization rate; ep, entry points; sp, spore density; hy, hyphal density.
ns, non-significant effect of treatment at a value of $P > 0.05$.
Means in each row followed by different letters are significantly different according to Single Classification ANOVA at a value of $P < 0.05$.

desiccation on hyphae (Juniper and Abbott, 1993); and (ii) indirect effects due to changes in net primary productivity and carbon allocation from the host plant to the AM fungi. Mycorrhizal hyphal length in soil in a natural grassland has been shown to decline markedly during the summer, indicating that there may be a rapid turnover of the AM-dominated hyphal carbon pool under extreme drought (Querejeta *et al.*, 2007).

The same host plant inoculated with different *Glomus* species can show different levels of mycorrhizal development under conditions of water deficit (Wu *et al.*, 2007). Greater numbers of intra-radical structures, such as arbuscules and vesicles, and entry points in the roots were seen in roots inoculated with *G. mosseae* and *G. versiforme* than with other *Glomus* species under drought stress conditions.

The effect of drought stress and microbial treatment on mycorrhizal and rhizobial symbiosis development was reviewed by Ruiz-Lozano *et al.* (2001), and they suggested that an alleviation of oxidative damage is strongly involved in AM protection against nodule senescence. In dual colonization experiments the colonization frequency (F) in soybean roots was similar during all the mycorrhizal treatments used, and this was not affected by the presence or absence of *Bradyrhizobium japonicum* or by drought stress (see Table 2.2).

Table 2.2. Effect of drought stress and microbial treatment on mycorrhizal and rhizobial symbiosis development (from Ruiz-Lozano *et al.*, 2001).

Treatment[1]	Water regime	F[2] (%)	M[2] (%)	a[2] (%)	A[2] (%)	NDW[2]
Ni	Well-watered	0^{b}	0^{c}	0^{c}	0^{c}	0^{d}
	Droughted	0^{b}	0^{c}	0^{c}	0^{c}	0^{d}
Br	Well-watered	0^{b}	0^{c}	0^{c}	0^{c}	71^{ab}
	Droughted	0^{b}	0^{c}	0^{c}	0^{c}	48^{c}
Gm	Well-watered	92.3^{a}	46.5^{b}	76.5^{b}	37.1^{b}	0^{d}
	Droughted	89.0^{a}	38.5^{b}	91.7^{a}	35.2^{b}	0^{d}
Gi	Well-watered	88.0^{a}	41.5^{b}	73.8^{b}	31.0^{b}	0^{d}
	Droughted	87.0^{a}	55.0^{ab}	93.3^{a}	52.0^{ab}	0^{d}
Gm + Br	Well-watered	96.0^{a}	61.5^{ab}	78.3^{b}	48.0^{ab}	80^{a}
	Droughted	94.5^{a}	48.5^{b}	93.0^{a}	46.0^{ab}	65^{b}
Gi + Br	Well-watered	96.0^{a}	76.5^{a}	89.3^{a}	68.0^{a}	63^{b}
	Droughted	86.0^{a}	52.0^{ab}	98.0^{a}	52.0^{ab}	36^{c}

[1] Ni, non-inoculated controls; Br, *Bradyrhizobium japonicum*; Gm, *Glomus mosseae*; Gi, *G. intraradices*; Gi + Br, *G. intraradices* + *Bradyrhizobium.* Plants were either well watered or drought stressed for 10 d.
[2] F, colonization frequency; M, colonization intensity; a, arbuscule richness in the mycorrhizal root fraction; A, arbuscule abundance in the whole root system; NDW, nodule dry weight (mg/plant).
Within each parameter, means followed by the same letter are not significantly different ($P < 0.05$), as determined by Duncan's Multiple Range test ($n = 4$).

Colonization intensity (M) and arbuscule abundance (A) were also not significantly affected by drought stress. Only well-watered plants colonized with both *Glomus intraradices* and *B. japonicum* gave higher M and A values than well-watered plants colonized by either single AM fungus. Roots colonized by *G. mosseae*, either alone or in dual combination with *B. japonicum*, produced more arbuscules as a consequence of drought. Under water stress conditions, plants colonized by *G. mosseae* had higher nodule dry weight than plants colonized only by *B. japonicum*. Each AM species showed differential influences in avoiding drought-induced nodule senescence (see Table 2.2).

Excess Soil Water and Mycorrhizal Symbiosis

Many flood-tolerant plants respond to the oxygen deficiency brought on by flooding through development of specialized tissue called aerenchyma. These facilitate internal oxygen transport between shoots and roots and can lead to oxygenation of the rhizosphere (Tornbjerg *et al.*, 1994). Miller and Bever (1999) identified mainly two mechanisms by which AM fungi could survive in anoxic conditions. First, certain AM fungi may require less oxygen than has previously been thought. Second, the fungus could be concentrated near the plant root, obtaining oxygen either directly from the root or as the oxygen diffuses from the root into the rhizosphere.

Flooding affects the length and morphology of root colonization, and this depends on the extent to which the association had already been established when flooding occurred. It has been shown that, when the fungi were established in the roots prior to the flooding, they were able to maintain and expand with the growing root system (Miller and Sharitz, 2000; Neto *et al.*, 2006). However, the plant's response to AM fungal colonization depends on the severity, periodicity and duration of a specific stress condition. Several studies have shown that the number of AM spores in soil were not affected by extended flooding, indicating that mycorrhizal formation is related to plant growth (Miller and Sharitz, 2000).

The effect of excess water on mycorrhizal associations cannot be generalized. Some studies have found an overall increase in mycorrhizal colonization intensity with increased water levels (Brown and Bledsoe, 1996), while others have reported an adverse effect on the AM fungi (Mendoza *et al.*, 2005; García *et al.*, 2008).

Excess water can reduce coils, arbuscules and vesicle formation in *Panicum hemitomon* roots, and in *Typha latifolia* roots colonization was virtually eliminated by flooding (Ipsilantis and Sylvia, 2007). Ray and Inouye (2006) observed a positive relationship between the duration of the non-flooded period and the levels of arbuscular and hyphal colonization in *T. latifolia* roots, while moderate to high levels of both forms of colonization were also observed during flooded periods (see Fig. 2.2). Furthermore, sustained flooding was accompanied by increasing levels of hyphal colonization in some instances, but overall the relationship was not significant. These

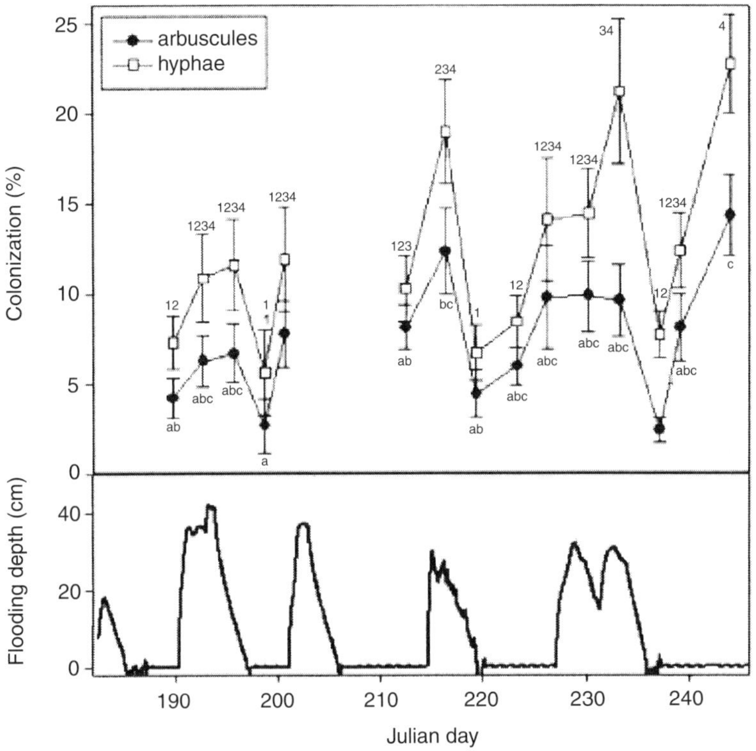

Fig. 2.2. Mean (± se) colonization levels (top) and water depth (bottom) plotted against Julian day. Points labelled with the same number (1, 2, 3, 4: hyphae) or letter (a, b, c: arbuscules) do not differ significantly (Tukey's *post hoc* analysis at $P < 0.05$). Water depth was recorded at 10-min intervals (from Ray and Inouye, 2006).

findings were in agreement with those made in a previous study by Miller and Sharitz (2000). *Typha latifolia* has been shown to be extremely effective in its ability to supply oxygen to the rhizosphere (Tornbjerg *et al.*, 1994), and this may explain why AM colonization was detected during both the flooded and non-flooded periods (Ray and Inouye, 2006).

Beck-Nielsen and Madsen (2001) found no mycorrhizal colonization in submerged plants, and very low colonization levels when the same species were not submerged. The potential for mycorrhizal colonization could increase due to a temporarily oxygenated rhizosphere and extensive aerenchyma tissue (Nielsen *et al.*, 2004). It has been suggested that AM morphology may partially depend on air spaces in the cortex, with arbuscules developing when large spaces are available and with coils developing when the air spaces are limited. However, no relationship between AM colonization and aerenchyma has been demonstrated, although Cornwell *et al.* (2001) reported higher colonization in plants with less aerenchyma. Conversely, Nielsen *et al.* (2004) presumed that an extensive aerenchyma promoted the development of AM fungi in aquatic plants.

Another example of the negative effects of soil waterlogging on AM symbiosis is the change in the morphology of colonization seen in waterlogged *L. tenuis* plants, where the number of arbuscules decreased and the number of vesicles increased in colonized roots of waterlogged plants (see Table 2.1; Mendoza *et al.*, 2005; García *et al.*, 2008). The arbuscule:vesicle ratio was 120 times lower in the roots of the waterlogged plants, suggesting a strong negative effect on the symbiotic efficiency in terms of nutrient uptake. A loss of arbuscules in the roots of colonized waterlogged plants may represent a benefit to the fungus and a C cost to the plant.

Waterlogged soils can contain a higher spore density at the end of the experiment in comparison with non-waterlogged soil. One field study has found spore counts of 239 and 195 spores/g of dry soil in two consecutive years (early autumn) (Escudero and Mendoza, 2005). These values were fairly similar to the spore density figures obtained in waterlogged and control soils at the end of the study (Mendoza *et al.*, 2005). The authors proposed that differences in spore density between treatments may be due to differences in spore germination and their loss, rather than being due to sporulation. As a result of the adverse soil condition, spores in waterlogged soil did not germinate. The higher number of entry points per mm of colonized root seen in roots of non-waterlogged plants may support this explanation (García *et al.*, 2008). Other studies have suggested that salt marsh soils contain high numbers of AM fungi spores, and that the reduction in spore germination at water levels above field capacity may be related to the low tolerance of AM fungi to hypoxic conditions (Carvalho *et al.*, 2004).

Flooding can also restrict the extension of the extra-radical mycelium, but a considerable hyphal network can remain and this may help maintain the mycorrhizal colonization (García *et al.*, 2008). Beck-Nielsen and Madsen (2001) found that hyphal length in AM fungi in wetlands was reduced under the low redox state that resulted from inundation. Extra-radical hyphae of the AM fungal assemblages were restricted to 2.5 cm by flooding, although there were differences among the assemblages and the largest hyphal extension recorded was 16.5 cm (Ipsilantis and Sylvia, 2007).

Effect of Soil Salinity on Mycorrhizal Symbiosis

Salinity is reported to affect the formation and function of mycorrhizal symbiosis (Abbott and Robson, 1991; Juniper and Abbott, 1993, 2006; McMillen *et al.*, 1998). Some research has shown that salinity may reduce mycorrhizal colonization by inhibiting spore germination and hyphal growth in soil, which in turn will inhibit the extension and spread of the hyphae after the initial infection (McMillen *et al.*, 1998). Salinity may also change the morphology of the colonization by reducing the number of arbuscules formed (Pfeiffer and Bloss, 1988). Decreased mycorrhizal formation in saline soils may be due to the effects of soil salinity on the fungus or the plant singly, or on both of the symbiotic partners. In contrast, some studies have reported that colonization was not reduced by salinity (Hartmond *et al.*, 1987).

Delay or prevention of all or any of the phases of spore germination by dissolved salts in the soil could delay or prevent hyphal growth and therefore could affect the colonization of plant roots and the establishment of the symbiosis (Juniper and Abbott, 1993). Spore germination of some AM fungi is inhibited by increasing concentrations of NaCl, and can range from no germination at all to maximum germination in 300 mM NaCl (Juniper and Abbott, 2006).

The capacity for hyphae of *Gigaspora decipiens* to grow through soil from germinated spores and colonize plant roots was reduced in soil containing increased concentrations of NaCl (McMillen *et al.*, 1998), a finding that has also been reported for other AM fungi (Juniper and Abbott, 1993). Ruiz-Lozano and Azcón (2000) showed that hyphal growth and/or viability of an isolate of *Glomus* sp. were more affected by soil salinity levels than were those properties of *G. deserticola*, and that the mycelial network generated by *G. deserticola* was more infective than that of *Glomus* sp. Hyphal growth in saline soil is dependent on the maintenance of the ionic balance in the mycelium and an internal water potential sufficient to maintain turgor, and these are both processes that require energy (Cook and Whipps, 1993). McMillen *et al.* (1998) suggested that the capacity for hyphal growth was reduced more rapidly in soil with added NaCl than in soil without added NaCl, due to the increased energy requirement for hyphal growth through saline soil.

Mycorrhizal colonization may be reduced by NaCl early after inoculation, but not at later observation times (Juniper and Abbott, 1993; McMillen *et al.*, 1998). Increasing concentrations of NaCl can also inhibit the spread of colonization after the initial infection, and this is probably due to NaCl in the soil inhibiting hyphal growth and possibly also affecting the supply of carbohydrates from the plant to the fungus (McMillen *et al.*, 1998). Tian *et al.* (2004) showed that, although mycorrhizal colonization was reduced with increasing NaCl levels, the dependency of cotton plants on AM fungi was increased. It appears that the symbiotic relationship between the AM fungi and the cotton plants was strengthened in a saline environment once the association had been established. The authors suggested that this may be an indication of the ecological importance of AM associations for plant growth and survival under saline stress. In *L. tenuis*, salinity reduced mycorrhization in a sensitive genotype but not in a tolerant one, suggesting that the latter plant genotype offered better protection to the fungal partner and better chances of growth within host tissues than were available from the salt-sensitive plants (Sannazzaro *et al.*, 2006).

Glomus clarum and *G. etunicatum* have been shown to give a higher colonization of banana roots than *Acaulospora scrobiculata* in salt-stressed soil, except at a soil conductivity of 7.39 dS/m (see Fig. 2.3). The increase of soil electrical conductivity (EC) promoted colonization by *A. scrobiculata* but, for *G. clarum* and *G. etunicatum*, colonization reached its maximum at EC values of 4 and 5 dS/m, respectively (Yano-Melo *et al.*, 2003). There was a high correlation between root colonization and spore density in the soil, with sporulation inhibited only at > 5 dS/m. *Glomus clarum* and *G. etunicatum* were

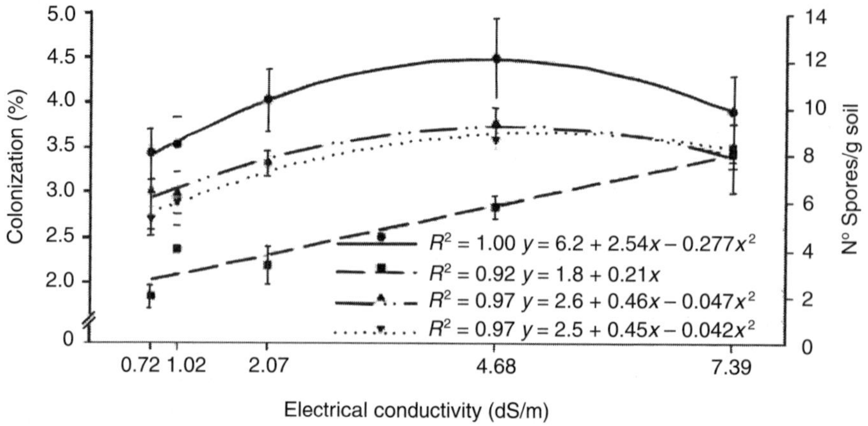

Fig. 2.3. Effect of increasing salt levels on root colonization of banana plantlets (cv. Pacovan) by *A. scrobiculata* (■), *Glomus clarum* (▲) and *G. etunicatum* (▼), and spore density (•). Data transformed to $\sqrt{x + 2.5}$ (spores) and $\ln(x + 2.5)$ (colonization); mean ± se ($n = 4$); $p \leq 0.05$ (from Yano-Melo *et al.*, 2003)

able to support an excess of NaCl and may be useful for soil recovery in saline soils. Increased AM fungal sporulation and colonization under salt-stress conditions has also been reported by Aliasgharzadeh *et al.* (2001), whereas Juniper and Abbott (1993) reported that a higher salt content of the soil reduced the number of spores.

Nutrient Acquisition by Mycorrhizal Plants under Stress

Water excesses or deficits and salinity cause changes in the physicochemical characteristics of the soil that can affect plant nutrient uptake and plant growth, and these in turn can also affect AM fungi and plant symbiosis. The decline of moisture in soil results in a decrease in the diffusion rate of nutrients (particularly P) from the soil matrix to the absorbing root surface; conversely, AM association enhances P uptake by plant roots under either drought (Ruiz-Lozano *et al.*, 1995) or non-drought (Hetrick *et al.*, 1996) conditions.

Improving host P nutrition by AM fungi during a stress period has been suggested as a primary mechanism for enhancing plant drought tolerance (Ruiz-Lozano *et al.*, 1995). Stress tolerance and P nutrition are both linked to the capacity of AM plants to maintain higher (less negative) leaf water potentials, despite a more negative soil water potential. Koide (1993) suggested that increased stomatal conductance and transpiration in AM plants may be due to P-mediated improvement in photosynthetic capacity. Mycorrhizal symbiosis can also assist the host plant in utilizing forms of N that are unavailable to non-mycorrhizal plants (Tobar *et al.*, 1994). The external mycelium may be able to transport nearly 40% of added N under a moderate drought condition (Subramanian and Charest, 1999). The capacity of AM fungi to increase plant tolerance to drought stress may be due to both improved nutrient uptake and an increase in N assimilation through nitrate reductase activity. Nitrate reductase activity decreases in plants exposed to water limitation because of a lower flux of nitrate from the soil to the root (Azcón *et al.*, 1996), but this reduction is much less in mycorrhizal plants than in those without mycorrhizae (Caravaca *et al.*, 2005).

In a 2-year field study, *Glomus intraradices* inoculation was found to improve the drought tolerance of tomato plants as a secondary effect of the enhanced nutritional status of the host, especially in relation to P and N (Subramanian *et al.*, 2006). AM colonization significantly increased the P contents of the tomato plants, and roots had 30–40% higher P levels than those of non-AM roots under varying intensities of drought stress (severe, moderate and mild drought). Furthermore, higher N contents were seen in both roots and shoots under various levels of water deficit, and these responses were not evident in the non-AM-treated plants.

In addition to an enhanced uptake of P and N in watermelon plants, mycorrhizal colonization of water-stressed plants was found to restore macro-(K, Ca and Mg) and micro-(Zn, Fe and Mn) nutrient concentrations in leaves, in most cases to levels comparable to those found in well-watered plants (Kaya *et al.*, 2003). These findings are similar to those reported by Al-Karaki and Clark (1998) in wheat plants, and by Wu and Xia (2006) in *Citrus tangerine* plants. The improved biomass production of mycorrhizal plants compared with non-mycorrhizal water-stressed plants has been related to the increased P and K uptake due to the fungi. Potassium plays a key role in plant water stress and has been found to be the cationic solute responsible for stomatal movement in response to changes in bulk leaf water status (Ruiz-Lozano *et al.*, 1995).

Soil salinity significantly reduces the absorption of mineral nutrients by plants, and this is particularly the case for P as phosphate ions precipitate with Ca^{2+} ions in salt-stressed soil and become unavailable to plants. AM fungi have been shown to have a positive influence on the composition of mineral nutrients of plants grown in salt-stressed conditions, particularly for nutrients with poor mobility such as P (Al-Karaki, 2006). Mycorrhizal plants have been reported to have higher shoot and root P contents than non-mycorrhizal plants, regardless of salinity level (Giri and Mukerji, 2004; Giri *et al.*, 2007), and the P concentration of AM plants has been shown to decrease with increasing NaCl levels (Tian *et al.*, 2004; Giri *et al.*, 2007).

AM plants can also have a significantly greater concentration of N than non-mycorrhizal plants (Giri and Mukerji, 2004). Lettuce plants colonized by *Glomus* sp. have been shown to accumulate N and P more efficiently at higher salt levels than at lower salt levels (Ruiz-Lozano and Azcón, 2000). Conversely, N and P accumulation under increasing salinity in mycorrhizal lettuce plants colonized by *G. deserticola* was not significantly different from that of non-mycorrhizal control plants (Ruiz-Lozano and Azcón, 2000). The difference in the effectiveness of *G. deserticola* and *Glomus* sp. on N and P uptake, and its translocation to shoots, was clearest at the two lower salt levels (1.1 and 1.4

dS/m), where *G. deserticola* gave the highest uptake (see Fig. 2.4). Higher Cu, Fe and Zn contents under salt stress were also noted in AM plants compared with non-AM plants (Al-Karaki, 2006; Giri *et al.*, 2007). The higher mineral nutrient acquisition in AM compared with non-AM plants may be due to either increased availability or increased transport (absorption and/or translocation) by AM fungal hyphae under saline conditions.

Increased N concentration under salt stress may help to decrease Na uptake, and this may be indirectly related to maintaining the chlorophyll content of the plant. Chlorophyll content was found to decline drastically under salt stress in leaves of non-mycorrhizal as compared with mycorrhizal *Acacia auriculiformis*, resulting in leaf chlorosis that was not seen in mycorrhizal plants (Giri *et al.*, 2003). Although there may be several reasons for low chlorophyll contents in plants growing under salinity stress, the antagonistic effect of Na on Mg uptake is well established (Alam, 1994). AM fungi have shown greater absorption of Mg, and this may suppress the antagonistic effect of Na on chlorophyll synthesis (Giri and Mukerji, 2004).

The detrimental effect of Na is a result of its ability to compete with K for binding sites essential for various cellular functions. Potassium activates a range of enzymes, and Na cannot substitute in this role. Thus, high levels of Na or high Na/K ratios can disrupt various enzymatic processes in the cytoplasm. Mycorrhizal *Acacia* and *Sesbania* plants have been found to have a higher concentration of K in root and shoot tissues at all salinity levels (Giri and Mukerji, 2004; Giri *et al.*, 2007). Similar results were reported by Sannazzaro *et al.* (2006) in *L. tenuis* plants colonized by *G. intraradices*. It seems that higher K accumulation by AM plants under salt-stress conditions

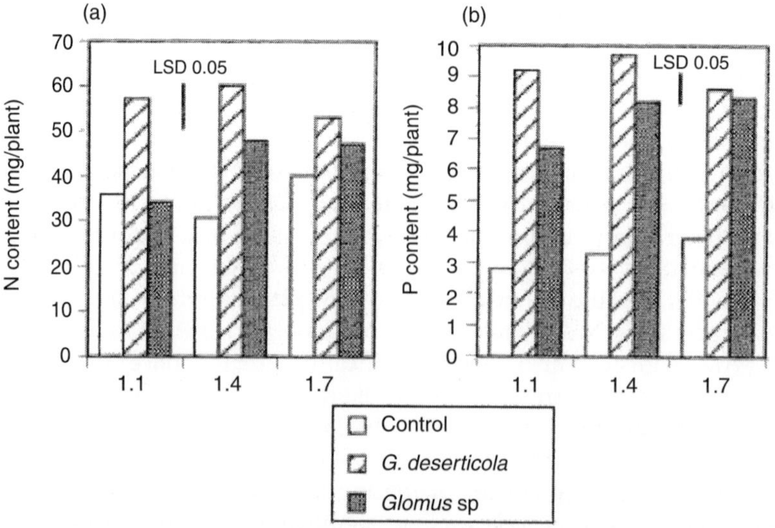

Fig. 2.4. Shoot N (a) and P (b) contents (mg/plant) in non-mycorrhizal (control) and mycorrhizal (*Glomus* sp. and *G. deserticola*) lettuce plants grown at three levels of salinity (1.1, 1.4 and 1.7 dS/m) (from Ruiz-Lozano and Azcón, 2000).

may help in maintaining a high K/Na ratio, thus preventing the disruption of various enzymatic processes and inhibition of protein synthesis (Giri *et al.*, 2007).

Waterlogged soil results in an increased uptake of P and N in mycorrhizal plants (Mendoza *et al.*, 2005; Fougnies *et al.*, 2007). Mendoza *et al.* (2005) reported that in *L. tenuis* plants the effect of excess water was more marked for P than for N. As waterlogging can decrease the biomass of roots more than that of shoots, waterlogged plants may compensate by absorbing much more P or N per unit of root biomass. The increase in shoot/root ratio shows a higher relative demand by the shoot for nutrients in waterlogged plants. In addition, waterlogging can increase the availability of P and N, and this may lead to an increased supply of nutrients per unit of root biomass for waterlogged plants. The roots of waterlogged plants have been reported to take up to 2.57 times more P and 1.44 times more N per unit of root biomass than non-waterlogged plants (Mendoza *et al.*, 2005).

Neto *et al.* (2006) found that flooding adversely affected N uptake by *Aster tripolium* plants. Nitrogen uptake was higher in AM plants than in non-mycorrhizal plants only under tidal flooding conditions. The reduced N acquisition caused by flooding is certainly related to the lower energy status of flooded roots and the energy costs of N uptake and assimilation. In addition, the altered N uptake may be a result of reduced transpiration, as reduced xylem flow is known to repress N uptake in herbaceous plants and trees (Geßler *et al.*, 2003). García *et al.* (2008) reported that in waterlogged *L. tenuis* plants the total N uptake per unit of total biomass was reduced and the proportion allocated to the roots remained constant. This suggests that N availability in soil (or as fixed atmospheric N) is insufficient to sustain growth demands in waterlogged plants, particularly those created by shoots. The total N absorbed per root nodule in waterlogged plants, and the clustering of a substantial proportion of the nodules at the base of the main root near the soil surface, suggest that most of the nodules are inefficient in fixing atmospheric N_2, which is consistent with the conclusion that N may not be sufficient for plant growth (García *et al.*, 2008).

Flooding has been reported to reduce slightly the total plant C content in *A. tripolium*, particularly due to reductions in shoots (Neto *et al.*, 2006). Mycorrhizal colonization increases the C sink strength in roots due to the net movement of C from the root into the fungus (Redman *et al.*, 2001). Therefore, flooding and mycorrhiza increase C demand in the roots, and this generally has a positive effect on C fixation by the shoots (Jackson, 2002).

Conclusions and Perspectives

Further studies on the dynamics of AM fungal colonization and plant nutrient uptake in stressed natural environments are required. There is a variety of information available on the effects of changes in soil fertility, salinity and moisture content on AM fungi and plant symbiosis in controlled conditions, but more field research is needed in order to understand more about the ecological implication of the symbiotic associations that help plants cope with stressed environments. Non-mycorrhizal plants are anomalous in nature, and so instead of comparing AM plants with non-AM plants, it may be more productive to examine the behaviour of colonized plants under different soil conditions, and to relate this to plant growth variables. Although it is known that grassland plants can be colonized by AM fungi over a wide range of soil conditions, it is difficult to relate changes in nutrient uptake or plant growth to changes in AM root colonization. When soil conditions change, plant communities may also change, and it can be difficult to justify investigating cause and effect relationships between nutrient uptake and AM fungal variables as a result of changes in a specific soil property. However, environmental factors seem mainly to influence the extent of AM colonization, while the form of the AM colonization seems to be mainly influenced by the plant species and plant phenological event.

There can be seasonal effects on the symbiosis between the plant and the AM fungi in a variety of soil environments. Nevertheless, changes in soil fertility, salinity and/or moisture content can partially inhibit AM fungal colonization, and so the symbiosis may be functionally adapted to a wide range of soils. It may be that the primary flood tolerance strategy for AM fungi would be to develop mainly intra-radical structures, and that the response to drought would be the development of extra-radical mycelium and propagules in soil.

References

Abbott, L. and Robson, A. (1991) Factors influencing the occurrence of vesicular arbuscular mycorrhizas. *Agriculture, Ecosystems and Environment* 35, 121–150.

Alam, S.M. (1994) Nutrient uptake by plants under stress condition. In: Pessakakli, M. (ed.) *Handbook of Plant Stress*. Dekker, New York, pp. 227–246.

Aliasgharzadeh, N., Saleh Rastin, N., Towfighi, H. and Alizadeh, A. (2001) Occurrence of arbuscular mycorrhizal fungi in saline soils of the Tabriz Plain of Iran in relation to some physical and chemical properties of soil. *Mycorrhiza* 11, 119–122.

Al-Karaki, G.N. (2006) Nursery inoculation of tomato with arbuscular mycorrhizal fungi and subsequent performance under irrigation with saline water. *Scientia Horticulturae* 109, 1–7.

Al-Karaki, G.N. and Clark, R.B. (1998) Growth, mineral acquisition, and water use by mycorrhizal wheat grown under water stress. *Journal of Plant Nutrition* 21, 263–276.

Allen, M.F., Klironomos, J.N., Treseder, K.K. and Oechel, W.C. (2005) Responses of soil biota to elevated CO_2 in a chaparral ecosystem. *Ecological Applications* 15, 1710–1711.

Augé, R.M. (2001) Water relations, drought and vesicular–arbuscular mycorrhizal symbiosis. *Mycorrhiza* 11, 3–42.

Azcón, R., Gomez, M. and Tobar, R. (1996) Physiological and nutritional responses by *Lactuca sativa* L. to nitrogen sources and mycorrhizal fungi under drought conditions. *Biology and Fertility of Soils* 22, 156–161.

Beck-Nielsen, D. and Madsen, T.V. (2001) Occurrence of vesicular–arbuscular mycorrhiza in aquatic macrophytes from lakes and streams. *Aquatic Botany* 71, 141–148.

Bohrer, K.E., Friese, C.F. and Amon, J.P. (2004) Seasonal dynamics of arbuscular mycorrhizal fungi in differing wetland habitats. *Mycorrhiza* 14, 329–337.

Brown, A.M. and Bledsoe, C. (1996) Spatial and temporal dynamics of mycorrhizas in *Jaumea carnosa*, a tidal saltmarsh halophyte. *Journal of Ecology* 84, 703–715.

Brundrett, M.C (1991) Mycorrhizas in natural ecosystems. *Advances in Ecological Research* 21, 171–313.

Caravaca, F., Alguacil, M.M., Hernández, J.A. and Rodán, A. (2005) Involvement of antioxidant enzyme and nitrate reductase activities during water stress and recovery of mycorrhizal *Myrtus communis* and *Phillyrea angustifolia* plants. *Plant Science* 169, 191–197.

Carvalho, L.M., Correia, P.M. and Martins-Loução, M.A. (2004) Arbuscular mycorrhizal fungal propagules in a salt marsh. *Mycorrhiza* 14, 165–170.

Cook, R.C. and Whipps, J.M. (1993) *Ecophysiology of Fungi*. Blackwell Scientific Publications, Oxford, UK.

Cornwell, W.K., Bedford, B.L. and Chapin, C.T. (2001) Occurrence of arbuscular mycorrhizal fungi in a phosphorus-poor wetland and mycorrhizal response to phosphorus fertilization. *American Journal of Botany* 88, 1824–1829.

Davies Jr., F.T., Olalde-Portugal, V., Aguilera-Gomez, L., Alvarado, M.J., Ferrera-Cerrato, R.C. and Boutton, T.W. (2002) Alleviation of drought stress of Chile ancho pepper (*Capsicum annuum* L. cv. San Luis) with arbuscular mycorrhiza indigenous to Mexico. *Scientia Horticulturae* 92, 347–359.

Escudero, V.G. and Mendoza, R.E. (2005) Seasonal variation of arbuscular mycorrhizal fungi in temperate grasslands along a wide hydrologic gradient. *Mycorrhiza* 15, 291–299.

Fougnies, L., Renciot, S., Muller, F., Plenchette, C., Prin, Y., de Faria, S.M., Bouvet, J.M., Sylla, S., Dreyfus, B. and Bâ, A.M. (2007) Arbuscular mycorrhizal colonization and nodulation improve flooding tolerance in *Pterocarpus officinalis* Jacq. seedlings. *Mycorrhiza* 17, 159–166.

García, I. and Mendoza, R. (2008) Relationships among soil properties, plant nutrition and arbuscular mycorrhizal fungi-plant symbioses in a temperate grassland along hydrologic, saline and sodic gradients. *FEMS Microbiology Ecology* 63, 359–371.

García, I., Mendoza, R. and Pomar, M.C. (2008) Deficit and excess of soil water impact on plant growth of *Lotus tenius* by affecting nutrient uptake and arbuscular mycorrhizal symbiosis. *Plant and Soil* 304, 117–131.

Geßler, A., Weber, P., Schneider, S. and Rennenberg, H. (2003) Bidirectional exchange of amino compounds between phloem and xylem during long-distance transport in Norway spruce trees (*Picea abies* (L.) Karst). *Journal of Experimental Botany* 54, 1389–1397.

Giri, B. and Mukerji, K.G. (2004) Mycorrhizal inoculat alleviates salt stress in *Sesbania aegyptiaca* and *Sesbania grandiflora* under field conditions: evidence for reduced sodium and improved magnesium uptake. *Mycorrhiza* 14, 307–312.

Giri, B., Kapoor, R. and Mukerji, K.G. (2003) Influence of arbuscular mycorrhizal fungi and salinty on growth, biomass, and mineral nutrition of *Acacia auriculiformis*. *Biology and Fertility of Soils* 38, 170–175.

Giri, B., Kapoor, R. and Mukerji, K.G. (2007) Improved tolerance of *Acacia nilotica* to salt stress by arbuscular mycorrhiza, *Glomus fasciculatum* may be partly related to elevated K/Na ratios in root and shoot tissues. *Microbial Ecology* 54, 753–760.

Hartmond, U., Schaesberg, N.V., Graham, J.H. and Syvertsen, J.P. (1987) Salinity and flooding stress effects on mycorrhizal and non-mycorrhizal citrus rootstock seedlings. *Plant and Soil* 104, 37–43.

Hetrick, B.A.D., Wilson, G.W.T. and Todd, T.C. (1996) Mycorrhizal response in wheat cultivars: relationship to phosphorus. *Canadian Journal of Botany* 74, 19–25.

Ipsilantis, I. and Sylvia, D.M. (2007) Interactions of assemblages of mycorrhizal fungi with two Florida Wetland plants. *Applied Soil Ecology* 35, 261–271.

Jackson, M.B. (2002) Long-distance signalling from roots to shoots assessed: the flooding story. *Journal of Experimental Botany* 53, 175–181.

Jacobson, K.M. (1997) Moisture and substrate stability determine VA-mycorrhizal fungal community distribution and structure in an arid grassland. *Journal of Arid Environments* 35, 59–75.

Juniper, S. and Abbott, L. (1993) Vesicular–arbuscular mycorrhizas and soil salinity. *Mycorrhiza* 4, 45–57.

Juniper, S. and Abbott, L. (2006) Soil salinity delays germination and limits growth of hyphae from propagules of arbuscular mycorrhizal fungi. *Mycorrhiza* 16, 371–379.

Kaya, C., Higgs, D., Kirnak, H. and Tas, I. (2003) Mycorrhizal colonisation improves fruit yield and water use efficiency in watermelon (*Citrullus lanatus* Thunb.) grown under well-watered and water-stressed conditions. *Plant and Soil* 253, 287–292.

Koide, R.T. (1993) Physiology of the mycorrhizal plants. In: *Advances in Plant Pathology Vol. 9*. Academic Press, London, pp. 33–54.

Lugo, M.A. and Cabello, M.N. (2002) Native arbuscular mycorrhizal fungi (AMF) from mountain grassland (Córdoba, Argentina) I: Seasonal variation of fungal spore diversity. *Mycologia* 94, 579–586.

McMillen, B.G., Juniper, S. and Abbott. L.K. (1998) Inhibition of hyphal growth of a vesicular–arbuscular mycorrhizal fungus in soil containing sodium chloride limits the spread of infection from spores. *Soil Biology and Biochemistry* 30, 1639–1646.

Mendoza, R., Goldmann, V., Rivas, J., Escudero, V., Pagani, E., Collantes, M. and Marbán, L. (2002) Arbuscular mycorrhizal fungi populations in relationship with soil properties and host plant in grasslands of Tierra del Fuego. *Ecologia Austral* 12, 105–116.

Mendoza, R., Escudero, V. and García, I. (2005) Plant growth, nutrient acquisition and mycorrhizal symbioses of a waterlogging tolerant legume (*Lotus glaber* Mill.) in a saline–sodic soil. *Plant and Soil* 275, 305–315.

Miller, S.P. and Bever, J. (1999) Distribution of arbuscular mycorrhizal fungi in stands of the wetland grass *Panicum hemitomon* along a wide hydrologic gradient. *Oecologia* 119, 586–592.

Miller, S.P. and Sharitz, R.R. (2000) Manipulation of flooding and arbuscular mycorrhiza formation influences growth and nutrition of two semiaquatic grass species. *Functional Ecology* 14, 738–748.

Mullen, R.B. and Schmidt, S.K. (1993) Mycorrhizal infection, phosphorus uptake and phenology in *Ranunculus adoneus*: implications for the functioning of mycorrhizae in alpine systems. *Oecologia* 94, 229–234.

Muthukumar, T. and Udaiyan, K. (2002) Seasonality of vesicular–arbuscular mycorrhizae in sedges in a semi-arid tropical grassland. *Acta Oecologica* 23, 337–347.

Neto, D., Carvalho, L.M., Cruz, C. and Matins-Loução, M.A. (2006) How do mycorrhizas affect C and N relationships in flooded *Aster tripolium* plants? *Plant and Soil* 279, 51–63.

Nielsen, K.B., Kjøller, R., Olsson, P.A., Schweiger, P.F., Andersen, F.O. and Rodendahl, S. (2004) Colonization and molecular diversity of arbusuclar mycorrhizal fungi in the aquatic plants *Littorella uniflora* and *Lobelia dortmana* in southern Sweden. *Mycological Research* 108, 616–625.

Pfeiffer, C.M. and Bloss, H.E. (1988) Growth and nutrition of Guayule (*Parthenium argentatum*) in a saline soil as influenced by vesicular–arbuscular mycorrhiza and phosphorus fertilization. *New Phytologist* 108, 315–321.

Pringle, A. and Bever, J.D. (2002) Divergent phenologies may facilitate the coexistence of arbuscular mycorrhizal fungi in a North Carolina grassland. *American Journal of Botany* 89, 1439–1446.

Querejeta, J.E., Egerton-Warburton, L.M. and Allen, M.F. (2007) Hydraulic lift may buffer rhizosphere hyphae against the negative effects of severe soil drying in a California Oak savanna. *Soil Biology and Biochemistry* 39, 409–417.

Ray, A.M. and Inouye, R.S. (2006) Effects of water-level fluctuations on the arbuscular mycorrhizal colonization of *Typha latifolia* L. *Aquatic Botany* 84, 210–216.

Redman, R.S., Dunigan, D.D. and Rodriguez, R.J. (2001) Fungal symbiosis from mutualism to parasitism: who control the outcome, host or invader? *New Phytologist* 151, 705–716.

Rilling, M.C. (2004) Arbuscular mycorrhizae and terrestrial ecosystem processes. *Ecology Letters* 7, 740–754.

Ruiz-Lozano, J.M. and Azcón, R. (2000) Symbiotic efficiency and infectivity of an autochthonous arbuscular mycorrhizal *Glomus* sp. from saline soils and *Glomus deserticola* under salinity. *Mycorrhiza* 10, 137–143.

Ruiz-Lozano, J.M., Azcón, R. and Gomez, M. (1996) Alleviation of salt stress by arbuscular-–mycorrhizal *Glomus* species in *Lactuca sativa* plants. *Physiologia Plantarum* 98, 767–772.

Ruiz-Lozano, J.M., Gomez, M. and Azcón, R. (1995) Influence of different *Glomus* species on the time-course of physiological plant responses of lettuce to progressive drought stress periods. *Plant Science* 110, 37–44.

Ruiz-Lozano, J.M., Collados, C., Barea, J.M. and Azcón, R. (2001) Arbuscular mycorrhizal symbiosis can alleviate drought-induced nodule senescence in soybean plants. *New Phytologist* 151, 493–502.

Sannazzaro, A.I., Ruiz, O.A., Albertó, E.O. and Menéndez, A.B. (2006) Alleviation of salt stress in *Lotus glaber* by *Glomus intraradices*. *Plant and Soil* 285, 279–287.

Schellenbaum, L., Müller, J., Boller, T., Wiemken, A. and Schüepp (1998) Effects of drought on non-mycorrhizal and mycorrhizal maize: changes in the pools of non-structural carbohydrates in the activities of invertase and trehalase, and in the pools of amino acids and imino acids. *New Phytologist* 138, 59–66.

Schüßler, A., Schwarzott, D. and Walker, C. (2001) A new fungal phylum, the Glomeromycota: phylogeny and evolution. *Mycological Research* 105, 1413–1421.

Smith, S.E. and Read, D.J. (1997) *Mycorrhizal Symbiosis*. 2nd edn., Academic Press Inc., Cambridge, Massachusetts, 605 pp.

Subramanian, K.S. and Charest, C. (1999) Acquisition of N by external hyphae of an arbuscular mycorrhizal fungus and its impact on physiological responses in maize under drought-stressed and well-watered conditions. *Mycorrhiza* 9, 69–75.

Subramanian, K.S., Santhanakrishnan, P. and Balasubramanian, P. (2006) Responses of field-grown tomato plants to arbuscular mycorrhizal fungal colonization under varying intensities of drought stress. *Scientia Horticulturae* 107, 245–253.

Taylor, T.N., Remy, W., Hass, H. and Kerp, H. (1995) Fossil arbuscular mycorrhizae from the early Devonian. *Mycologia* 87, 560–573.

Tian, C.Y., Feng, G., Li, X.L. and Zhang, F.S. (2004) Different effects of arbuscular mycorrhizal fungal isolates from saline or non-saline soil on salinity tolerance of plants. *Applied Soil Ecology* 26, 143–148.

Tobar, R.M., Azcón, R. and Barea, J.M. (1994) The improvement of plant N acquisition from an ammonium-treated, drought-stressed soil by the fungal symbiont in arbuscular mycorrhizae. *Mycorrhiza* 4, 105–108.

Tornbjerg, T., Bendix, M. and Brix, H. (1994) Internal gas transport in *Typha latifolia* L. and *Typha angustifolia* L. 2. Convective throughflow pathways and ecological significance. *Aquatic Botany* 49, 91–105.

Vandenkoornhuyse, P., Husband, R., Daniell, T.J., Watson, I.J., Duck, J.M., Fitter, A.H. and Young, P.W. (2002) Arbuscular mycorrhizal community composition associated with two plant species in a grassland ecosystem. *Molecular Ecology* 11, 1555–1564.

Wu, Q-S. and Xia, R-X. (2006) Arbuscular mycorrhizal fungi influence growth, osmotic adjustment and photosynthesis of citrus under well-watered and water stress conditions. *Journal of Plant Physiology* 163, 417–425.

Wu, Q-S., Zou, Y.N., Xia, R.X. and Wang, M.Y. (2007) Five *Glomus* species affect water relations of *Citrus tangerine* during drought stress. *Botanical Studies* 48, 147–158.

Yano-Melo, A.M., Saggin Jr., O.J. and Costa Maia, L. (2003) Tolerance of mycorrhized banana (*Musa* sp. cv. Pacovan) plantlets to saline stress. *Agriculture, Ecosystems and Environment* 95, 343–348.

3 An Overview of Ochratoxin Research

JÁNOS VARGA[1, 2], SÁNDOR KOCSUBÉ[1], ZSANETT PÉTERI[1] AND ROBERT A. SAMSON[2]

[1]*Department of Microbiology, University of Szeged, Faculty of Science and Informatics, Hungary;* [2]*CBS Fungal Biodiversity Centre, The Netherlands*

Introduction

Mycotoxins are a diverse group of fungal secondary metabolites, which are harmful to animals and humans. Several fungi are able to produce mycotoxins. Most of the mycotoxin-producing species are filamentous ascomycetes, basidiomycetes or imperfect fungi, with *Aspergillus, Fusarium, Penicillium* and *Alternaria* usually being considered the most important mycotoxin-producing genera. Postharvest spoilage caused by these and other fungi is the most important source of mycotoxins in foods and feeds. However, many fungi produce toxins while interacting with living plants as pathogens or as endophytes (preharvest spoilage).

Ochratoxin A (OTA) is a pentaketide mycotoxin that contaminates different plant products, including cereals, coffee beans, nuts, cocoa, pulses, beer, wine, spices and dried vine fruits (see Fig. 3.1; Varga *et al.*, 2001). Ochratoxins are cyclic pentaketides, and are dihydroisocoumarin derivatives linked to an L-phenylalanine moiety. OTA was first discovered in 1965 in an isolate of *Aspergillus ochraceus* (van der Merwe *et al.*, 1965). OTA was shown to exhibit nephrotoxic, immunosuppressive, teratogenic and carcinogenic properties.

Several nephropathies affecting animals as well as humans have been attributed to OTA; and for example, it is the aetiological agent of Danish porcine nephropathy, and renal disorders observed in other animals (Smith and Moss, 1985). In humans, OTA is frequently cited as the possible causative agent of Balkan endemic nephropathy, a syndrome characterized by contracted kidneys with tubular degeneration, interstitial fibrosis and hyalinization of glomeruli (Krogh *et al.*, 1977). In 1993, the International Agency for Research on Cancer (IARC) classified OTA as a possible human carcinogen (Group 2B) and concluded that there was sufficient evidence in experimental animals for the carcinogenicity of OTA, but inadequate

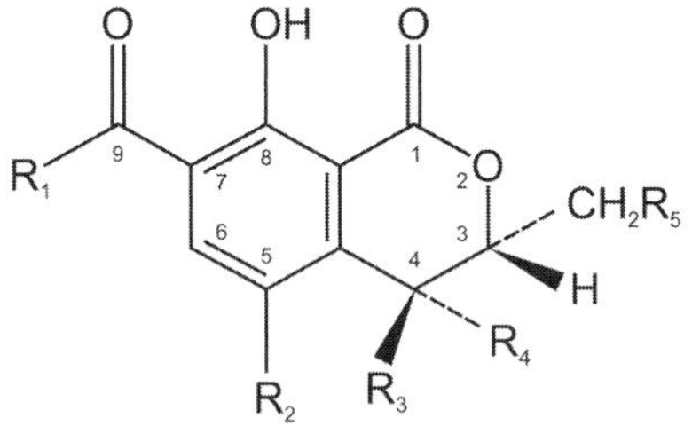

R_1	R_2	R_3	R_4	R_5	
Phenylalanyl	Cl	H	H	H	Ochratoxin A
Phenylalanyl	H	H	H	H	Ochratoxin B
Phenylalanyl ethyl ester	Cl	H	H	H	Ochratoxin C
Phenylalanyl methyl ester	Cl	H	H	H	Ochratoxin A methyl ester
Phenylalanyl methyl ester	H	H	H	H	Ochratoxin B methyl ester
Phenylalanyl ethyl ester	H	H	H	H	Ochratoxin B ethyl ester
OH	Cl	H	H	H	Ochratoxin α
OH	H	H	H	H	Ochratoxin β
Phenylalanyl	Cl	H	OH	H	4R-Hydroxyochratoxin A
Phenylalanyl	Cl	OH	H	H	4S-Hydroxyochratoxin A
Phenylalanyl	Cl	H	H	OH	10-Hydroxyochratoxin A

Fig. 3.1. Chemical structures of ochratoxin derivatives.

evidence for carcinogenicity in humans (IARC, 1993). Recently, it has been suggested that OTA may have a role in chronic karyomegalic interstitial nephropathy and chronic interstitial nephropathy in Tunisia (Maaroufi *et al.*, 1999), urothelial tumours (end-stage renal disease) in Egypt (Wafa *et al.*, 1998) and testicular cancer (Schwartz, 2002).

Several reviews have recently been published on the chemistry, molecular biology and mode of action of ochratoxins (Abarca *et al.*, 2001; O'Brien and Dietrich, 2005; O'Callaghan and Dobson, 2005; Bayman and Baker, 2006; Pfohl-Leszkowitz and Manderville, 2007). In this chapter we will give a general overview of recent progress in ochratoxin research, with emphasis on OTA-producing organisms, biosynthesis, molecular detection methods and control strategies.

Occurrence and Legislation of Ochratoxins in Agricultural Products

Ochratoxins have been detected in several agricultural products including cereals, coffee beans, cocoa, spices, soybeans, peanut, rice, maize, figs and grapes (Varga *et al.*, 2001). Ochratoxin-producing organisms are usually considered as causing postharvest spoilage. However, *Penicillium verrucosum* is often isolated from surface-sterilized cereals examined at harvest, indicating that preharvest infestation by OTA-producing fungi is at least partly responsible for OTA contamination (Miller, 1995). Indeed, OTA contamination of green coffee beans has recently been suggested to occur before storage (Bucheli *et al.*, 1998). Mantle (2000) examined the uptake of radiolabelled OTA from the soil by coffee plants, and suggested that OTA contamination of coffee beans is partly due to fungal activity in soil rather than to fungi growing on coffee beans. Further studies are necessary to clarify the significance of these observations.

OTA-producing *Penicillium* spp. are most frequently associated with stored cereals and fermented products, whereas aspergilli mainly contaminate coffee beans, spices, cocoa beans, soybeans, groundnuts, rice and maize (Moss, 1996; Table 3.1). OTA contamination of cereals, green coffee beans and foods such as mouldy cheese, fish, milk powder, bread and spices is a serious health hazard throughout the world. Contamination of green coffee beans is especially important, as some studies have found that OTA is not completely degraded during roasting (Tsubouchi *et al.*, 1987). In naturally contaminated coffee samples a 69–96% reduction in OTA content could be achieved by roasting (Blanc *et al.*, 1998; van der Stegen *et al.*, 2001), and different roasting times and temperatures affected the amount of OTA reduction (Viani, 2002). OTA has also been detected in beer (Mateo *et al.*, 2007), in wine and in other grape-derived products (Varga and Kozakiewicz, 2006), and in the body fluids and kidneys of animals and humans (Solti *et al.*, 1997).

The potential danger presented by OTA to humans and animals has led many countries to take measures to restrict OTA contamination of foods and feedstuffs. In 2003, about 100 countries had in place some regulations regarding mycotoxins (FAO, 2004) and, of these, 37 had regulations for OTA levels in agricultural products. Proposed limits for OTA range from 1–5 μg/kg for infant foods, 2–50 μg/kg for foods and 5–300 μg/kg for animal feeds. Maximum levels for OTA in cereals have been set by the European Commission Regulation No. 472/2002, and the sampling method has been regulated by the European Commission Directives 2002/26/EC and 2002/27/EC. Maximum tolerable levels of ochratoxins in foodstuffs in the European Union (EU) are given in Table 3.2.

Table 3.1. Fungal species responsible for ochratoxin contamination of different agricultural products (after Frisvad *et al.*, 2004; Samson *et al.*, 2004; Varga and Kozakiewicz, 2006).

Agricultural products	Species responsible
Cereals	*Penicillium verrucosum*, *Aspergillus ochraceus*
Fermented products (meat, cheese)	*P. nordicum*
Grapes, wine	*A. niger*, *A. carbonarius*
Coffee, cocoa, spices	*A. ochraceus*, *A. steynii*, *A. westerdijkiae*, *A. niger*, *A. carbonarius*
Figs	*A. alliaceus*, *A. niger*

Table 3.2. Legal limits for ochratoxin A in different food products as set by the European Commission (Commission Regulation (EC) No.123/2005 of 26 January 2005).

Products	Maximum level (μg/kg)
Raw cereal grains (including raw rice and buckwheat)	5.0
All products derived from cereals (including processed cereal products and cereal grains intended for direct human consumption)	3.0
Dried vine fruits (currants, raisins and sultanas)	10.0
Roasted coffee beans and ground roasted coffee, with the exception of soluble coffee	5.0
Soluble coffee (instant coffee)	10.0
Wine (red, white and rosé) and other wine and/or grape must-based drinks	2.0
Grape juice, grape juice ingredients in other beverages, including grape nectar and concentrated grape juice as reconstituted	2.0
Grape must and concentrated grape must as reconstituted, intended for direct human consumption	2.0
Baby foods and processed cereal-based foods for infants and young children	0.5
Dietary foods for special medical purposes intended specifically for infants	0.5

Following evaluations of OTA carried out in the 1990s, the Joint FAO/WHO Expert Committee on Food Additives (JECFA) has established a provisional tolerable weekly intake of 100 ng/kg body weight (b.w.) per week for OTA (JECFA, 2001). The World Health Organisation (WHO) has proposed a provisional tolerable daily intake of 16 ng/kg b.w., and this was lowered to 5 ng/kg by the EC Scientific Committee on Food in 1998 in view of the carcinogenicity of OTA. More recently, the Panel on Contaminants in the Food Chain of the European Food Safety Authority (EFSA) concluded that, according to the estimated exposure levels (2–8 ng/kg b.w./day), exposure to OTA varies between 15 and 60 ng/kg b.w./week. Based on the evaluation of the toxicity of OTA, the EFSA established a tolerable weekly intake of 120 ng/kg b.w. for OTA (EFSA, 2006). The main contributors to OTA intake in the EU are cereals and cereal products and, to a lesser degree, wine (see Chapter 9, this volume), coffee, beer, pork, pulses and spices (EFSA, 2006).

Ochratoxin-producing Organisms

Ochratoxins are produced mostly by *Penicillium* spp. in colder temperate climates, and by a number of *Aspergillus* spp. in warmer and tropical parts of the world. *Aspergillus* isolates usually produce both OTA and ochratoxin B (a dechlorinated analogue of OTA), while penicillia produce only OTA. In *Aspergillus* section *Circumdati*, the species *A. cretensis*, *A. flocculosus*, *A. pseudoelegans*, *A. roseoglobulosus*, *A. westerdijkiae*, *A. sulphureus* and *Neopetromyces muricatus* produce consistently high amounts of OTA (Frisvad *et al.*, 2004; Table 3.3). Two other species, *A. ochraceus* and *A. sclerotiorum*, produce large or small amounts of OTA, but less consistently, while four further species have been reported to produce OTA inconsistently and in trace amounts: *A. melleus*, *A. ostianus*, *A. petrakii* and *A. persii* (see Table 3.3).

The most important species regarding potential OTA production in coffee, rice, beverages and other foodstuffs are *A. ochraceus*, *A. westerdijkiae* and *A. steynii*. In *Aspergillus* section *Flavi*, *Petromyces albertensis* was found to produce the highest amounts of OTA in synthetic media (Varga *et al.*, 1996). Another important OTA-producing group are the black aspergilli (*Aspergillus* section *Nigri*). The most important OTA-producing species in this section is *A. carbonarius*, which is unable to produce ochratoxin B, in contrast to *A. niger*, which produces both analogues. While most *A. carbonarius* isolates are able to produce OTA, different surveys found that only 5–15% of *A. niger* isolates were OTA producers. Two other black *Aspergillus* species, *A. lacticoffeatus* and *A. sclerotioniger*, are also able to produce OTA (Samson *et al.*, 2004). Black aspergilli have been implicated in OTA contamination of coffee beans, grapes and spices (see Table 3.3).

A number of species in the genus *Penicillium* were described as OTA producers in the early 1970s (Ciegler *et al.*, 1972). Some of these, such as the OTA-producing *P. viridicatum* and *P. cyclopium* isolates, have since been placed in *P. verrucosum* (Pitt, 1987). *P. nordicum* has also been shown to produce OTA (Larsen *et al.*, 2001). Although Pitt (1987) suggested that *P. verrucosum* was the only OTA-producing species in *Penicillium* subgenus *Penicillium*, Bridge *et al.* (1989) subsequently found that some isolates of *P. expansum*, *P. aurantiogriseum* (*P. solitum*) and *P. atramentosum* were also able to

Table 3.3. Ochratoxin-producing *Aspergillus* species.

Species	Section	Reference(s)
A. auricomus	*Circumdati*	Varga *et al.*, 1996
A. melleus	*Circumdati*	Ciegler, 1972
A. muricatus	*Circumdati*	Frisvad and Samson, 2000
A. ochraceus	*Circumdati*	van der Merwe *et al.*, 1965
A. ostianus	*Circumdati*	Ciegler, 1972
A. petrakii	*Circumdati*	Ciegler, 1972
A. sclerotiorum	*Circumdati*	Ciegler, 1972
A. sulphureus	*Circumdati*	Ciegler, 1972
A. persii	*Circumdati*	Zotti and Corte, 2002
A. westerdijkiae	*Circumdati*	Frisvad *et al.*, 2004
A. steynii	*Circumdati*	Frisvad *et al.*, 2004
A. cretensis	*Circumdati*	Frisvad *et al.*, 2004
A. flocculosus	*Circumdati*	Frisvad *et al.*, 2004
A. pseudoelegans	*Circumdati*	Frisvad *et al.*, 2004
A. roseoglobulosus	*Circumdati*	Frisvad *et al.*, 2004
A. alliaceus	*Flavi*	Ciegler, 1972
A. albertensis	*Flavi*	Varga *et al.*, 1996
A. carbonarius	*Nigri*	Horie 1995; Téren *et al.*, 1996
A. niger	*Nigri*	Abarca *et al.*, 1994
A. lacticoffeatus	*Nigri*	Samson *et al.*, 2004
A. sclerotioniger	*Nigri*	Samson *et al.*, 2004

produce OTA. The OTA-producing abilities of these species have not yet been confirmed by other laboratories. Recently, OTA production has also been claimed for some endophytes of coffee, including one isolate each of *P. brevicompactum, P. crustosum, P. olsonii and P. oxalicum* (Vega *et al.*, 2006), and for some grape-derived isolates related to *P. radicum* and *P. rugulosum* (Torelli *et al.*, 2006).

Chemistry and Biosynthesis of Ochratoxins

OTA contains a 7-carboxy-5-chloro-8-hydroxy-3,4-dihydro-3R-methylisocoumarin ring linked through the 7-carboxy group to L-β-phenylalanine by an amide bond (see Fig. 3.1). The isocoumarin ring of OTA is a pentaketide skeleton, to which a chlorine atom is introduced, and a C1 unit is added from methionine which is subsequently oxidized to carboxyl. An intact L-phenylalanine is linked to the isocoumarin ring through this carboxyl group (Moss, 1996). Several related compounds are also synthesized in OTA-producing organisms: these are the dechlorinated analogue (ochratoxin B), the isocoumarin nucleus of OTA (ochratoxin α), its dechlorinated analogue (ochratoxin β), methyl and ethyl esters including ochratoxin C, which is an ethyl ester derivative of OTA, and several amino acid analogues (see Fig. 3.1; Moss, 1996).

Early studies on OTA biosynthesis clarified that phenylalanine was incorporated into OTA, whereas ochratoxin α was constructed from five acetate units with the carbon at C-8 obtained from methionine. Wei *et al.* (1971) demonstrated incorporation of ^{36}Cl into OTA, possibly due to the action of a chloroperoxidase enzyme. Several schemes have been proposed for the OTA biosynthetic pathway, but the steps involved in OTA production are not well established (Huff and Hamilton, 1979; Harris and Mantle, 2001). The isocoumarin group is a pentaketide skeleton formed from acetate and malonate via a polyketide pathway. A chlorine atom is incorporated most probably through the action of a chloroperoxidase to form the isocoumarin portion of OTA. Phenylalanine that is derived from the shikimic acid pathway is linked to OTA through the additional carboxyl group.

Precursor feeding experiments carried out by Harris and Mantle (2001) did not support an intermediary role for mellein in OTA biosynthesis, but could not rule out the role of 7-methyl-mellein. Their results indicated that chlorination of both ochratoxin β and OTB can give rise to OTA. However, there was no evidence for the intermediate role of the ester ochratoxin C as proposed by Huff and Hamilton (1979). These results suggest that one biosynthetic pathway involves ochratoxins β → α → OTA, with an additional branch for ochratoxins β → OTB. A scheme of the OTA biosynthetic pathway, adapted from Huff and Hamilton (1979), Harris and Mantle (2001) and O'Callaghan and Dobson (2005) is shown in Fig. 3.2.

Various enzymes and genes are involved in OTA biosynthesis. Polyketide synthase, chloroperoxidase, cytochrome P450-containing enzymes and a so-called 'OTA synthase', which catalyses the incorporation of phenyl-

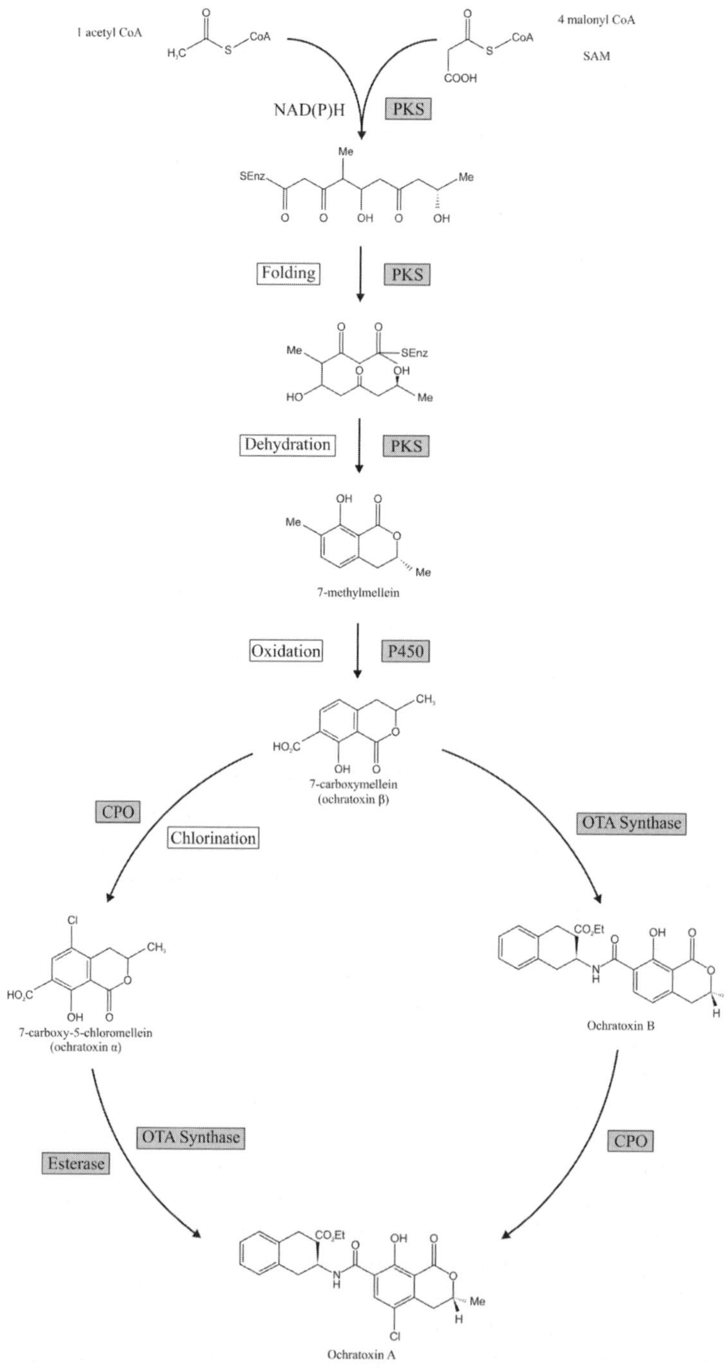

Fig. 3.2. A proposed pathway of ochratoxin biosynthesis (adapted from Huff and Hamilton, 1979; Harris and Mantle, 2001; O'Callaghan and Dobson, 2005). SAM, S-adenosyl methionine; PKS, polyketide synthase; P450, cytochrome P450-containing enzyme; CPO, chloroperoxidase.

alanine into OTA, have all been reported in the biosynthetic pathway. The latter enzyme is possibly a non-ribosomal peptide synthetase. Genes coding for some of these enzymes have been identified in some species, including *P. nordicum* and some aspergilli (Dobson and O'Callaghan, 2004; Geisen *et al.*, 2004, 2006), and part of the gene cluster responsible for OTA production in *P. nordicum* has been elucidated. This includes two putative genes, namely the OTA polyketide synthase (*otapks*PN) and a non-ribosomal peptide synthetase (*otanps*PN). Recently, two additional open reading frames have also been identified from *P. nordicum* with homology to a chloroperoxidase (*otachl*PN) and a putative OTA transporter gene (*otatra*PN; Geisen, 2007).

Molecular Detection of Ochratoxin A-producing Fungi

Several approaches have been used for the molecular detection of ochratoxin-producing fungi. Schmidt *et al.* (2004a,b) developed species-specific primers for the detection of *A. ochraceus* and *A. carbonarius* based on amplified fragment length polymorphisms (AFLP) profiles, and used these primers for the quantitative detection of these fungi in coffee beans. Fungaro *et al.* (2004) designed PCR primers based on RAPD sequences for the detection of *A. carbonarius*. Perrone *et al.* (2004) developed PCR primers based on the calmodulin gene sequence for the detection of *A. carbonarius* and *A. japonicus*. Mule *et al.* (2006) have also developed a real-time PCR approach for the identification of *A. carbonarius* on grapes using species-specific primer pairs based on partial sequences of the calmodulin gene. Morello *et al.* (2007) developed a real-time PCR method based on amplification of partial β-tubulin sequences for the detection of *A. westerdijkiae* in coffee beans and, recently, Susca *et al.* (2007) developed a PCR-single-stranded conformational polymorphism (SSCP) screening method based on the detection of sequence variation in part of the calmodulin gene. A low-complexity oligonucleotide microarray (OLISA) has also been developed based on oligonucleotide probes obtained from sequences of the calmodulin gene for the detection of black aspergilli (*A. carbonarius, A. ibericus* and *A. aculeatus*/*A. japonicus*) from grapes (Bufflier *et al.*, 2007).

O'Callaghan *et al.* (2003) identified polyketide synthases (PKSs) involved in the biosynthesis of OTA in *A. ochraceus.* The PKS sequences were used to design primers for the detection of OTA-producing fungi, including *A. ochraceus, A. alliaceus, A. carbonarius, A. niger* and some penicillia (Dobson and O'Callaghan, 2004). Polyketide synthase-specific primers have also been developed by other authors for the detection of OTA-producing *Aspergillus* species (Tóth *et al.*, 2004; Dao *et al.*, 2005). Atoui *et al.* (2007) developed a real-time PCR approach to identify *A. carbonarius* on grapes. The latter two groups have developed species-specific primer pairs designed from the acyltransferase domain of the polyketide synthase sequence.

Part of the gene cluster responsible for OTA production in *P. nordicum* has been elucidated (Geisen, 2007). Specific primers have been developed for either *P. nordicum* or *P. nordicum* and *P. verrucosum* and used to detect the presence of *P. nordicum* on cured meats (Bogs *et al.*, 2005). Recently, Schmidt-Heydt and Geisen (2007) developed a microarray method for the detection of various mycotoxin-producing fungi based on sequences of genes involved in the aflatoxin, trichothecene, fumonisin, patulin and ochratoxin biosynthetic pathways.

Controlling Ochratoxin Levels in Foods and Feeds

Several strategies have been investigated for lowering mycotoxin contamination of feeds and foods. The prevention of growth and mycotoxin production by fungi on plants and in feedstuffs is usually considered as the best approach to limit the harmful effects of mycotoxins on animal and human health. Knowledge of the key critical control points during harvesting, drying and storage stages in the cereal production chain is essential in developing effective pre- and postharvest prevention strategies. Ecological studies on the effects of environmental factors on growth and mycotoxin production have been identified for *P. verrucosum* and *A. ochraceus* during cereal production, and for *A. carbonarius* in grape and wine production (Magan, 2006). Critical control points have been identified at different stages of the coffee, cereal and wine production chains, and a Hazard Analysis and Critical Control Points (HACCP) approach was successfully used to control OTA levels in these agricultural products (Frank, 1999; Magan and Aldred, 2007; Lopez-Garcia *et al.*, 2008; Visconti *et al.*, 2008).

Plant breeding is traditionally used to improve the resistance of host plants to fungal infection. Kernels of several varieties of wheat, rye and barley have been found to have varying levels of resistance to fungal attack and OTA accumulation, and so varieties with stronger resistance to fungal invasion during storage could be selected (Chelkowski *et al.*, 1981). However, to our knowledge, breeding has not been used to increase the resistance of host plants to OTA accumulation, and treating field crops with fungicides remains the traditional technique for lowering preharvest contamination. Several studies have examined the effects of fungicide treatments on the OTA content of wine. Lo Curto *et al.* (2004) reported that application of some pesticides such as Azoxystrobin (a strobilurin derivative) or Dinocap (a dinitrophenyl derivative) in combination with sulfur effectively decreased the OTA content of wines. Greek studies have found pesticides such as Carbendazim and Chorus to be ineffective in controlling sour rot caused by aspergilli, while the application of another pesticide, Switch, led to a significant decrease in the incidence of aspergilli on grapes (Tjamos *et al.*, 2004).

Although preventing mycotoxin contamination in the field is the main goal of the agricultural and food industries, the contamination of various commodities with Aspergillus or Penicillium isolates and their mycotoxins is unavoidable under certain environmental conditions. For postharvest control, storage conditions should be improved to minimize the mycotoxin content of foods and feeds. Mycotoxin production is dependent on a number of factors, including the water activity of the stored product, temperature, gas composition, the presence of chemical preservatives and microbial interactions. An integrated approach for controlling several of these factors could give much more effective control of deterioration, without requiring the extreme control of any one factor.

Decontamination/detoxification procedures are useful in order to recondition mycotoxin-contaminated commodities. While certain treatments have been found to reduce levels of specific mycotoxins, no single method has been developed that is equally effective against the wide variety of mycotoxins that may co-occur in different commodities. Several strategies are available for the detoxification or decontamination of commodities containing mycotoxins. These can be classified as physical, physicochemical, chemical and (micro)biological approaches. OTA is a moderately stable molecule which can survive most food processing such as roasting, brewing and baking to some extent (Scott, 1996). Ensiling was found to reduce the OTA content of barley (Rotter *et al.*, 1990), and several chemical and physical methods such as ammoniation (Chelkowski *et al.*, 1982) and heat treatment (Boudra *et al.*, 1995) have been developed to detoxify OTA in animal feeds. These methods have, however, achieved only varying degrees of success, and none of them have been recommended for practical detoxification of OTA-contaminated grains and feeds (Scott, 1996).

Alternatively, microbes or their enzymes could also be applied for mycotoxin detoxification, and such biological approaches have been widely studied (Sweeney and Dobson, 1998). There are several reports of OTA-degrading activities in the microbial flora of the mammalian gastrointestinal tract, including ruminal microbes in cows and sheep (Hult *et al.*, 1976). The species responsible for OTA detoxification have not yet been identified, although protozoa were suggested as taking part in the biotransformation process in ruminants (Ozpinar *et al.*, 2002). Numerous bacteria, protozoa and fungi have been found that are able to degrade OTA (see Table 3.4).

Some enzymes, such as carboxypeptidase A (Pitout, 1969), a lipase from *A. niger* (Stander *et al.*, 2000) and some commercial proteases (Abrunhosa *et al.*, 2006) have been identified as being able to convert OTA to ochratoxin α (Fig. 3.3). This reduces the toxicity of OTA as ochratoxin α is almost non-toxic (Creppy *et al.*, 1995). *A. niger* can also degrade ochratoxin α to an unknown compound, but the pathway leading to the opening of the isocoumarin ring is unknown (Varga *et al.*, 2000). Bacteria such as *Lactobacillus plantarum* and *Oenococcus oeni*, and application of absorbents such as charcoal, liquid gelatine, yeast cell wall preparations or conidia of aspergilli have all been found to reduce the OTA content of wine (Bejaoui *et al.*, 2004, 2005).

Prevention of the Toxic Effects of Ochratoxins

In some instances it is possible to prevent the toxic effects of OTA after it has been ingested. Sodium bicarbonate can be effective in preventing absorption from the stomach (Creppy *et al.*, 1995). The phenylalanine analogue

Table 3.4. Microbes and enzymes able to degrade ochratoxin A.

Agents	Reference
Bacteria	
Ruminal microbes	Hult *et al.*, 1976
Butyrivibrio fibrisolvens	Westlake *et al.*, 1987
Lactobacillus, Streptococcus, Bifidobacterium spp.	Skrinjar *et al.*, 1996
Bacillus subtilis, B. licheniformis	Böhm *et al.*, 2000
Acinetobacter calcoaceticus	Hwang and Draughon, 1994
Phenylobacterium immobile	Wegst and Lingens, 1983
Nocardia corynebacterioides, Rhodococcus erythropolis, Mycobacterium spp.	Holzapfel *et al.*, 2002
Lactobacillus spp.	Piotrowska and Zakowska, 2000
Eubacterium callenderi, E. ramulus, *Streptococcus pleomorphus, Lactobacillus vitullinus, Sphingomonas paucimobilis, S. saccharolytica, Stenotrophomonas nitritreducens, Ralstonia eutropha, R. basilensis, Ochrobactrum* spp., *Agrobacterium* spp.	Schatzmayr *et al.*, 2006
Protozoa	Ozpinar *et al.*, 2002
Fungi	
Aspergillus niger, A. fumigatus	Varga *et al.*, 2000
Aspergillus niger, A. versicolor, A. wentii, A. ochraceus	Abrunhosa *et al.*, 2002
Aspergillus niger, A. japonicus	Bejaoui *et al.*, 2006
Pleurotus ostreatus	Engelhardt, 2002
Saccharomyces cerevisiae	Piotrowska and Zakowska 2000
Saccharomyces cerevisiae, S. bayanus	Bejaoui *et al.*, 2004
Rhizopus stolonifer, R. microsporus, R. homothallicus, R. oryzae	Varga *et al.*, 2005
Trichosporon mycotoxinivorans	Molnar *et al.*, 2004
Phaffia rhodozyma	Péteri *et al.*, 2007
Kloeckera apiculata	Angioni *et al.*, 2007
Enzymes	
Carboxypeptidase A	Pitout, 1969
Commercial proteases	Abrunhosa *et al.*, 2006
Commercial hydrolases	Stander *et al.*, 2000
A. niger enzyme	Abrunhosa and Venancio, 2007

aspartame, and piroxicam, a non-steroidal anti-inflammatory drug have been reported effectively to prevent OTA-induced toxic effects in rats (Creppy *et al.*, 1998).

Adsorbents have also been tested as potential binding agents for lowering OTA levels in body fluids. Cholestyramine, a resin used for pharmaceutical purposes in decreasing total and LDL cholesterol, can adsorb zearalenone, aflatoxins, OTA and fumonisins in *in vivo* experiments (Kerkadi *et al.*, 1999).

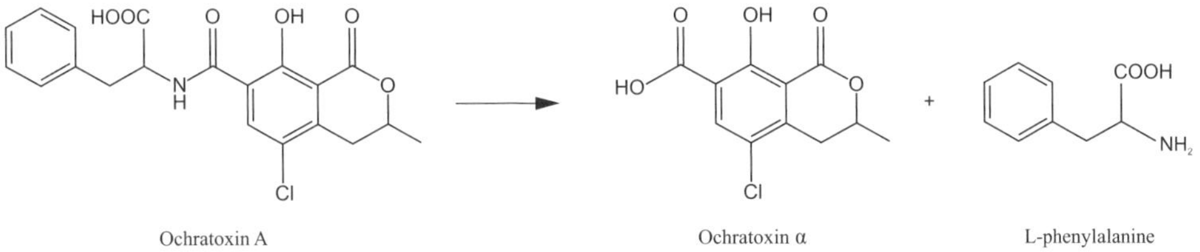

Fig. 3.3. Degradation of ochratoxin A by carboxypeptidase A.

However, it is too expensive to be used on farms efficiently. Activated carbons have been shown to have high *in vitro* affinity to OTA (Galvano *et al.*, 1998), although addition of charcoal to the diet of chicken did not significantly reduce OTA toxicity (Rotter *et al.*, 1989). However, although several adsorbents have been developed for lowering the toxic effects of mycotoxins as feed additives, most of them have been found to be ineffective against OTA-induced toxicity in *in vivo* experiments (Garcia *et al.*, 2003).

Future Prospects

Ochratoxins are among the most economically important mycotoxins. However, our knowledge of the occurrence, biosynthesis, mode of action and control of OTA is still fragmentary. Understanding the biosynthesis and effects of ochratoxins could facilitate the safeguarding of human and animal health. To achieve this goal, further research is needed to clarify the biosynthetic pathway of ochratoxins, to identify the significance of preharvest contamination of agricultural products and to determine the role of ochratoxins in animal and human pathogenesis. Further efforts are needed in order to reduce OTA contamination of foods. Projects that can further describe known sources of exposure and that can identify potential emerging sources of OTA are also required.

References

Abarca, M.L., Bragulat, M.R., Castella, G. and Cabanes, F.J. (1994) Ochratoxin A production by strains of *Aspergillus niger* var. *niger*. *Applied and Environmental Microbiology* 60, 2650–2652.

Abarca, M.L., Accensi, F., Bragulat, M.R. and Cabanes, F.J. (2001) Current importance of ochratoxin A-producing *Aspergillus* species. *Journal of Food Protection* 64, 903–906.

Abrunhosa, L. and Venancio, A. (2007) Isolation and purification of an enzyme hydrolyzing ochratoxin A from *Aspergillus niger*. *Biotechnology Letters* 29, 1909–1914.

Abrunhosa, L., Serra, R. and Venancio, A. (2002) Biodegradation of ochratoxin A by fungi isolated from grapes. *Journal of Agricultural and Food Chemistry* 50, 7493–7496.

Abrunhosa, L., Santos, L. and Venancio, A. (2006) Degradation of ochratoxin A by proteases and by a crude enzyme of *Aspergillus niger*. *Food Biotechnology* 20, 231–242.

Angioni, A., Caboni, P., Garau, A., Farris, A., Orro, D., Budroni, M. and Cabras, P. (2007) *In vitro* interaction between ochratoxin A and different strains of *Saccharomyces cerevisiae* and *Kloeckera apiculata*. *Journal of Agricultural and Food Chemistry* 55, 2043–2048.

Atoui, A., Mathieu, F. and Lebrihi, A. (2007) Targeting a polyketide synthase gene for *Aspergillus carbonarius* quantification and ochratoxin A assessment in grapes using real-time PCR. *International Journal of Food Microbiology* 115, 313–318.

Bayman, P. and Baker, J.L. (2006) Ochratoxins: a global perspective. *Mycopathologia* 162, 215–223.

Bejaoui, H., Mathieu, F., Taillandier, P. and Lebrihi, A. (2004) Ochratoxin A removal in synthetic and natural grape juices by selected oenological *Saccharomyces* strains. *Journal of Applied Microbiology* 97(5), 1038–1044.

Bejaoui, H., Mathieu, F., Taillandier, P. and Lebrihi, A. (2005) Conidia of black Aspergilli as new biological adsorbents for ochratoxin A in grape juices and musts. *Journal of Agricultural and Food Chemistry* 53, 8224–8229.

Bejaoui, H., Mathieu, F., Taillandier, P. and Lebrihi, A. (2006) Biodegradation of ochratoxin A by *Aspergillus* section *Nigri* species isolated from French grapes: a potential means of ochratoxin A decontamination in grape juices and musts. *FEMS Microbiology Letters* 255(2), 203–208.

Blanc, M., Pittet, A., Munoz-Box, R. and Viani, R. (1998) Behavior of ochratoxin A during green coffee roasting and soluble coffee manufacture. *Journal of Agricultural and Food Chemistry* 46, 673–675.

Bogs, C., Battilani, P. and Geisen, R. (2005) Development of a molecular detection and differentiation system for ochratoxin A producing *Penicillium* species and its application to analyse the occurrence of *P. nordicum* in cured meats. *International Journal of Food Microbiology* 107, 39–47.

Böhm, J., Grajewski, J., Asperger, H., Cecon, B., Rabus, B. and Razzazi, E. (2000) Study on biodegradation of some A- and B-trichothecenes and ochratoxin A by use of probiotic microorganisms. *Mycotoxin Research* 16A, 70–74.

Boudra, H., Le Bars, P. and Le Bars, J. (1995) Thermostability of ochratoxin A in wheat under two moisture conditions. *Applied and Environmental Microbiology* 61, 1156–1158.

Bridge, P.D., Hawksworth, D.L., Kozakiewicz, Z., Onions, A.H.S., Paterson, R.R.M., Sackin, M.J. and Sneath, P.H.A. (1989) A reappraisal of terverticillate Penicillia using biochemical, physiological and morphological features. I. Numerical taxonomy. *Journal of General Microbiology* 135, 2941–2966.

Bucheli, P., Meyer, I., Pittet, A., Vuataz, G. and Viani, R. (1998) Industrial storage of green robusta coffee under tropical conditions and its impact on raw material quality and ochratoxin A content. *Journal of Agriculture and Food Chemistry* 46, 4507–4511.

Bufflier, E., Susca, A., Baud, A., Mule, G., Brengel, K. and Logrieco, A. (2007) Detection of *Aspergillus carbonarius* and other black Aspergilli from grapes by DNA OLISA™ microarray. *Food Additives and Contaminants* 24, 1138–1147.

Chelkowski, J., Dopierala, G., Godlewska, B., Radomyska, W. and Szebiotko, K. (1981) Mycotoxins in cereal grain. Part III. Production of ochratoxin A in different varieties of wheat, rye and barley. *Food* 25, 625–629.

Chelkowski, J., Szebiotko, K., Golinski, P., Buchowski, M., Godlewska, B., Radomyrska, W. and Wiewiorowska, M. (1982) Mycotoxins in cereal grain. Part V. Changes of cereal grain biological value after ammoniation and mycotoxin (ochratoxin) inactivation. *Nahrung* 26, 1–7.

Ciegler, A. (1972) Bioproduction of ochratoxin A and penicillic acid by members of the *Aspergillus ochraceus* group. *Canadian Journal of Microbiology* 18, 631–636.

Ciegler, A., Fennell, D.J., Mintzlaff, H.J. and Leistner, L. (1972) Ochratoxin synthesis by *Penicillium* species. *Naturwissenschaften* 59, 365–366.

Creppy, E.E., Baudrimont, I. and Betbeder, A.-M. (1995) Intoxication by ochratoxin A (OTA): prevention of toxic effects. *Toxicon* 33, 1115–1116.

Creppy, E.E., Baudrimont, I. and Betbeter, A.-M. (1998) How aspartame prevents the toxicity of ochratoxin A. *Journal of Toxicological Sciences* 23(Suppl. 2), 165–172.

Dao, H.P., Mathieu, F. and Lebrihi, A. (2005) Two primer pairs to detect OTA producers by PCR method. *International Journal of Food Microbiology* 104, 61–67.

Dobson, A. and O'Callaghan, J. (2004) Detection of ochratoxin A-producing fungi. Patent No. WO 2004/072224 A2.

EFSA (2006) Opinion of the Scientific Panel on contaminants in the food chain on a request from the Commission related to ochratoxin A in food. *The EFSA Journal* 365, 1–56.

Engelhardt, G. (2002) Degradation of ochratoxin A and B by the white rot fungus *Pleurotus ostreatus*. *Mycotoxin Research* 18, 37–43.

FAO (2004) *Worldwide Regulations for Mycotoxins in Food and Feed in 2003.* FAO, Rome.

Frank, M. (1999) HACCP and its mycotoxin control potential: an evaluation of ochratoxin A in coffee production. Paper presented at the *Third Joint FAO/WHO/UNEP International Conference on Mycotoxins*, Tunis (ftp://ftp.fao.org/es/esn/food/myco6c.pdf).

Frisvad, J.C. and Samson, R.A. (2000) *Neopetromyces* gen. nov. and an overview of teleomorphs of *Aspergillus* subgenus *Circumdati*. *Studies in Mycology* 45, 201–207.

Frisvad, J.C., Frank, J.M., Houbraken, J.A.M.P., Kuijpers, A.F.A. and Samson, R.A. (2004) New ochratoxin A-producing species of *Aspergillus* section *Circumdati*. *Studies in Mycology* 50, 23–43.

Fungaro, M.H., Vissotto, P.C., Sartori, D., Vilas-Boas, L.A., Furlaneto, M.C. and Taniwaki, M.H. (2004) A molecular method for detection of *Aspergillus carbonarius* in coffee beans. *Current Microbiology* 49, 123–127.

Galvano, F., Pietri, A., Bertuzzi, T., Piva, A., Chies, L. and Galvano, M. (1998) Activated carbons: *in vitro* affinity for ochratoxin A and deoxynivalenol and relation of adsorption ability to physicochemical parameters. *Journal of Food Protection* 61, 469–475.

García, A.R., Avila, E., Rosiles, R. and Petrone, V.M. (2003) Evaluation of two mycotoxin binders to reduce toxicity of broiler diets containing ochratoxin A- and T-2 toxin-contaminated grain. *Avian Disease* 47, 691–699.

Geisen, R. (2007) Molecular detection and monitoring. In: Dijksterhuis, J. and Samson, R.A. (eds) *Food Mycology: A Multifaceted Approach to Fungi and Food*. CRC Press, Boca Raton, Florida, pp. 255–278.

Geisen, R., Mayer, Z., Karolewiez, A. and Färber, P. (2004) Development of a real-time PCR system for detection of *Penicillium nordicum* and for monitoring ochratoxin A production in foods by targeting the ochratoxin polyketide synthase gene. *Systematic and Applied Microbiology* 27, 501–507.

Geisen, R., Schmidt-Heydt, M. and Karolewiez, A. (2006) A gene cluster of ochratoxin A biosynthetic genes in *Penicillium*. *Mycotoxin Research* 22, 134–141.

Harris, J.P. and Mantle, P.G. (2001) Biosynthesis of ochratoxins by *Aspergillus ochraceus*. *Phytochemistry* 58, 709–716.

Holzapfel, W., Brost, I., Farber, P., Geisen, R., Bresch, H., Jany, K.D., Mengu, M., Jakobsen, M., Steyn, P.S., Teniola, D. and Addo, P. (2002) *Actinomycetes* for breaking down aflatoxin B1, ochratoxin A, and/or zearalenon. International patent No. WO 02/099142 A3.

Horie, Y. (1995) Productivity of ochratoxin A of *Aspergillus carbonarius* in *Aspergillus* section *Nigri*. *Nippon Kingakkai Kaiho* 36, 73–76.

Huff, W.E. and Hamilton, P.B. (1979) Mycotoxins – their biosynthesis in fungi: ochratoxins – metabolites of combined pathways. *Journal of Food Protection* 42, 815–820.

Hult, K., Teiling, A. and Gatenbeck, S. (1976) Degradation of ochratoxin A by a ruminant. *Applied and Environmental Microbiology* 32, 443–444.

Hwang, C.-A. and Draughon, F.A. (1994) Degradation of ochratoxin A by *Acinetobacter calcoaceticus*. *Journal of Food Protection* 57, 410–414.

IARC (1993) Ochratoxin A. In: *IARC Monographs on the Evaluation of Carcinogenic Risk of Chemicals to Humans*, Vol. 56. IARC Press, Lyon, France, pp. 489–521.

JECFA (2001) Ochratoxin A. Available from: http://www.inchem.org/documents/jecfa/jecmono/v47je04.htm

Kerkadi, A., Barriault, C., Marquardt, R.R., Frohlich, A.A., Yousef, I.M., Zhu, X.X. and Tuchweber, B. (1999) Cholestyramine protection against ochratoxin A toxicity: role of ochratoxin A sorption by the resin and bile acid enterohepatic circulation. *Journal of Food Protection* 62, 1461–1465.

Krogh, P., Hald, B., Plestina, R. and Ceovic, S. (1977) Balkan (endemic) nephropathy and food-borne ochratoxin A: preliminary results of a survey of foodstuff. *Acta Pathologica et Microbiologica Scandinavica Section B* 85, 238–240.

Larsen, T.O., Svendsen, A. and Smedsgaard, J., (2001) Biochemical characterization of ochratoxin A-producing strains of the genus *Penicillium*. *Applied and Environmental Microbiology* 67, 3630–3635.

Lo Curto, R., Pellicano, T., Vilasi, F., Munafo, P. and Dugo, G. (2004) Ochratoxin A occurrence in experimental wines in relationship with different pesticide treatments of grapes. *Food Chemistry* 84, 71–75.

Lopez-Garcia, R., Mallmann, C.A. and Pineiro, M. (2008) Design and implementation of an integrated management system for ochratoxin A in the coffee production chain. *Food Additives and Contaminants* 25, 231–240.

Maaroufi ,K., Zakhama, A., Baudrimont, I., Achour, A., Abid, S., Ellouz, F., Dhouib, S., Creppy, E.E. and Bacha, H. (1999) Karyomegaly of tubular cells as early stage marker of the nephrotoxicity induced by ochratoxin A in rats. *Human and Experimental Toxicology* 18, 410–415.

Magan, N. (2006) Mycotoxin contamination of food in Europe: early detection and prevention strategies. *Mycopathologia* 162, 245–253.

Magan, N, and Aldred, D. (2007) Post-harvest control strategies: minimizing mycotoxins in the food chain. *International Journal of Food Microbiology* 119, 131–139.

Mantle, P.G. (2000) Uptake of radiolabelled ochratoxin A from soil by coffee plants. *Phytochemistry* 53, 377–378.

Mateo, R., Medina, A., Mateo, E.M., Mateo, F. and Jiménez, M. (2007) An overview of ochratoxin A in beer and wine. *International Journal of Food Microbiology* 119, 79–83.

Miller, J.D. (1995) Fungi and mycotoxins in grain: implications for stored product research. *Journal of Stored Products Research* 31, 1–16.

Molnar, O., Schatzmayr, G., Fuchs, E. and Prillinger, H. (2004) *Trichosporon mycotoxinivorans* sp. *nov.*, a new yeast species useful in biological detoxification of various mycotoxins. *Systematic and Applied Microbiology* 27(6), 661–671.

Morello, L.G., Sartori, D., de Oliveira Martinez, A.L., Vieira, M.L.C., Taniwaki, M.H. and Fungaro M.H.P. (2007) Detection and quantification of *Aspergillus westerdijkiae* in coffee beans based on selective amplification of β-tubulin gene by using real-time PCR. *International Journal of Food Microbiology* 119, 270–276.

Moss, M.O. (1996) Mode of formation of ochratoxin A. *Food Additives and Contaminants* 13(Suppl.), 5–9.

Mule, G., Susca, A., Logrieco, A., Stea, G. and Visconti, A. (2006) Development of a quantitative real-time PCR assay for the detection of *Aspergillus carbonarius* in grapes. *International Journal of Food Microbiology* 111(Suppl. 1), S28–S34.

O'Brien, E. and Dietrich, D.R. (2005) Ochratoxin A: the continuing enigma. *Critical Reviews in Toxicology* 35, 33–60.

O'Callaghan, J. and Dobson, A.D. (2005) Molecular characterization of ochratoxin A biosynthesis and producing fungi. *Advances in Applied Microbiology* 58, 227–243.

O'Callaghan, J., Caddick, M.X. and Dobson, A.D. (2003) A polyketide synthase gene required for ochratoxin A biosynthesis in *Aspergillus ochraceus*. *Microbiology* 149, 3485–3491.

Ozpinar, H., Bilal, T., Abas, I. and Kutay, C. (2002) Degradation of ochratoxin A in rumen fluid *in vitro*. *Medical Biology* 9, 66–69.
Perrone, G., Susca, A., Stea, G. and Mule, G. (2004) PCR assay for identification of *Aspergillus carbonarius* and *Aspergillus japonicas*. *European Journal of Plant Pathology* 110, 641–649.
Péteri, Z., Téren, J., Vágvölgyi, C. and Varga, J. (2007) Ochratoxin degradation and adsorption caused by astaxanthin-producing yeasts. *Food Microbiology* 24, 205–210.
Pfohl-Leszkowicz, A. and Manderville, R.A. (2007) Ochratoxin A: an overview on toxicity and carcinogenicity in animals and humans. *Molecular Nutrition and Food Research* 51(1), 61–99.
Piotrowska, M. and Zakowska, Z. (2000) The biodegradation of ochratoxin A in food products by lactic acid bacteria and baker's yeast. In: Bielecki, S., Tramper, J. and Polak, J. (eds) *Food Biotechnology*. Elsevier, Amsterdam, pp. 307–310.
Pitout, M.J. (1969) The hydrolysis of ochratoxin A by some proteolytic enzymes. *Biochemical Pharmacology* 18, 485–491.
Pitt, J.I. (1987) *Penicillium viridicatum, Penicillium verrucosum*, and production of ochratoxin A. *Applied and Environmental Microbiology* 53, 266–269.
Rotter, R.G., Frohlich, A.A. and Marquardt, R.R. (1989) Influence of dietary charcoal on ochratoxin A toxicity in Leghorn chicks. *Canadian Journal of Veterinary Research* 53, 449–453.
Rotter, R.G., Marquardt, R.R., Frohlich, A.A. and Abramson, D. (1990) Ensiling as a means of reducing ochratoxin A concentrations in contaminated barley. *Journal of the Science of Food and Agriculture* 50, 155–160.
Samson, R.A., Houbraken, J.A.M.P., Kuijpers, A.F.A., Frank, J.M. and Frisvad, J.C. (2004) New ochratoxin- or sclerotium-producing species in *Aspergillus* section *Nigri*. *Studies in Mycology* 50, 45–61.
Schatzmayr, G., Zehner, F., Taubel, M., Schatzmayr, D., Klimitsch, A., Loibner, A.P. and Binder, E.M. (2006) Microbiologicals for deactivating mycotoxins. *Molecular Nutrition and Food Research* 50, 543–551.
Schmidt, H., Bannier, M., Vogel, R.F. and Niessen, L. (2004a) Detection and quantification of *Aspergillus ochraceus* in green coffee by PCR. *Letters in Applied Microbiology* 38, 464–469.
Schmidt, H., Taniwaki, M.H., Vogel, R.F. and Niessen, L. (2004b) Utilization of AFLP markers for PCR-based identification of *Aspergillus carbonarius* and indication of its presence in green coffee samples. *Journal of Applied Microbiology* 97, 899–909.
Schmidt-Heydt, M. and Geisen, R. (2007) A microarray for monitoring the production of mycotoxins in food. *International Journal of Food Microbiology* 117, 131–140.
Schwartz, G.G. (2002) Hypothesis: does ochratoxin A cause testicular cancer? *Cancer Causes Control* 13(1), 91–100.
Scott, P.M. (1996) Effects of processing and detoxification treatments on ochratoxin A: introduction. *Food Additives and Contamination* 13(Suppl.), 19–21.
Skrinjar, M., Rasic, J.L. and Stojicic, V. (1996) Lowering ochratoxin A level in milk by yoghurt bacteria and bifidobacteria. *Folia Microbiologica* 41, 26–28.
Smith, J.E. and Moss, M.O. (1985) *Mycotoxins. Formation, Analysis and Significance*, John Wiley & Sons, Chichester, UK.
Solti, L., Salamon, F., Barna-Vetró, I., Gyöngyösi, A., Szabó, E. and Wolfing, A. (1997) Ochratoxin content of human sera determined by a sensitive ELISA. *Journal of Analytical Toxicology* 21, 44–48.
Stander, M.A., Bornscheurer, U.T., Henke, E. and Steyn, P.S. (2000) Screening of commercial hydrolases for the degradation of ochratoxin A. *Journal of Agriculture and Food Chemistry* 48, 5736–5739.
Susca, A., Stea, G. and Perrone, G. (2007) Rapid polymerase chain reaction (PCR)-single-stranded conformational polymorphism (SSCP) screening method for the identification

of *Aspergillus* section *Nigri* species by the detection of calmodulin nucleotide variation. *Food Additives and Contaminants* 24, 1148–1153.

Sweeney, M.J. and Dobson, A.D.W. (1998) Mycotoxin production by *Aspergillus, Fusarium* and *Penicillium* species. *International Journal of Food Microbiology* 43, 141–158.

Téren, J., Varga, J., Hamari, Z., Rinyu, E. and Kevei, F. (1996) Immunochemical detection of ochratoxin A in black *Aspergillus* strains. *Mycopathologia* 134, 171–176.

Tjamos, S.E., Antoniou, P.P., Kazantzidou, A., Antonopoulos, D.F., Papageorgiou, I. and Tjamos, E.C. (2004) *Aspergillus niger* and *A. carbonarius* in Corinth raisin- and wine-producing vineyards in Greece: population composition, ochratoxin A production and chemical control. *Journal of Phytopathology* 152, 250–255.

Torelli, E., Firrao, G., Locci, R. and Gobbi, E. (2006) Ochratoxin A-producing strains of *Penicillium* spp. isolated from grapes used for the production of 'passito' wines. *International Journal of Food Microbiology* 106, 307–312.

Tóth, B., Kiss, R., Kocsubé, S., Téren, J. and Varga, J. (2004) Real-time PCR detection of fungi and their toxins in wheat. In: *Proceedings of the 2nd Central European Food Congress*, Budapest, Hungary.

Tsubouchi, H., Yamamoto, K., Hisada, K., Sakabe, Y. and Udagawa, S. (1987) Effect of roasting on ochratoxin A level in green coffee beans inoculated with *Aspergillus ochraceus*. *Mycopathologia* 97, 111–115.

van der Merwe, K.J., Steyn, P.S., Fourie, L., Scott, D.B. and Theron, J.J. (1965) Ochratoxin A, a toxic metabolite produced by *Aspergillus ochraceus* Wilh. *Nature* 205, 1112–1113.

van der Stegen, G.H., Essens, P.J. and van der Lijn, J. (2001) Effect of roasting conditions on reduction of ochratoxin A in coffee. *Journal of Agricultural and Food Chemistry* 49, 4713–4715.

Varga, J. and Kozakiewicz, Z. (2006) Ochratoxin A in grapes and grape-derived products. *Trends in Food Science and Technology* 17, 72–81.

Varga, J., Kevei, É., Rinyu, E., Téren, J. and Kozakiewicz, Z. (1996) Ochratoxin production by *Aspergillus* species. *Applied and Environmental Microbiology* 62, 4461–4464.

Varga, J., Rigó, K. and Téren, J. (2000) Degradation of ochratoxin A by *Aspergillus* species. *International Journal of Food Microbiology* 59, 1–7.

Varga, J., Rigó, K., Téren, J. and Mesterházy, Á. (2001) Recent advances in ochratoxin research I. Production, detection and occurrence of ochratoxins. *Cereal Research Communications* 29, 85–92.

Varga, J., Péteri, Z., Tábori, K., Téren, J. and Vágvölgyi, C. (2005) Degradation of ochratoxin A and other mycotoxins by *Rhizopus* isolates. *International Journal of Food Microbiology* 99, 321–328.

Vega, F., Posada, F., Peterson, S.W., Gianfagna, T. and Chaves, F. (2006) *Penicillium* species endophytic in coffee plants and ochratoxin A production. *Mycologia* 98, 31–42.

Viani, R. (2002) Effect of processing on ochratoxin A content of coffee, *Advances in Experimental Medicine and Biology* 504, 189–193.

Visconti, A., Perrone, G., Cozzi, G. and Solfrizzo, M. (2008) Managing ochratoxin A risk in the grape–wine food chain. *Food Additives and Contaminants* 25, 193–202.

Wafa, E.W., Yahya, R.S., Sobh, M.A., Eraky, I., El Baz, H., El Gayar, H.A.M., Betbeder, A.M. and Creppy, E.E. (1998) Human ochratoxicosis and nephropathy in Egypt: a preliminary study. *Human and Experimental Toxicology* 17, 124–129.

Wegst, W. and Lingens, F. (1983) Bacterial degradation of ochratoxin A. *FEMS Microbiology Letters* 17, 341–344.

Wei, R.-D., Strong, F.M. and Smalley, E.B. (1971) Incorporation of chlorine-36 into ochratoxin A. *Applied Microbiology* 22, 276–277.

Westlake, K., Mackie, R.I. and Dutton, M.F. (1987) Effects of several mycotoxins on specific growth rate of *Butyrivibrio fibrisolvens* and toxin degradation *in vitro*. *Applied and Environmental Microbiology* 53, 613–614.

Zotti, M. and Corte, A.M. (2002) *Aspergillus persii*: a new species in section *Circumdati*. *Mycotaxon* 83, 269–278.

4 Improvement of Controlled Mycorrhizal Usage in Forest Nurseries

R. Duponnois[1], D. Diouf[2], A. Galiana[3] and Y. Prin[3]

[1]*IRD, UMR 113 CIRAD/INRA/IRD/SUPAGRO/UM2, Laboratoire des Symbioses Tropicales et Méditerranéennes (LSTM), Montpellier, France;*
[2]*Université Cheikh Anta Diop (UCAD), Faculté de Sciences et Techniques, Département de Biologie Végétale, Dakar, Sénégal;*
[3]*CIRAD, UMR 113 CIRAD/INRA/IRD/SUPAGRO/UM2, Laboratoire des Symbioses Tropicales et Méditerranéennes (LSTM), Montpellier, France*

Introduction

Mycorrhizal fungi are ubiquitous components of most ecosystems throughout the world, and are considered key ecological factors in governing the cycles of major plant nutrients and in sustaining the vegetation cover (Schreiner *et al.*, 2003). Two major morphological forms of mycorrhizas are usually distinguished, namely arbuscular mycorrhizas (AM) and ectomycorrhizas (ECM). AM symbiosis is the most widespread mycorrhizal association and is found on average in 80–90% of land plants in natural, agricultural and forest ecosystems (Brundrett, 2002). ECMs are observed on trees, on woody shrubs and on a small number of herbaceous plants (Smith and Read, 1997), on gymnosperms and angiosperms and usually result from the association between Homobasidiomycetes and about 20 families of mainly woody plants (Smith and Read, 1997).

Mycorrhizal fungi enhance the uptake of low-mobility minerals such as phosphorus and micronutrients, and aid water absorption (Smith and Read, 1997). Additionally they can improve plant health by providing protection against pathogens (Smith and Read, 1997) and heavy metal pollution (Chen *et al.*, 2003). Hyphae of mycorrhizal fungi also play a role in the formation and stability of soil aggregates (Caravaca *et al.*, 2002), and influence the composition of plant community structures (van der Heijden *et al.*, 1998; Simard and Durall, 2004). Hence, these symbiotic microorganisms have been found to be essential components of sustainable soil–plant systems (van der Heijden *et al.*, 1998; Dunabeita *et al.*, 2004).

In Mediterranean and tropical ecosystems, desertification has occurred following several decades of scarce and irregular rainfall, together with the over-exploitation of natural resources (Francis and Thornes, 1990). Desertification mainly results from the degradation of natural plant communities (population structure, succession pattern and species diversity) and the degradation of physicochemical and biological soil properties (nutrient availability, microbial activity, soil structure, etc.) (Requena *et al.*, 2001). These disturbances generally induce a loss or a reduction of mycorrhizal propagules in the soil and, consequently, decrease the mycorrhizal potential in the degraded areas (Requena *et al.*, 2001). Failures in artificial or natural processes of re-vegetation have been hypothesized as resulting from the absence or diminution of mycosymbiont propagules (Marx, 1980; Requena *et al.*, 2001).

To overcome this problem, it has already been shown that mycorrhizal inoculation of seedlings can be very efficient in establishing plants at outplanting on disturbed soils (Duponnois *et al.*, 2007). The practice of controlled mycorrhization is based on the use of mycorrhizal strains best suited to host plant species that rapidly colonize their root systems, and that are well adapted to the environmental conditions of the planting site (Duponnois *et al.*, 2007). This practice needs high quantities of mycorrhizal fungal inoculum, and the biotechnological processes required for this are generally associated with a large investment of resources. Such an investment limits the use of mycorrhization in forest nurseries, and especially those in developing countries.

It has been suggested that, in order to reduce the quantities of fungal inoculum required, some rhizosphere bacteria (Mycorrhiza Helper Bacteria, or MHB) could improve the establishment of the fungal inoculant and consequently, minimize the volume of fungal inoculum added into the soil (Duponnois *et al.*, 1993). It has recently been demonstrated that the termite mounds of *Macrotermes subhyalinus* (a litter-foraging termite commonly found in tropical areas) contain a specific microflora that enhanced ectomycorrhizal fungal development (Duponnois *et al.*, 2006, Duponnois, 2007). Therefore *M. subhyalinus* mound amendment may be a tool that could improve the efficiency of controlled mycorrhization in forest nurseries and minimize the technical and financial investment required.

The aims of this chapter are to present innovative processes of mycorrhizal inoculation using a 'catalyst' of MHB or termite mounds of *M. subhyalinus* for mycorrhizal establishment in order to minimize the volume of fungal inoculum required. The potential of MHB to improve the efficiency of an ectomycorrhizal fungal inoculum was tested with a northern hemisphere tree species (*Pseudotsuga menziesii*) and an ectomycorrhizal fungus, *Laccaria bicolor*, in nursery conditions. The effect of the termite mound on mycorrhizal formation was explored on an Australian acacia, *Acacia holosericea* A. Cunn. ex G. Don (a fast-growing leguminous tree species frequently used in afforestation programmes in West Africa) with an ECM fungus, *Pisolithus albus* IR 100, and the AM fungus *Glomus intraradices*.

Mycorrhiza Helper Bacteria

Assessment of the existence of MHB

The existence of a group of mycorrhiza helper bacteria was first suggested by Garbaye and Bowen (1989). They showed that some bacteria could interact with the host fungus, either through the fungal mycelium or inside the mycorrhiza at the interface between the root and the fungal mantle. They surface-sterilized ectomycorrhizas in order to isolate only bacteria located internally. Most of the bacteria isolated from *Rhizopogon luteolus*/*Pinus radiata* ectomycorrhizas were identified as being in the fluorescent pseudomonads group. About 80% of the bacterial strains tested in their experiment stimulated mycorrhiza formation, whereas 20% had either neutral or negative effects. Fourteen of the bacteria isolated from sporocarps of *L. bicolor* and ectomycorrhizas of Douglas fir with *L. bicolor* significantly stimulated the ectomycorrhizal colonization of the Douglas fir root systems by a *L. bicolor* strain after 4 months' culture in glasshouse conditions (Duponnois and Garbaye, 1991). These authors did not find any relationships between the bacterial effect on mycorrhizal formation and either the origin of the bacteria or their taxonomic placement.

A strain of *Bacillus subtilis*, isolated from Douglas fir/*L. bicolor* ectomycorrhizas, has been shown to increase the ectomycorrhizal rate (percentage of short roots mycorrhized with *L. bicolor*) to 97.3%, as compared with 67.3%, in the control treatment (inoculated fungus without bacteria). A *Pseudomonas fluorescens* strain isolated from a *L. bicolor* sporocarp showed a similar positive effect on mycorrhizal formation. The MHB effect has been demonstrated in both glasshouse and nursery conditions, which shows the high competitiveness of the bacterial inoculant towards the native microflora, and in axenic conditions that showed that the stimulation is an intrinsic property of each bacterial strain and did not result from interactions within the microbial community in the rhizosphere (Duponnois, 1992).

Mycorrhiza Helper Bacteria, defined as telluric bacteria promoting the development of mycorrhizal symbiosis, have been isolated from varying plant–fungal combinations with AM fungi (herbaceous plants) or ECM fungi (trees). To date, there have been many reports in the literature of various bacterial strains that can promote the establishment of either arbuscular or ectomycorrhizal symbiosis (see Table 4.1) (Frey-Klett *et al.*, 2007). These results show that MHB is a generic concept that is not dependent on the type of mycorrhizal symbiosis or the taxonomy of the bacterial strains (Frey-Klett *et al.*, 2007).

The MHB effect on ectomycorrhizal symbiosis has been investigated mainly with northern hemisphere tree species, including *Pseudotsuga menziesii*, *Quercus robur* (Duponnois and Garbaye, 1991), *Picea abies*, *Pinus sylvestris* (Schrey *et al.*, 2005), *Pinus radiata* (Garbaye and Bowen, 1989) and some ectomycorrhizal fungi, principally *L. bicolor* (Frey-Klett *et al.*, 2007) and *Suillus luteus* (Bending *et al.*, 2002). It has, however, been shown that some fluorescent pseudomonads could help in ectomycorrhizal formation in tropical conditions with Australian *Acacia* species (Founoune *et al.*, 2002a,b).

Use of MHBs in forest nurseries

Mycorrhiza Helper Bacteria belonging to the fluorescent pseudomonad group have mainly been studied for their effect on ectomycorrhizal formation in glasshouse conditions. For example, the inoculation of a strain of a *Pseudomonas monteilii* MHB has been investigated on the mycorrhization of: (i) an Australian *Acacia*, *A. holosericea*, by several ectomycorrhizal fungi or one endomycorrhizal fungus, *Glomus intraradices*; and (ii) several Australian *Acacia* species by a *Pisolithus albus* strain under glasshouse conditions (Duponnois and Plenchette, 2003). The promotion of the ectomycorrhizal establishment by the *P. monteilii* strain was seen with all of the fungal isolates used (strains of *Pisolithus* and *Scleroderma*; see Fig. 4.1). This MHB also significantly increased the ectomycorrhizal colonization of all of the *Acacia* species tested in this experiment (*A. auriculiformis*, *A. eriopoda*, *A. holosericea*, *A. mangium* and *A. platycarpa*) (see Fig. 4.2).

To date, MHB have been tested for their influence on ectomycorrhizal formation between Douglas fir and a *L. bicolor* strain in only nursery conditions. One bacterial isolate (a strain of *P. fluorescens*) increased mycorrhizal formation from around 60% (percentage of short roots mycorrhized with *L. bicolor*) without bacterial inoculation to 87%. This experiment was performed with a peat–vermiculite fungal inoculum. Similar positive effects were recorded when this bacterial strain was immobilized with the fungus in alginate beads (Duponnois, 1992). After 4 months' culture, the *P. fluorescens* strain increased ectomycorrhizal infection from 42 to 75%. Frey-Klett *et al.* (1999) showed that, five months after inoculation and sowing, bacterial inoculation significantly increased the percentage of ectomycorrhizal short roots on plants inoculated with two doses of fungal inoculum of *L. bicolor* (50 and 100 mg/m^2 dry weight mycelium on alginate beads mixed into the soil at a constant dose of 1 l/m^2) in a nursery (see Fig. 4.3). The lowest bacterial dose increased mycorrhizal colonization from 45 to 70% in plants inoculated with the lowest amount of fungal inoculum, and from 64 to 77% in plants inoculated with the highest amount. The lowest bacterial dose increased mycorrhizal colonization more than did the highest bacterial dose (see Fig. 4.3).

Use of Termite Mound Powder to Promote Mycorrhizal Formation

Termite mounds of *Macrotermes subhyalinus*: islands of microbial fertility in tropical forest ecosystems

Termite (*Isoptera*) mounds are ubiquitous features of tropical ecosystems, more particularly in savannah environments. Termite activities result in the

Table 4.1. Studies on the effects of MHB on mycorrhizal formation (from Frey-Klett *et al.*, 2007).

Mycorrhizal fungi	MHB species	Host plant	Reference
Ectomycorrhizal fungi			
Amanita muscaria, *Suillus bovinus*	*Streptomyces*	*Picea abies*, *Pinus sylvestris*	Schrey *et al.* (2005)
Hebelome crustuliniforme	Unidentified bacteria	*Fagus sylvatica*	de Oliveira (1988)
Laccaria bicolor	*Pseudomonas fluorescens*, *Pseudomonas* sp., *Bacillus* sp.	*Pseudotsuga menziesii*	Duponnois and Garbaye (1991)
Laccaria fraterna, *L. laccata*	*Bacillus* sp., *Pseudomonas* sp.	*Eucalyptus diversicolor*	Dunstan *et al.* (1998)
Lactarius rufus	*Paenibacillus* sp., *Burkholderia* sp.	*Pinus sylvestris*	Poole *et al.* (2001)
Pisolithus albus	*Pseudomonas monteilii* *Pseudomonas resinovorans*	*Acacia holosericea*	Founoune *et al.* (2002a)
Pisolithus sp.	Fluorescent Pseudomonads	*Acacia holosericea*	Founoune *et al.* (2002b)
Rhizopogon luteolus	Unidentified bacteria	*Pinus radiata*	Garbaye and Bowen (1989)
Rhizopogon vinicolor, *Laccaria laccata*	*Arthrobacter* sp.	*Pinus sylvestris*	Rozycki *et al.* (1994)
Scleroderma spp., *Pisolithus* spp.	*Pseudomonas monteilii*	Australian *Acacia* species	Duponnois and Plenchette (2003)
Suillus luteus	*Bacillus* sp.	*Pinus sylvestris*	Bending *et al.* (2002)
Arbuscular mycorrhizal fungi			
Endogone sp.	*Pseudomonas* sp.	*Trifolium* spp., *Cucumis sativum*, *Allium cepa*	Mosse (1962)
Gigaspora margarita	*Azospirillum brasilense*	*Pennidetum americanum*	Rao *et al.* (1985)
Glomus clarum	*Azotobacter diazotrophicus*, *Klebsiella* sp.	*Ipomoea batatas*	Paula *et al.* (1992)
G. deserticola	*Klebsiella pneumoniae*, *Alcaligenes denetrificans*	*Unicola paniculata*	Will and Sylvia (1990)
G. fasciculatum	*Azotobacter chroococcum*	*Lycopersicum esculentum*	Bagyaraj and Menge (1978)
G. fasciculatum, *G. mosseae*	*Rhizobium meliloti*	*Medicago sativa*	Azcón *et al.* (1991)
G. fasciculatum, *G. mosseae*, *G. caledonium*	*Bacillus coagulans*	*Morus alba*, *Carica papaya*	Mamatha *et al.* (2002)

G. fistulosum	*Pseudomonas putida*	*Zea mays*, *Solanum tuberosum*	Vósatka and Gryndler (1999)
G. intraradices	*Bacillus subtilis*, *Enterobacter* sp.	*Allium cepa*	Toro *et al.* (1997)
G. intraradices	*Pseudomonas monteilii*	*Acacia holosericea*	Duponnois and Plenchette (2003)
G. intraradices	*Rhizobium*	*Anthyllis cytisoides*	Requena *et al.* (1997)
G. intraradices	*Agrobacterium rhizogenes*, *Pseudomonas fluorescens*, *Rhizobium leguminosarum*	*Hordeum vulgare*, *Triticum aestivum*	Fester *et al.* (1999)
G. intraradices	*Streptomyces coelicolor*	Sorghum	Abdel-Fattah and Mohamedin (2000)
G. mosseae	*Paenibacillus* sp.	*Sorghum bicolor*	Budi *et al.* (1999)
G. mosseae	*Pseudomonas* sp.	*Lycopersicum esculentum*	Barea *et al.* (1998)
G. mosseae	*Bradhyrhizobium japonicum*	*Glycine max*	Xie *et al.* (1995)
G. mosseae	*Pseudomonas fluorescens*	*Lycopersicum esculentum*	Gamalero *et al.* (2004)
G. mosseae	*Brevibacillus* sp.	*Trifolium pratense*	Vivas *et al.* (2003)
G. mosseae, *G. intraradices*	*Paenibacillus brasilensis*	*Trifolium*	Artursson (2005)
Complex of AMF[a]	*Pseudomonas putida*	*Trifolium*	Meyer and Linderman (1986)
Complex of indigenous AMF	*Pseudomonas* sp.	*Triticum aestivum*	Babana and Antoun (2005)
Complex of AMF	*Bacillus mycoides*	Herbaceous plant species	von Alten *et al.* (1993)

[a]AMF, arbuscular mycorrhizal fungi.

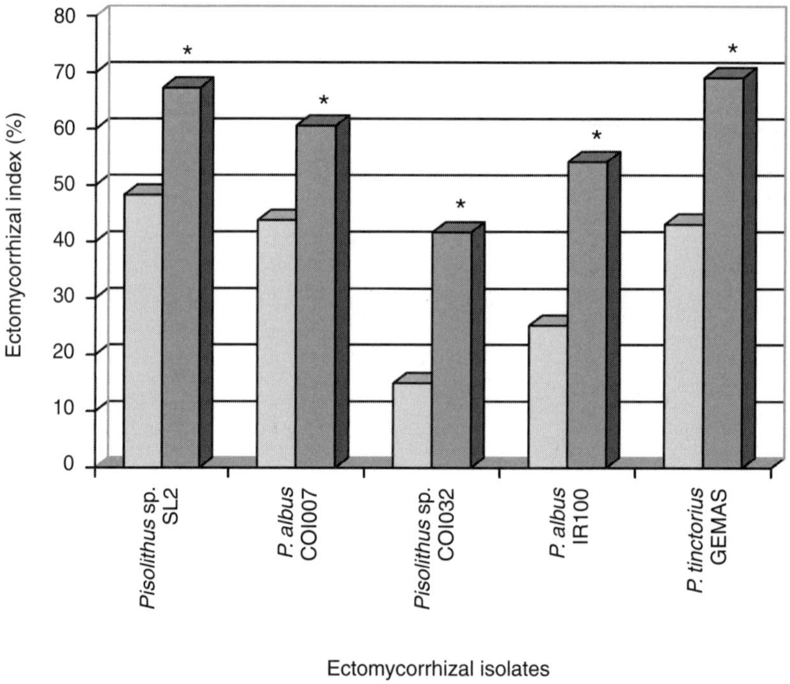

Fig. 4.1. Effect of *Pseudomonas monteilii* on ectomycorrhizal formation of *Acacia holosericea* with ectomycorrhizal fungi after 4 months' culture under glasshouse conditions (from Duponnois and Plenchette, 2003). * Significant effect of *P. monteilii* isolate on ectomycorrhizal formation for each fungal strain. ▪ Fungal strain inoculated alone; ▪ fungal strain inoculated with *P. monteilii* HR13.

translocation of large amounts of soil from various depths of the soil profile (Holt and Lepage, 2000). Termite mounds strongly influence their environment and have a considerable impact on the physical and chemical properties of the soil (Holt and Lepage, 2000), giving the termites a role as ecosystem engineers.

Previous microbiological studies of termite mounds have been carried out to compare the microbial communities in grass-, litter- and soil-feeding termite mounds (Duponnois *et al.*, 2005), and fluorescent pseudomonads have been detected only in powder from *Macrotermes subhyalinus* mounds (see Table 4.2). Phylogenetic analysis of these fluorescent pseudomonads showed that most were placed in *Pseudomonas monteilii* (Duponnois *et al.*, 2006). A dual inoculant comprising termite mound and ectomycorrhizal fungus (*Scleroderma* sp. or *S. dictyosporum*) significantly improved the root growth of *A. holosericea* seedlings after 4 months' culture in greenhouse conditions (see Table 4.3).

In addition, termite mound amendment significantly increased ectomycorrhizal formation of *A. holosericea* with both of the fungal isolates tested in the study (see Table 4.3). Another isolate of *P. monteilii* has also been

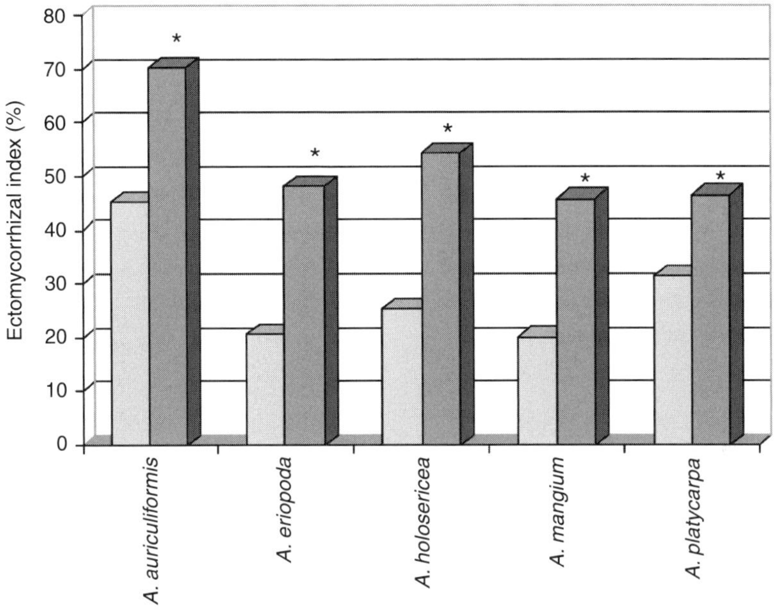

Fig. 4.2. Effect of *Pseudomonas monteilii* on ectomycorrhiza formation with *Pisolithus albus* isolate IR100 of five Australian *Acacia* species after 4 months' culture under glasshouse conditions (from Duponnois and Plenchette, 2003). For graph legend, see Fig. 4.1.

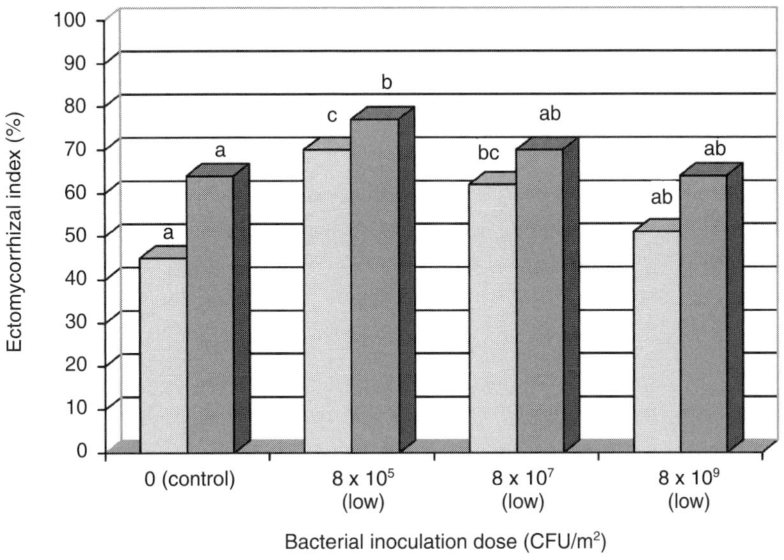

Fig. 4.3. Dose effect of *Pseudomonas fluorescens* on the percentage of Douglas fir short roots mycorrhiza with *Laccaria bicolor* S238, 23 weeks following fungal and bacterial inoculations. ■ Fungal inoculation dose, 50 mg dry weight mycelium/m^2; ■ fungal inoculation dose, 100 mg dry weight mycelium/m^2. For each fungal inoculation dose, columns indexed by the same letters are not significantly different according to a Scheffe test (from Frey-Klett *et al.*, 1999).

Table 4.2. Biological and chemical characteristics of powders from termite mounds (from Duponnois *et al.*, 2005).

Characteristic	*Cubitermes* sp.	*Macrotermes subhyalinus*	*Trinervitermes* sp.
NH_4^+ (μg N/g of dry mound powder)	40.9[b1]	9.4[a]	37.1[b]
NO_3^- (μg N/g of dry mound powder)	206.9[a]	3408.9[b]	42.3[a]
Available P (μg/g of dry mound powder)	7.7[b]	3.5[a]	10.2[c]
Microbial biomass (μg C/g of dry mound powder)	17.0[a]	22.5[ab]	36.5[b]
Fluorescent pseudomonads (x 10^2 CFU/g of dry mound powder)	< 1[a]	79.3[b]	< 1[a]
Actinomycetes (x 10^2 CFU/g of dry mound powder)	37.3[b]	39.5[b]	22.5[a]
Ergosterol (μg/g of dry mound powder)	1.381[b]	0.316[a]	1.717[b]

CFU, colony-forming unit.

[1]Data in the same line followed by the same letter are not significantly different, according to Student's t-test ($p < 0.05$).

Table 4.3. Effects of fungal inoculation and *Macrotermes subhyalinus* amendment on growth of *Acacia holosericea* and ectomycorrhizal colonization after 4 months' culture in greenhouse conditions (from Duponnois *et al.*, 2006).

Treatment	Shoot biomass (mg dry weight)	Root biomass (mg dry weight)	Ectomycorrhizal colonization (%)
Control	261[a1]	33[a]	0
Scleroderma sp. IR408	1458[c]	318[bc]	12.8[a]
S. disctyosporum IR412	964[b]	190[ab]	14.2[a]
Macrotermes subhyalinus (Ms)	1288[bc]	238[b]	0
IR408 + Ms	1051[b]	432[cd]	23.6[b]
IR412 + Ms	1140[bc]	606[d]	25.8[b]

[1]Data in the same column followed by the same letters are not significantly different, according to the Student's t-test ($p < 0.05$).

shown to stimulate ectomycorrhizal formation between *A. holosericea* and different fungal isolates (*S. dictyosporum, S. verrucosum, P. albus* and *P. tinctorius*) (Founoune *et al.*, 2002a; Duponnois and Plenchette, 2003), and to stimulate AM establishment between this *Acacia* species and *G. intraradices* (Duponnois and Plenchette, 2003).

More recently, a further strain of *P. monteillii* isolated from a *M. subhyalinus* termite mound has been found to stimulate ectomycorrhizal formation between *S. dictyosporum* and *A. holosericea* and to also stimulate the growth of the host plant and rhizobial development (Duponnois *et al.*, 2006 ; Table 4.4). This last result suggests that termite mound powder, which naturally contains some MHB, can provide a beneficial inoculum where the bacterial

Table 4.4. Effect of a *Pseudomonas monteilii* strain and/or *Scleroderma dictyosporum* strain on mycorrhizal formation, rhizobial development and growth of *Acacia holosericea* after 4 months' culture under glasshouse conditions (from Duponnois *et al.*, 2006).

	Treatments			
Effect	Control	*P. monteilii* KR9	*S. dictyosporum* IR412	KR9 + IR412
Shoot biomass (mg dry weight)	532[a1]	553[a]	1236[b]	1786[c]
Root biomass (mg dry weight)	184[a]	198[a]	536[b]	868[c]
Nodules per plant (*n*)	4.2[a]	4.6[a]	8.3[b]	12.4[c]
Total nodule weight per plant (mg)	6.8[a]	7.1[a]	15.9[b]	25.3[c]
Ectomycorrhizal colonization (%)	0	0	28.3[a]	48.5[c]

[1]Data in the same line followed by the same letter are not significantly different, according to the Student's t-test ($p < 0.05$).

cells may be protected from adverse environmental factors when the microbial inoculant is mixed to the soil.

Use of termite mound powder in controlled mycorrhization practice

As termite mound powder showed a significant promotion effect on mycorrhizal colonization and plant growth, a study has been conducted to test a new mycorrhizal inoculation process using *M. subhyalinus* termite mound powder as a 'catalyst' for mycorrhizal establishment, in order to minimize the volume of fungal inoculum required (Duponnois, 2007). The effects of the termite mound have been explored on mycorrhizal formation on an Australian acacia, *Acacia holosericea,* by an ECM fungus (*Pisolithus albus*) and an AM fungus (*Glomus intraradices*). *Acacia holosericea* is a fast-growing leguminous tree species frequently used in afforestation programmes in West Africa.

Seeds of *A. holosericea* were surface-sterilized with 95% sulphuric acid for 60 min and transferred aseptically to Petri dishes filled with 1% (w/v) agar/water medium. These plates were incubated at 25°C in the dark. The germinating seeds were used when rootlets were 1–2 cm long.

A disinfected (120°C, 40 min) sandy soil was mixed with 0, 1, 5 and 10% (v/v) of *M. subhyalinus* mound powder. Ectomycorrhizal and AM inoculation were performed by mixing the soils with fungal inoculum (10/1: v/v), with an autoclaved mixture of moistened vermiculite/peat moss at the same rate (ectomycorrhizal control) or with non-mycorrhizal millet roots and their rhizosphere soil (AM control). Plastic containers (30 x 30 x 5 cm) filled with soil mixtures were each planted with 100 pre-germinated seeds of *A. holosericea*. Seedlings were kept in a glasshouse under natural light (daylight approximately 12 h, mean daytime temperature 30°C) and watered daily without fertilizer.

After 2 months' culture, the seedlings were uprooted from the containers and transferred into 1 l pots (one seedling per pot) filled with the same sterilized soil but without termite mound powder amendment and fungal inoculation. Two months after the seedling transplantation, *A. holosericea* plants were uprooted and the root systems were gently washed. Root growth was measured, as well as the extent of ectomycorrhizal or arbuscular mycorrhizal colonization.

A conventional process used for controlled mycorrhization in forest nurseries was also undertaken by planting *A. holosericea* seedlings in 1 l pots filled with the same disinfected soil as before. For ectomycorrhizal inoculation, the soil was mixed with *P. albus* fungal inoculum (10/1: v/v), whereas the control treatment (without fungus) received an autoclaved mixture of moistened (MMN medium) vermiculite/peat moss at the same rate. For arbuscular mycorrhizal inoculation, a hole was made in each pot and filled with 1 g fresh millet root (mycorrhizal or not for the control treatment without fungus). The *A. holosericea* seedlings were kept in a glasshouse under natural light. After 4 months' culture, *A. holosericea* plants were uprooted when their shoots and shoot biomasses, and their mycorrhizal indexes, were measured as described before.

The results showed that termite mound amendment significantly enhanced the mycorrhizal formation by both types of fungi (see Tables 4.5 and 4.6). This stimulating effect could be attributed to: (i) the increase of plant growth induced by nutrients added with the termite mound powder (particularly root growth); and (ii) the introduction via the termite mound of a bacterial group (i.e. fluorescent pseudomonads) that could act as MHB (Duponnois and Plenchette, 2003). The data recorded from this study were similar to those obtained from the cultural practice using the conventional process of controlled mycorrhization (see Table 4.7).

One of the main problems encountered with the conventional controlled mycorrhization of forest planting stocks is the large quantity of fungal inoculum needed for the production of high-quality mycorrhizal plants in nursery conditions. In tropical and Mediterranean areas, the mycorrhizal inoculum dose added to the substrate per plant is usually of the order of 0.1 l/l of soil (ECM inoculation) and 1 g fresh weight of mycorrhizal root/l of soil (AM inoculation) (Duponnois *et al.*, 2007). With this innovative process, these quantities are significantly reduced and only 450 ml of ECM inoculum or AM inoculum (mixture of spores, mycorrhizal roots and rhizosphere soil) was required. About 100 mycorrhized *A. holosericea* plants can be produced with the same growth and mycorrhizal colonization compared with single mycorrhized plants produced by the conventional controlled mycorrhization procedure. Therefore, in tropical ecosystems where termite mounds are commonly found, the inoculation cost could be significantly reduced as only small amounts of fungal culture would be required.

Table 4.5. Effect of arbuscular mycorrhizal inoculation and termite mound powder amendment rates on *Acacia holosericea* growth and on mycorrhizal formation after 2 months' culture in 1 l pots (from Duponnois, 2007).

Factor[1]	Shoot biomass (mg dry weight)	Root biomass (mg dry weight)	Mycorrhizal colonization (%)
Fungal inoculum rate (%)			
0	190.1 (22.6) ^{1a2}	70.5 (8.5) ^{a}	0
1	783.5 (41.5) ^{b}	238.1 (18.4) ^{b}	41.8 (2.5) ^{a}
5	1236.1 (48.9) ^{c}	357.1 (20.3) ^{c}	70.1 (2.6) ^{c}
10	1510.1 (73.9) ^{d}	334.5 (35.9) ^{c}	55.5 (4.3) ^{b}
Termite mound amendment rate (%)			
0	709.6 (101.1) ^{a}	175.0 (25.6) ^{a}	31.0 (5.3) a
1	1000.5 (142.9) ^{b}	274.5 (43.3) ^{b}	47.3 (7.3) ^{b}
5	1019.5 (125.2) ^{b}	294.5 (30.7) ^{b}	41.5 (6.4) ^{b}
10	988.5 (113.4) ^{b}	256.1 (29.9) ^{ab}	47.5 (6.6) ^{b}
Fungal inoculum rate (FIR)	S	S	S
Termite mound amendment rate (TAR)	S	S	S
FIR x TAR	NS	S	S

S, significant ($p < 0.05$); NS, not significant ($p < 0.05$).

[1]Values are means of 20 replicates for fungal inoculum and termite mound amendment rates. Fungal inoculum rate factor is for all termite mound amendment rate treatments combined; the termite amendment factor is for all fungal inoculum rate treatments combined.

[2]Standard error of the mean.

[3]Data in the same column and for each factor followed by the same letter are not significantly different, according to the Newman–Keuls test ($p < 0.05$).

Table 4.6. Effect of ectomycorrhizal inoculation and termite mound powder amendment rates on *Acacia holosericea* growth and on mycorrhizal formation after 2 months' culture in 1 l pots (from Duponnois, 2007).

Factor[1]	Shoot biomass (mg dry weight)	Root biomass (mg dry weight)	Mycorrhizal colonization (%)
Fungal inoculum rate (%)			
0	191.0 (27.9)[2a3]	74.5 (12.9)[a]	0
1	556.5 (61.4)[b]	166.1 (15.2)[b]	57.6 (2.1)[a]
5	745.5 (43.6)[c]	221.1 (17.4)[c]	61.1 (1.5)[a]
10	1197.5 (96.6)[d]	364.5 (27.1)[d]	61.6 (1.9)[a]
Termite mound amendment rate (%)			
0	501.0 (77.1)[a]	167.0 (27.4)[a]	41.1 (5.5)[a]
1	612.2 (78.1)[ab]	195.0 (30.2)[ab]	41.6 (5.6)[a]
5	858.6 (96.9)[c]	250.5 (27.8)[b]	51.2 (6.8)[c]
10	719.1 (135.5)[bc]	213.5 (33.6)[b]	46.3 (6.4)[b]
Fungal inoculum rate (FIR)	S	S	S
Termite mound amendment rate (TAR)	S	S	S
FIR x TAR	S	S	S

For table legend see Table 4.5.

Table 4.7. Effect of ectomycorrhizal or AM inoculations on growth of *Acacia holosericea* and mycorrhizal formation after 4 months' culture under glasshouse conditions with the conventional process of controlled mycorrhization (from Duponnois, 2007).

Treatment	Shoot biomass (mg dry weight)	Root biomass (mg dry weight)	Mycorrhizal colonization (%)
AM inoculation			
Non-inoculated control	648.1[a1]	312.2[a]	0
Glomus intraradices	1834.1[b]	546.3[b]	59.3
ECM inoculation			
Non-inoculated control	550[a]	290[a]	0
Pisolithus albus IR100	1120.1[b]	560.2[b]	35.6

[1]For each type of fungal inoculation, data in the same column followed by the same letter are not significantly different, according to the Newman–Keuls test ($p < 0.05$).

Conclusion

Although controlled mycorrhization is acknowledged as a beneficial tool for improving the survival and productivity of tree species in degraded areas, the practice is poorly used in forest nurseries for the reasons described above. The microbial inoculation process could be significantly improved when the fungal inoculation is associated with Mycorrhiza Helper Bacteria or powder from *M. subhyalinus* termite mounds. These innovative procedures reduce the amount of fungal inoculum required while promoting the same mycorrhizal formation that is usually recorded with the conventional process of controlled mycorrhization. This process could be facilitated and encouraged in forest nurseries, but care must be taken as so far most of these results have been obtained in experimental conditions and their relevance in field conditions has yet to be assessed through long-term tree plantations.

References

Abdel-Fattah, G.M. and Mohamedin, A.H. (2000) Interactions between a vesicular–arbuscular mycorrhizal fungus (*Glomus intraradices*) and *Streptomyces coelicolor* and their effects on sorghum plants grown in soil amended with chitin of brawn scales. *Biology and Fertility of Soils* 32, 401–409.

Artursson, V. (2005) Bacterial–fungal interactions highlighted using microbiomics: potential application for plant growth enhancement. PhD thesis, University of Uppsala, Uppsala, Sweden.

Azcón, R., Rubio, R. and Barea, J.M. (1991) Selective interactions between different species of mycorrhizal fungi and *Rhizobium meliloti* strains, and their effects on growth, N_2-fixation (^{15}N) and nutrition of *Medicago sativa* L. *New Phytologist* 117, 399–404.

Babana, A.H. and Antoun, H. (2005) Biological system for improving the availability of Tilemsi phosphate rock for wheat (*Triticum aestivum* L.) cultivated in Mali. *Nutrient Cycling in Agroecosystems* 72, 147–157.

Bagyaraj, D.J. and Menge, J.A. (1978) Interaction between a VA mycorrhiza and *Azotobacter* and their effects on the rhizosphere microflora and plant growth. *New Phytologist* 80, 567–573.

Barea, J.M., Andrade, G., Bianciotto, V., Dowling, D., Lohrke, S., Bonfante, P., O'Gara, F., and Azcón-Aguilar, C. (1998) Impact on arbuscular mycorrhiza formation of *Pseudomonas* strains used as inoculants for biocontrol of soil-borne fungal plant pathogens. *Applied Environmental Microbiology* 64, 2304–2307.

Bending, G.D., Poole, E.J., Whipps, J.M. and Read, D.J. (2002) Characterisation of bacteria from *Pinus sylvestris – Suillus luteus* mycorrhizas and their effects on root–fungus interactions and plant growth. *FEMS Microbiology and Ecology* 39, 219–227.

Brundrett, M.C. (2002) Coevolution of roots and mycorrhizas of land plants. *New Phytologist* 154, 275–304.

Budi, S.W., van Tuinen, D., Martinotti, G. and Gianinazzi, S. (1999) Isolation from the *Sorghum bicolor* mycorrhizosphere of a bacterium compatible with arbuscular mycorrhiza development and antagonistic towards soilborne fungal pathogens. *Applied and Environmental Microbiology* 65, 5148–5150.

Caravaca, F., Garcia, C., Hernandez, M.T. and Roldan, A. (2002) Aggregate stability changes after organic amendment and mycorrhizal inoculation in the afforestation of a semiarid site with *Pinus halepensis*. *Applied Soil Ecology* 19, 199–208.

Chen, B.D., Tao, H.Q., Christie, P. and Wong, M.H. (2003) The role of arbuscular mycorrhiza in zinc uptake by red clover growing in a calcareous soil spiked with various quantities of zinc. *Chemistry* 50, 839–846.

de Oliveira, V. (1988) Interactions entre les micro-organismes du sol et l'établissement de la symbiose ectomycorhizienne chez le hêtre (*Fagus sylvatica* L.) avec *Hebeloma crustuliniforme* (Bull. ex Saint-Amans) Quél. et *Paxillus involutus* (Batsch ex Fr.). PhD thesis, University of Nancy, Nancy, France.

Dunabeita, M., Rodriguez, N., Salcedo, I. and Sarrionandia, E. (2004) Field mycorrhization and its influence on the establishment and development of the seedlings in a broadleaf plantation in the Basque Country. *Forest Ecology and Management* 195, 129–139.

Dunstan, W.A., Malajczuk, N. and Dell, B. (1998) Effects of bacteria on mycorrhizal development and growth of container-grown *Eucalyptus diversicolor* F. Muell. seedlings. *Plant and Soil* 201, 241–249.

Duponnois, R. (1992) Les bactéries auxiliaires de la mycorhization du Douglas (*Pseudotsuga menziesii* (Mirb.) Franco) par *Laccaria laccata* souche S238. PhD Thesis, University of Nancy, Nancy, France.

Duponnois, R. (2007) Nouvelles compositions d'inocula fongiques, leur procédé de préparation et leur application à l'amélioration de la croissance des cultures. Patent no PCT/FR2007/000828.

Duponnois, R. and Garbaye, J. (1991) Techniques for controlled synthesis of the Douglas fir–*Laccaria laccata* ectomycorrhizal symbiosis. *Annals of Forest Sciences* 48, 239–251.

Duponnois, R. and Plenchette, C. (2003) A mycorrhiza helper bacterium enhances ectomycorrhizal and endomycorrhizal symbiosis of Australian *Acacia* species. *Mycorrhiza* 13, 85–91.

Duponnois, R., Garbaye, J., Bouchard, D. and Churin, J.L. (1993) The fungus-specificity of mycorrhization helper bacteria (MHBs) used as an alternative to soil fumigation for ectomycorrhizal inoculation of bare-root Douglas fir planting stocks with *Laccaria laccata*. *Plant and Soil* 157, 257–262.

Duponnois, R., Paugy, M., Thioulouse, J., Masse, D. and Lepage, M. (2005) Functional diversity of soil microbial community, rock phosphate dissolution and growth of *Acacia seyal* as influenced by grass-, litter- and soil-feeding termite nest structure amendments. *Geoderma* 124, 349–361.

Duponnois, R., Assigbetse, K., Ramanankierana, Kisa, M., Thioulouse, J. and Lepage, M. (2006) Litter-forager termite mounds enhance the ectomycorrhizal symbiosis between *Acacia holosericea* A. Cunn. Ex G. Don and *Scleroderma dictyosporum* isolates. *FEMS Microbiology Ecology* 56, 292–303.

Duponnois, R., Plenchette, C., Prin, Y., Ducousso, M., Kisa, M., Bâ, A.M. and Galiana, A. (2007) Use of mycorrhizal inoculation to improve reafforestation process with Australian Acacia in Sahelian ecozones. *Ecological Engineering* 29, 105–112.

Fester, T., Maier, W. and Strack, D. (1999) Accumulation of secondary compounds in barley and wheat roots in response to inoculation with an arbuscular mycorrhizal fungus and co-inoculation with rhizosphere bacteria. *Mycorrhiza* 8, 241–246.

Founoune H., Duponnois R., Meyer J.M., Thioulouse J., Masse D., Chotte J.L. and Neyra M. (2002a) Interactions between ectomycorrhizal symbiosis and fluorescent pseudomonads on *Acacia holosericea* : isolation of mycorrhiza helper bacteria (MHB) from a soudano-Sahelian soil. *FEMS Microbiology Ecology* 41, 37–46.

Founoune, H., Duponnois, R., Bâ, A.M., Sall, S., Branget, I., Lorquin, J., Neyra, M. and Chotte, J.L. (2002b) Mycorrhiza helper bacteria stimulated ectomycorrhizal symbiosis of *Acacia holosericea* with *Pisolithus alba*. *New Phytologist* 153, 81–89.

Francis, D.F. and Thornes, J.B. (1990) Matorral: erosion and reclamation. In: Albaladejo, J., Stocking, M.A. and Diaz, E. (eds) *Soil Degradation and Rehabilitation in Mediterranean Environmental Conditions*. CSIC, Murcia, Spain, pp. 87–115.

Frey-Klett, P., Churin, J.-L., Pierrat, J.-C. and Garbaye, J. (1999) Dose effect in the dual inoculation of an ectomycorrhizal fungus and a mycorrhiza helper bacterium in two forest nurseries. *Soil Biology and Biochemistry* 31, 1555–1562.

Frey-Klett, P., Garbaye, J. and Tarkka, M. (2007) The mycorrhiza helper bacteria revisited. *New Phytologist* 176, 22–36.

Gamalero, E., Trotta, A., Massa, N., Copetta, A., Martinotti, M.G. and Berta, G. (2004) Impact of two fluorescent pseudomonads and an arbuscular mycorrhizal fungus on tomato plant growth, root architecture and P acquisition. *Mycorrhiza* 14, 185–192.

Garbaye, J. and Bowen, G.D. (1989) Stimulation of ectomycorrhizal infection of *Pinus radiata* by some microorganisms associated with the mantle of ectomycorrhizas. *New Phytologist* 112, 383–388.

Holt, J.A. and Lepage, M. (2000) Termites and soil properties. In: Abe, T., Bignell, D.E. and Higashi, M. (eds) *Termites: Evolution, Sociality, Symbioses, Ecology.* Kluwer Academic Publishers, Dordrecht, Netherlands, pp. 389–407.

Mamatha, G., Bagyaraj, D.J. and Jaganath, S. (2002) Inoculation of field-established mulberry and papaya with arbuscular mycorrhizal fungi and a mycorrhiza helper bacterium. *Mycorrhiza* 12, 313–316.

Marx, D.H. (1980) Ectomycorrhiza fungus inoculations: a tool to improve forestation practices. In: Mikola, P. (ed.) *Tropical Mycorrhiza Research.* Oxford University Press, Oxford, UK, pp. 13–71.

Meyer, J.R. and Linderman, R.G. (1986) Response of subterranean clover to dual-inoculation with vesicular-arbuscular mycorrhizal fungi and a plant growth-promoting bacterium, *Pseudomonas putida. Soil Biology and Biochemistry* 18, 185–190.

Mosse, B. (1962) The establishment of vesicular–arbuscular mycorrhiza under aseptic conditions. *Journal of General Microbiology* 27, 509–520.

Paula, M.A., Urquiaga, S. and Sijqueira, J.O. (1992) Synergistic effects of vesicular–arbuscular mycorrhizal fungi and diazotrophicus bacteria on nutrition and growth of sweet potato (*Ipomoea batatas*). *Biology and Fertility of Soils* 14, 61–66.

Poole, E.J., Bending, G.D., Whipps, J.M. and Read, D.J.(2001) Bacteria associated with *Pinus sylvestris – Lactarius rufus* ectomycorrhizas and their effects on mycorrhiza formation *in vitro*. *New Phytologist* 151, 743–751.

Rao, N.S.S., Tilak, K.V.B.R. and Singh, C.S. (1985) Effect of combined inoculation of *Azospirillum brasilense* and vesicular–arbuscular mycorrhiza on pearl millet (*Pennisetum americanum*). *Plant and Soil* 84, 283–286.

Requena, N., Jimenez, I., Toro, M. and Barea, J.M. (1997) Interactions between plant growth-promoting rhizobacteria (PGPR), arbuscular mycorrhizal fungi and *Rhizobium* spp. in the rhizosphere of *Anthyllis cytisoides,* a model legume for revegetation in Mediterranean semi-arid ecosystems. *New Phytologist* 136, 667–677.

Requena, N., Perez-Solis, E., Azcón-Aguilar, C., Jeffries, P. and Barea, J.M. (2001) Management of indigenous plant–microbe symbioses aids restoration of desertified ecosystems. *Applied and Environmental Microbiology* 67, 495–498.

Rózycki, H., Kampert, M., Strzelczyk, E., Li, C.Y. and Perry, D.A. (1994) Effect of different soil bacteria on mycorrhizae formation in pine (*Pinus sylvestris* L.). *Folia Forestalia Polonica, Series A (Forestry)* 36, 91–102.

Schreiner, R.P., Mihara, K.L., McDaniel, K.L. and Bethlenfalvay, G.J. (2003) Mycorrhizal fungi influence plant and soil functions and interactions. *Plant Soil* 188, 199–209.

Schrey, S.D., Schellhammer, M., Ecke, M., Hampp, R. and Tarkka, M.T. (2005) Mycorrhiza helper bacterium *Streptomyces* AcH 505 induces differential gene expression in the ectomycorrhizal fungus *Amanita muscaria. New Phytologist* 168, 205–216.

Simard, S.W. and Durall, D.M. (2004) Mycorrhizal networks: a review of their extent, function, and importance. *Canadian Journal of Botany* 82, 1140–1165.

Smith, S. and Read, J. (1997) Mycorrhizal symbiosis, 2nd edn. Clarendon, Oxford, UK, pp 22–30.

Toro, M., Azcón, R. and Barea, J.M. (1997) Improvement of arbuscular mycorrhiza development by inoculation of soil with phosphate-solubilizing rhizobacteria to improve rock phosphate bioavailability (P_{32}) and nutrient cycling. *Applied and Environmental Microbiology* 63, 4408–4412.

van der Heijden, M.G.A., Klironomos, J.N., Ursic, M., Moutoglis, P., Streitwolf-Engel, R., Boller, T., Wiemken, A. and Sanders, I.R. (1998) Mycorrhizal fungal diversity determines plant biodiversity ecosystem variability and productivity. *Nature* 396, 69–72.
Vivas, A., Marulanda, A., Ruiz-Lozano, J.M., Barea, J.M. and Azcón R. (2003) Influence of a *Bacillus* sp. on physiological activities of two arbuscular mycorrhizal fungi and on plant responses to PEG-induced drought stress. *Mycorrhiza* 13, 249–256.
von Alten, H., Lindemann, A. and Schönbeck, F. (1993) Stimulation of vesicular-arbuscular mycorrhiza by fungicides or rhizosphere bacteria. *Mycorrhiza* 2, 167–173.
Vósatka, M. and Gryndler, M. (1999) Treatment with culture fractions from *Pseudomonas putida* modifies the development of *Glomus fistulosum* mycorrhiza and the response of potato and maize plants to inoculation. *Applied Soil Ecology* 11, 245–251.
Will, M.E. and Sylvia, D.M. (1990) Interaction of rhizosphere bacteria, fertilizer, and vesicular-arbuscular mycorrrhizal fungi sea oats. *Applied and Environmental Microbiology* 56, 2073–2079.
Xie, Z.P., Staehelin, C., Vierheilig, H., Wiemken, A., Jabbouri, S., Broughton, W.J., Vogeli-Lange, R. and Boller, T. (1995) Rhizobial nodulation factors stimulate mycorrhizal colonization of nodulating and non-nodulating soybeans. *Plant Physiology* 108, 1519–1525.

5 Fungi in the Tree Canopy: an Appraisal

K.R. Sridhar

Microbiology and Biotechnology, Department of Biosciences, Mangalore University, Mangalore, Karnataka, India

Introduction

The collection of above-ground plant organs within a community is called the canopy (Moffett, 2000). The forest canopy is an aggregate of all of the tree crowns in a forest stand, and is a combination of foliage, twigs, fine branches, epiphytes and atmosphere (Parker, 1995). It represents a mosaic of structurally complex and ecologically important subsystems conducive to the evolution of flora, fauna and microorganisms that are either rare or not encountered on the forest floor (Nadkarni *et al.*, 2001). Forest canopies constitute the interface between 90% of the earth's terrestrial biomass and the atmosphere and, as such, they are recognized as a cradle of biological diversity (Hammond *et al.*, 1977; Ozanne *et al.*, 2003). Canopies trap a considerable amount of organic matter (e.g. leaf litter, twigs, inflorescence) and transform it into 'crown humus', which supports innumerable life forms.

The trunks, branches and leaves of the canopy provide mechanical support for many epiphytes, including tree ferns (Nadkarni *et al.*, 2001). As well as non-vascular (cryptogamic) epiphytes (lichens, bryophytes and algae), canopies are known to support extensive numbers of vascular epiphytes (> 24,000 species), hemi-epiphytes and parasites (Kress, 1986; Nadkarni *et al.*, 2001).

The common terms for epiphytes present in forest canopies are: humus epiphytes, tank and trash-basket epiphytes, ant-associated epiphytes, bark epiphytes and atmospheric epiphytes (Nadkarni *et al.*, 2001). These epiphytes support a complex community of fauna and microorganisms, including micro- and macrofungi (Counts *et al.*, 2000). Organic soils trapped in canopy pockets are mainly derived from decomposing heartwood and phyllode litter, with high levels of exchangeable cations, total nitrogen and lower aluminum levels than the terrestrial soils. Interestingly, *Acacia koa* is known for canopy nodulation, which is a unique symbiosis between the adventitious roots and bradyrhizobia (Leary *et al.*, 2004). Such canopy nodulation helps to maintain symbiosis and atmospheric nitrogen fixation under stressed rhizosphere and terrestrial environments. Canopy fauna consists of a wide variety of arthropods, gastropods, annelids, amphibians, reptiles and mammals (Ellwood and Foster, 2004).

Although a conservative estimate of global fungal diversity is 1.5 million species, only about 5–6% of these are currently described (Hawksworth, 1991). Fungi are an important component of the biota of forest canopy habitats, but mycology has been largely neglected in the 20 or so years of canopy research (Unterseher *et al.*, 2005). A variety of fungi have been reported from the canopy in some forest pathology studies and biodiversity surveys. A few studies have been undertaken on the life style of conidial and water-borne fungi in tree canopies, but these have largely been limited to temperate regions (Gönczöl and Révay, 2006). This chapter offers a brief commentary on fungal association in canopies, with a major emphasis on the significance, occurrence, ecology and survival of water-borne conidial fungi.

Fungi in the Canopy

Forest canopies consist of many surfaces in complex structures, and these give rise to many microhabitats and microclimates for insects and fungi. For instance, tall, wet tropical forests develop long vertical gradients of temperature and relative humidity, which leads to the distribution and stratification of fungal species (Hedger, 1985). Based on abundance and partitioning of resources, Hammond (1992) hypothesized that the overall forest architecture may better predict the number of fungi and small animals present in a given area than plant species richness. Several studies have been carried out to assess fungi in canopies, although they have not been designed exclusively to investigate canopy fungi. Fungi have direct associations with host canopy through colonizing the foliar and twig surface as epiphytes, with internal foliage and bark/wood as endophytes, and with dead wood as decomposers (Stone *et al.*, 1996). A variety of fungi have been reported from the forest canopies, including endophytic, pathogenic, phylloplane and lignicolous fungi (Lodge and Cantrell, 1995). Other fungal groups found in canopies include aquatic/water-borne, aero-aquatic and dematiaceous fungi (see Tables 5.1, 5.2 and 5.3).

Endophytic fungi

Fungi that grow asymptomatically in any living part of a plant are referred to as endophytic fungi. Due to the wide distribution of endophytic fungi, they can become important cryptic guilds in canopies. Endophytes are well documented in the aerial parts of rain forest canopies (Laessøe and Lodge, 1994), in epiphytes such as Araceae, Bromeliaceae and Orchidaceae (Dreyfuss

Table 5.1. Water-borne conidial fungi recovered from tree canopy habitats.

Taxon	Canopy habitat	Reference
Alatosessilispora bibrachiata K. Ando & Tubaki	Tree leaves, Japan	2
Alatospora acuminata Ingold	Tree holes, Hungary	9
Alatospora prolifera Ingold	Tree holes, Hungary	9, 10
	Stemflow, Hungary	11
Anguillospora crassa Ingold	Fern leaves, roots and rhizome; trapped leaves, India	18
	Throughfall, Romania	12
A. longissima (Sacc. & P. Syd.) Ingold	Fern leaves and roots; trapped leaves, India	18
A. pseudolongissima Ranzoni	Tree leaves, Poland	5
	Tree snow, Poland	7
Arborispora palma K. Ando	Tree leaves, Japan	1
	Tree snow, Poland	7
A. paupera Marvanová and Bärl.	Stemflow, Germany and Hungary	11, 12
	Throughfall, Sweden	12
Articulospora proliferata A. Roldán & W.J.J. van der Merwe	Stemflow, Poland	8
	Tree leaves, Poland	6
Brachiosphaera jamaicensis (J.L. Crane and Dumont) Nawawi	Tree leaves, Poland	6
Campylospora chaetocladia Ranzoni	Tree leaves, Poland	5
	Fern leaves, India	18
Ceratosporium cornutum Matsush.	Tree leaves, Japan	3
	Throughfall, Hungary and Sweden	12
	Honeydew, Italy	16
Clavariopsis aquatica de Wild.	Stemflow and throughfall, Canada	15
C. brachycladia Tubaki	Tree leaves, Poland	5
Colispora cavincola J. Gönczöl & Révay	Tree holes, Hungary	9, 10
C. elongata Marvanová	Stemflow, Poland	8
Curucispora flabelliformis K. Ando	Stemflow, Germany and Hungary	12
C. ombrogena K. Ando & Tubaki	Tree leaves, Japan	2
	Tree leaves, Poland	6
C. ponapensis Matsush.	Tree leaves, Japan	2
	Stemflow, Gernany and Hungary	11, 12
	Throughfall, Sweden and Romania	12
Dactylella submersa (Ingold) Sv. Nilsson	Tree leaves, Poland	5
Dendrospora erecta Ingold	Tree holes, India	13
Dicranidion fissile K. Ando & Tubaki	Stemflow, Japan	3
Dimorphospora foliicola Tubaki	Tree leaves, Poland	5
	Tree holes, Hungary	10
	Tree holes, USA	14
Dwayaangam colodena Sokolski & Bérubé	Endophyte of Picea needles, Canada	17

Continued

Table 5.1. – *Continued.*

Taxon	Canopy habitat	Reference
D. cornuta Descals	Tree holes, Hungary	10
	Treeflow, Germany	12
D. dichotoma Nawawi	Stemflow, Hungary	11
	Stemflow, Germany and Hungary	12
	Honeydew, Italy	16
D. yakuensis (Matsush.) Matsush.	Tree leaves, Japan	2
	Stemflow, Germany and Hungary	11, 12
Flabellospora crassa Alas.	Tree leaves, Japan	2
	Fern leaves, India	18
	Tree holes, India	13
F. multiradiata Nawawi	Tree holes, India	13
F. verticillata Alas.	Tree leaves, Japan	2
	Fern roots; trapped leaves, India	18
	Tree holes, India	13
Flagellospora curvula Ingold	Fern leaves, roots; trapped leaves, India	18
	Tree holes, India	13
F. penicillioides Ingold	Fern leaves, rhizome; trapped leaves, India	18
	Tree holes, India	13
Gyoerffyella biappendiculata (G.R.W. Arnold) Ingold	Stemflow, Canada	4
	Stemflow and throughfall, Canada	15
G. gemellipara Marvanová	Stemflow and trapped leaves, Canada	4
	Stemflow and throughfall, Canada	15
G. myrmecophagiformis Melnik & Dudka	Honeydew, Croatia	16
G. tricapillata Ingold	Tree leaves, Poland	5
Heliscus lugdunensis Sacc. & Théry	Tree leaves, Poland	5
Hymenoscyphus tetracladius Abdullah (Descals & J. Webster)	Tree holes, Hungary	9
Ingoldiella hamata D.E. Shaw	Tree holes, India	13
Isthmolongispora minima Matsush.	Tree holes, Hungary	9
Isthmotricladia gombakiensis Nawawi	Tree holes, India	13
I. laeensis Matsush.	Tree holes, Hungary	11
	Tree holes, India	13
Lateriramulosa uni-inflata Matsush.	Stemflow and tree holes, Hungary	11
	Stemflow and throughfall, Canada	15
Lemonniera aquatica de Wild.	Tree leaves, Poland	5
L. cornuta Ranzoni	Stemflow, Hungary	11
L. terrestris Tubaki	Stemflow, Hungary	11
Lunulospora curvula Ingold	Tree leaves, Poland	5
	Tree holes, India	13
L. cymbiformis K. Miura	Tree holes, India	13
Margaritispora aquatica Ingold	Tree leaves, Poland	5
Microstella pluvioriens K. Ando & Tubaki	Tree leaves, Japan	2
Miladina lecithina (Cooke) Svrček)	Tree leaves, Poland	5
Mirandiana corticola Arnaud	Tree holes, Hungary	10
	Tree snow, Poland	7

Table 5.1. – *Continued.*

Taxon	Canopy habitat	Reference
Mycocentrospora aquatica S.H. Iqbal	Tree snow, Poland	7
Ordus tribrachiatus K. Ando & Tubaki	Tree leaves, Japan	2
	Tree leaves, Poland	6
Phalangispora constricta Nawawi & J. Webster	Tree holes, India	13
Pleuropedium tricladioides Marvanoá & S.H. Iqbal	Stemflow, Poland	8
Pyramidospora casuarinae Sv. Nilsson	Tree leaves, Poland	5
P. fluminea Miura & K.I. Kudo	Tree leaves, Poland	5
Retiarius bovicornutus D.L. Olivier	Stemflow, Hungary	11, 12
	Stemflow, Germany	12
	Throughfall, Hungary, Romania and Sweden	12
	Fern leaves; trapped leaves, India	18
	Honeydew, Grece and Italy	16
Seuratia millardetii (Racib.) Meeker)	Throughfall, Romania	12
Sigmoidea prolifera (R.H. Peterson) J.L. Crane	Stemflow, Poland	8
Speiropsis pedatospora Tubki	Throughfall, Romania	12
Tetracladium marchalianum de Wild.	Stemflow and tree holes, Hungary	11
T. maxilliforme (Rostr.) Ingold	Tree leaves, Poland	6
	Stemflow and tree holes, Hungary	11
T. setigerum (Grove) Ingold	Tree holes, Hungary	11
Titaea clarkeae Ellis & Everh.	Tree leaves, Poland	6
	Stemflow, Hungary	11, 12
	Stemflow, Germany	12
	Throughfall, Romania	12
T. complexa K. Matsush. and Matsush.	Stemflow, Germany and Hungary	12
	Throughfall, Romania	12
Tricellula aquatica J. Webster	Tree leaves, Poland	6
	Stemflow and tree holes, Hungary	11
	Stemflow, Hungary	12
T. aurantiaca (Haskins) Arx	Tree holes, Hungary	11
T. inaequalis Beverw.	Tree leaves, Poland	6
Tricladiella pluvialis K. Ando & Tubaki	Stemflow, Japan	3
	Tree leaves, Poland	6
	Throughfall, Romania	12
Tricladium angulatum Ingold	Tree holes, India	13
T. castaneicola B. Sutton	Tree holes, Hungary	9, 10
	Stemflow, Hungary	12
T. patulum Marvanová	Tree leaves, Poland	15
	Tree holes, USA	14
T. splendens Ingold	Tree holes, USA	14
	Tree holes, India	13

Continued

Table 5.1. – *Continued.*

Taxon	Canopy habitat	Reference
Trifurcospora irregularis (Matsush.) K. Ando & Tubaki	Tree snow, Poland	7
	Tree leaves, Poland	6
	Stemflow, Hungary	11, 12
	Stemflow, Germany	12
	Throughfall, Romania and Sweden	12
Trinacrium parvisporum Matsush.	Honeydew, Italy and Greece	16
T. robustum Tzean & J.L. Chen	Stemflow, Hungary	11
	Stemflow, Germany and Hungary	12
	Throughfall, Romania	12
	Honeydew, Italy	16
	Tree holes, India	13
T. subtile Riess	Tree snow, Poland	7
	Treeflow, Germany	12
	Throughfall, Hungary and Romania	12
	Fern leaves, India	18
	Honeydew, Italy	16
	Tree holes, India	13
Tripospermum acerinum P. Syd.	Tree snow, Poland	7
T. camelopardus Ingold, Dunn & P.J. McDougall	Stemflow, Hungary	11, 12
	Tree leaves, Poland	5
	Stemflow, Germany	12
	Throughfall, Romania and Sweden	12
	Honeydew, Italy and Greece	16
	Endophyte of *Picea* needles, Canada	17
	Tree holes, India	13
T. gardneri (Berk.) Speg.	Tree leaves, Poland	6
T. inflacatum K. Ando & Tubaki	Stemflow, Japan	3
	Tree leaves, Poland	6
	Fern leaves, India	18
T. myrti (Lind.) S. Hughes	Tree snow, Poland	7
	Tree leaves, Poland	6
	Tree holes, Hungary	10, 11
	Tree holes, India	13
	Stemflow, Hungary	11, 12
	Stemflow, Germany	12
	Throughfall, Hungary	12
	Floral honey, South Africa	16
	Honeydew, Croatia, Greece and Italy	16
	Endophyte of *Picea* needles, Canada	17
T. porosporiferum Matsush.	Tree leaves, Poland	6
Triscelophorus acuminatus Nawawi	Fern leaves, roots; trapped leaves, India	18
	Tree holes, India	13
T. konajensis K.R. Sridhar & Kaver.	Fern leaves, roots; trapped leaves, India	18
	Tree holes, India	13
T. monosporus Ingold	Tree holes, India	13

Table 5.1. – *Continued.*

Taxon	Canopy habitat	Reference
Trisulcosporium acerinum Hudson & Sutton	Throughfall, Romania	12
Tumularia aquatica (Ingold) Descals & Marvanová	Tree holes, USA	14
Vargamyces aquaticus (Dudka) Tóth	Stemflow, Hungary	5
	Tree snow, Poland	7
	Tree holes, Hungary	10
Varicosporium elodeae W. Kegel	Trapped leaves, Canada	4
	Tree holes, Hungary	10, 11
	Tree holes, India	13
	Stemflow, Hungary	11
Volucrispora aurantiaca Haskins	Tree leaves, Poland	6
Wiesneriomyces conjunctosporus Kuthub. & Nawawi	Tree leaves, Poland	6
Ypsilina graminea (Ingold, P.J. McDougall & Dann) Descals, J. Webster & Marvanová	Tree holes, Hungary	10
	Tree holes, India	13

References: 1, Ando and Kawamoto, 1986; 2, Ando and Tubaki, 1984a; 3, Ando and Tubaki, 1984b; 4, Bandoni, 1981; 5, Czeczuga and Orłowska, 1994; 6, Czeczuga and Orłowska, 1998a; 7, Czeczuga and Orłowska, 1998b; 8, Czeczuga and Orłowska, 1999; 9, Gönczöl, 1976; 10, Gönczöl and Révay, 2003; 11, Gönczöl and Révay, 2004; 12, Gönczöl and Révay, 2006; 13, Karamchand and Sridhar, 2008; 14, Kaufman *et al.*, 2008; 15, Mackinnon, 1982; 16, Magyar *et al.*, 2005; 17, Sokolski *et al.*, 2006; 18, Sridhar *et al.*, 2006.

and Petrini, 1984), in medicinal plants (Raviraja *et al.*, 2006), in tropical forest trees (Suryanarayanan *et al.*, 2003; Murali *et al.*, 2006) and in mangrove plants and mangrove associates (Ananda and Sridhar, 2002; Maria and Sridhar, 2003). Several endophytic fungal associations with plants are antagonistic to insect herbivores, and the canopy endophytes may protect plant species from herbivores (Carroll, 1991; Azevedo *et al.*, 2000). Many canopy fungi produce secondary metabolites and the bark endophyte, *Taxomyces andreanae,* has been considered for *in vitro* production of the anticancer agent taxol (Stierle *et al.*, 1993; Strobel *et al.*, 1993).

McCutcheon *et al.* (1993) have conservatively estimated that up to 10^{11} individuals of the endophytic fungus, *Rhabdocline parkeri,* may be involved in colonizing a single tree. Zonation has also been seen in endophytic microfungal distribution in canopies (Unterseher *et al.*, 2007). A species of *Phoma* was reported to prefer significantly shaded leaves from the lower canopy layer of *Acer* and *Quercus,* while *Sordaria fimicola* preferred the sun-exposed leaves of the upper tree crowns. Seasonal patterns appear to occur in some endophytic fungi, and *Apiognomonia errabunda* was found to be common in young leaves of *Acer* and *Quercus* in the spring, but was almost completely absent in aged autumn leaves (Unterseher *et al.*, 2007). The endophytic fungal host range, specificity, species complexities, modes of

Table 5.2. Aero-aquatic and dematiaceous conidial fungi recovered from tree canopy habitats.

Taxon	Canopy habitat	Reference
Beverwykella cerebriformis Nawawi & Kuthub.	Stemflow, Poland	1
B. pulchrum Hol-Jech & Mercado	Stemflow, Poland	1
Blodgettia indica Subram.	Stemflow, Poland	1
Camposporium cambrense S. Hughes	Stemflow, Hungary	5
C. japonicum Ichinoe	Stemflow, Hungary	5
C. ontariense Matsush.	Stemflow, Hungary	5
C. pellucidum (Grove) S. Hughes	Stemflow, Hungary	5
	Tree snow, Poland	12
	Tree holes, Hungary	3
Camposporium spp.	Tree holes, Hungary	4
	Honeydew, Italy	8
	Stemflow, Hungary	6
Diplocladiella scalaroides G. Arnaud	Floral honey, South Africa	8
	Tree holes, Hungary	4
	Stemflow, Hungary	5
Helicodendron paradoxum Peyr.	Tree holes, Hungary	3
Helicodendron triglitziense (Jaap) Linder	Tree holes, Hungary	4
H. tubulosum (Riess) Linder	Tree holes, Hungary	4
H. westerdijkiae Beverw.	Tree holes, Hungary	3
	Tree holes, Hungary	4
Helicodendron sp.	Honeydew, Croatia, Greece and Italy	8
Helicoma sp.	Stemflow, Hungary	5
Helicomyces sp.	Stemflow, Hungary	5
Helicoon macrosporum Aa & Samson	Stemflow, Poland	1
Helicosporium sp.	Epipiphytic fern leaf, India	9
	Tree holes, India	7
Monodictys peruviana Matush.	Tree snow, Poland	2
Monodictys putredinis (Wallr.) S. Hughes	Tree holes, Hungary	4
Tetraploa aristata Berk. & Broome	Floral honey, South Africa	8
	Tree holes, India	7

References: 1, Czeczuga and Orłowska, 1998a; 2, Czeczuga and Orłowska, 1998b; 3, Gönczöl, 1976; 4, Gönczöl and Révay, 2003; 5, Gönczöl and Révay, 2004; 6, Gönczöl and Révay, 2006; 7, Karamchand and Sridhar, 2008; 8, Magyar *et al.*, 2005; 9, Sridhar *et al.*, 2006.

dispersal, secondary metabolites and physiological specializations for specific host require intensive investigation. From a biodiversity standpoint, work on endophytic fungi is still in the inventory and documentation stage.

Phylloplane fungi

A variety of phylloplane (surface of an unshed leaf) fungi have been obtained from forest canopies. A study on the distribution of phyllosphere fungi within the canopy of giant dogwood (*Swida controversa*) recovered between 13 and 33 fungi from the interior and surfaces of leaves (Osono and Mori, 2004). The fungal species composition was markedly different between the interior and

Table 5.3. Genera of unidentified species of water-borne conidial fungi recovered from tree canopy habitats (see also Table 5.2 for some aero-aquatic and dematiaceous genera).

Taxon	Canopy habitat	Reference
Arborispora spp.	Tree holes, Hungary	2
	Epipiphytic fern, India	9
Articulospora sp.	Honeydew, Italy	6
Atichia sp.	Honeydew, Italy and Greece	6
Cornutispora sp.	Stemflow, Hungary	3
Curucispora spp.	Honeydew, Italy	6
	Stemflow, Hungary	4
Cylindrocarpon spp.	Epiphytic fern leaf, rhizome and root, India	9
	Tree holes, India	5
Dactylaria sp.	Stemflow, Hungary	3
Dicranidion sp.	Honeydew, Croatia, Greece and Italy	6
Dwayaangam spp.	Honeydew, Italy	6
	Tree holes, Hungary	2
	Stemflow, Hungary	3
	Stemflow, Germany	4
	Throughfall, Hungary, Romania and Sweden	4
Flagellospora sp.	Leaf surface, Japan	7
Geniculospora spp.	Floral honey, Mexico	6
	Honeydew, Croatia	6
Isthmotricladia spp.	Honeydew, Italy	6
	Stemflow, Hungary	3
	Stemflow, Germany	4
	Stemflow, Hungary	4
Lemonniera spp.	Floral honey, Mexico	6
	Honeydew, Greece and Italy	6
Mirandiana spp.	Tree holes, Hungary	2
	Stemflow, Hungary	3
Mycocentrospora spp.	Honeydew, Italy	6
	Stemflow, Hungary	3
Retiarius spp.	Stemflow, Germany and Hungary	4
	Throughfall, Hungary, Romania and Sweden	4
	Endophyte of *Picea* needles, Canada	8
Titaea spp.	Tree holes, Hungary	2
	Stemflow, Germany and Hungary	4
	Throughfall, Romania	4
Tricellula sp.	Honeydew, Italy	6
Tricladium spp.	Honeydew, Greece	6
	Epipiphytic fern leaf, root, trapped leaf, India	9
	Stemflow, Hungary	3
	Treeflow, Germany	4
	Throughfall, Hungary	4
Tridentaria sp.	Stemflow, Hungary	3
Trifurcospora spp.	Honeydew, Italy	6
	Tree holes, Hungary	2
Trinacrium spp.	Honeydew, Italy	6
	Tree holes, Hungary	2
	Stemflow, Hungary	3
	Stemflow, Germany and Hungary	4
	Throughfall, Hungary	4
Tripospermum spp.	Floral honey, Italy	6
	Honeydew, Croatia, Greece and Italy	6
	Stemflow, Hungary	4
	Stemflow, Hungary	3
Varicosporium spp.	Stemflow, Canada	1
	Stemflow, Hungary	3
Volucrispora sp.	Stemflow, Canada	1

References: 1, Bandoni, 1981; 2, Gönczöl and Révay, 2003; 3, Gönczöl and Révay, 2004; 4, Gönczöl and Révay, 2006; 5, Karamchand and Sridhar, 2008; 6, Magyar *et al.*, 2005; 7, Osono and Mori, 2004; 8, Sokolski *et al.*, 2006; 9, Sridhar *et al.*, 2006.

the surface of an individual leaf. Similarly, species composition for both interior fungi and surface fungi also differed in five canopy positions. The fungi thus recovered differed according to the order of shoots (leaf age), the height of leaf layer and/or the distance from the trunk (sunlight intensity). Our knowledge of parasitic fungi in rain forest canopies is meagre, with the exception of the Meliolaceae (Laessøe and Lodge, 1994). Meliolaceous fungi have been reported from the canopies of trees and shrubs in Puerto Rico and Venezuela, as both hyperparasites and commensals (Dennis, 1970; Stevenson, 1975).

Mycena citricolor is an agaric that is one of the non-specialized parasites reported on a wide range of native and introduced hosts of neotropical rain forest understorey, including coffee, where it causes American leaf spot (Stevenson, 1975; Pegler, 1983). Similarly, another agaric parasite, *Crinipellis perniciosa,* causes Witches' Broom of cacao in neotropical forests (Hedger *et al.,* 1987). *Crinipellis* and *Marasmius* have been considered as canopy pathogens, and they fruit on dead plant parts where they are also assumed to be saprophytic (Pegler, 1983). The horsehair fungus, *Marasmius crinis-equi,* establishes rhizomorph nets in the understorey in rain forests (Hedger, 1990), but is not generally considered to be a pathogen (Weber, 1973). Studies on pathogens and symbiotic fungi in canopy and understorey in Australian rainforest revealed that they commonly disperse as air-borne spores (Gilbert and Reynolds, 2005). Spore abundance was 52-fold higher in understorey than in the canopy, and spores were between five- and 35-fold more abundant at night than during the day, possibly due to the environmental conditions at night being more conducive for spore germination and infection.

Lignicolous fungi

Sixty-two taxa of lignicolous fungi have been reported fruiting on twigs of canopy and understorey in Cameroon and West African rainforests (Ryvarden and Nuñez, 1992; Nuñez and Ryvarden, 1993). It has been suggested that the extremes of moisture and temperature experienced in the canopy are more selective for wood decomposers. Species of Aphyllophorales that occur in canopies have long-lived basidiocarps, while agarics rarely fruit in the canopy and any fruit bodies are confined to the wet season, such as in *Armillaria* sp. on orchids (Hedger, 1985). Wood-decomposing fungi are more common and diverse in the understorey than in the canopy of the rain forests. The canopy agaric communities usually have a distinct species composition that is different to that found on the forest floor, as they have adapted to different moisture regimes (Hedger, 1985; Hedger *et al.*, 1987).

Several phases of canopy wood decomposition have been recognized in mixed deciduous woodland (Swift *et al.*, 1976). Fungal decomposition leads to an average loss of about 8.4% of wood per annum in canopy as against 17.1% at the forest floor. The rates of wood decay differ between individual branches, although in a study on *Betula pendula pluspubescens*, *Corylus avellana*, *Fraxinus excelsior* and *Quercus robur pluspetraea,* no significant differences were found between branches of the different trees (Swift *et al.*, 1976).

Water-borne Fungi

Water-borne hyphomycetes (also known as aquatic fungi, freshwater hyphomycetes, amphibious hyphomycetes or Ingoldian fungi) are usually found associated with submerged dead leaves in streams, and are involved in the energy flow between leaf litter and invertebrates (Ingold, 1942; Baerlocher 1992). In addition to their occurrence in streams, they are also known from terrestrial habitats such as soil, leaf litter, roots and leaf surfaces (see Sridhar and Baerlocher, 1993). They are also reported from several canopy habitats.

Throughfall, stemflow and tree holes

Water from mist or rain dripping from foliage to the ground is known as 'throughfall', while 'stemflow' is water from mist or rain flowing to the ground along the outside of stems (Moffett, 2000). The 'tree holes' are formed due to damage to the branches, and part of the stem decomposes creating a hole, which allows the accumulation of water, sediment and plant litter and serves as another important canopy habitat for several organisms, including fungi. The specific features of canopy throughfall and stemflow habitats (e.g. temperature, humidity, nutrients) may be conducive for growth, survival and dissemination for various organisms, including water-borne fungi (see Table 5.1). Likewise, snow accumulating on canopies, and floral honey and honeydew excreted by aphids in canopies, are also sources of fungal flora.

Typical water-borne hyphomycetes have been reported in tree canopies (e.g. throughfall, stemflow and tree holes) from temperate habitats, including Canada, Hungary, Japan and Poland (Gönczöl 1976; Bandoni 1981; Ando and Tubaki, 1984a,b; Czeczuga and Orłowska, 1994, 1998a,b; Gönczöl and Révay, 2003, 2004). Tree holes in canopies constitute natural continuous-flow chambers, and they can support a variety of fungi as they trap leaf litter (Gönczöl and Révay, 2003; Karamchand and Sridhar, 2008). So far about 96 species of water-borne conidial fungi have been reported to occur in several habitats of tree canopies (see checklist in Table 5.1). Most of such reports are confined to temperate regions.

Several morpho-species (water-borne, aero-aquatic and dematiaceous fungi) recovered from the canopy habitats have not yet been identified to species level (see Tables 5.2, 5.3). In addition, the huge variety of conidia recovered in throughfall, stemflow and tree holes have not yet been fully identified, even to genus level (Gönczöl and Révay, 2003, 2004, 2006). This raises the possibility that these environments may harbour currently undescribed fungal taxa. These observations support the hypothesis that conidial fungi constitute a guild in the canopy and function like typical Ingoldian fungi in streams (Carroll, 1981). As water-borne fungi are at risk of extinction due to unidirectional flow of water in streams, they recover due to their ability to colonize stationary substrates in stream or stream banks (e.g. roots, wood). Similarly, habitats in the tree canopy (e.g. tree holes, epiphytes) also serve as potential refuges for these fungi.

It is interesting to note that 33 species of fungi – including many Ingoldian fungi – have been reported as occurring in rainwater dripping from six types of building roofs (copper, zinc, red tiles, asbestos, tar paper and thatch) in rural and urban region of Poland (Czeczuga and Orłowska, 1997). Czeczuga and Orłowska (1997) postulated that the irregular roof structure accumulates sediments and supports the survival of Ingoldian fungi. Several Ingoldian fungi that had colonized coffee and rubber leaves in a Western Ghat stream in India have been shown to survive for up to 360 days on exposure to terrestrial conditions (Sridhar and Kaveriappa, 1988). This supports the potential dissemination of Ingoldian fungi through dried-stream leaf litter. In addition to the typical Ingoldian fungi, several aero-aquatic and dematiaceous fungi are also inhabitants of canopy habitats (see Table 5.2).

Endophytes

Although there have been extensive studies on the floristic and ecological functions of water-borne fungi in woodland streams, their vertical distribution and ecological functions in riparian tree canopies are largely unknown. Interestingly, typical aquatic or water-borne hyphomycetes (e.g. *Dwayaangam colodena*, *Retiarius* sp., *Tripospermum camelopardus* and *T. myrti*) have recently been reported as endophytes in a canopy of black spruce needles (*Picea mariana*) in a mixed-wood forest of Canada (Sokolski *et al.*, 2006). These examples suggest that water-borne and aero-aquatic fungi have also adapted to canopy habitats.

Water-borne fungi in *Drynaria*

The Western Ghats and west coast of India experience heavy rain fall between May and September. The environmental conditions that result in the forest canopies are ideal for growth, sporulation and dispersal of water-borne hyphomycetes. *Drynaria quercifolia*, commonly known as 'oak-leaf basket fern', is a widely distributed pteridophyte in this area. It produces fragile, slender, short-lived nest leaves and tough, leathery, plate-like bracket leaves that resemble oak leaves and are attached to both living and dead stages of the creeping rhizome. The bracket leaves act as funnels and trap debris from the canopy water runoff.

Drynaria is widespread and occurs at all levels in the canopy, where it can grow to an exceptionally large size. Both living and dead tissues of *Drynaria* and trapped leaf litter are colonized by water-borne hyphomycetes (Sridhar *et al.*, 2006), and these persist until the rainy season and reproduce. This is analagous to the colonization of dry leaves on stream banks (Sanders and Webster, 1978; Sridhar and Kaveriappa, 1987; Sridhar and Baerlocher, 1993). Fewer conidia are found in the fern rhizome than in the roots, bracket leaves and trapped leaves (15 versus 101–1780/g), and this has been taken to indicate that water-borne fungi in the rhizome are present as endophytes (Sridhar *et al.*, 2006). Gönczöl and Révay (2003) also suggested that the occurrence of water-borne hyphomycetes in canopy could be epiphytic or endophytic in origin. The rachis of bracket leaves and nest leaves of *Drynaria* are more persistent compared with pinna, and may be suitable for investigating anamorph–teleomorph connections of water-borne hyphomycetes.

Dissemination and evolution

Apart from the movement and dispersal of water-borne fungal conidia in water films on wet canopy (Bandoni, 1974), tree hole-inhabiting insects (Kitching, 1971) – including mosquitoes (Kaufman *et al.*, 2008) – may be involved in the dissemination of spores or propagules of fungi within or across canopies. Such transmission of fungi may also occur through riparian canopy as a result of birds and other animals. The true spores of the perfect states of aquatic or water-borne hyphomycetes may also be aerially dispersed. Perfect states of some aquatic and water-borne hyphomycetes have been recorded on the plant detritus gathered from terrestrial habitats surrounding streams (Webster, 1992). The presence of the perfect state in the tree canopy might allow dissemination of propagules to streams or terrestrial habitats via throughfall, stemflow or invertebrates. The majority of species reported in Table 5.1 produce multi-radiate conidia, and the branched nature of these may be a factor in their anchorage on the canopy substrate surface (Kearns and Baerlocher, 2008).

Stone *et al.* (1996) suggested that canopy fungi may be early colonizers of live foliage and twigs, and act as a minor link between soil and aquatic foodwebs by completing their life cycles. Ingold (1942) first described the aquatic hyphomycetes in 1942, and hypothesized that the shape of their multi-radiate conidia might be a secondary adaptation for aquatic life and a product of multiple convergent evolutions. This hypothesis has recently been supported through molecular evidence that has shown that Ingoldian fungi have multiple origins (Belliveau and Baerlocher, 2005).

Recently, Selosse *et al.* (2008) proposed another hypothesis – that the shape of Ingoldian asexual conidia evolved for dispersal in water. Very large numbers of conidia can be found in air bubbles in foam in streams, and this would favour their dispersal through wind or aerosols. Therefore, Ingoldian fungal propagules in an aerial phase could be spread to tree canopies (e.g. leaves, bark, tree holes), where they could lead either a saprophytic or an endophytic lifestyle (see Sokolski *et al.*, 2006). Molecular studies have confirmed that some Ingoldian fungi have close terrestrial relatives (Kong *et al.*, 2000; Liew *et al.*, 2002). Anamorph–teleomorph studies of Ingoldian fungi have also shown that the majority of species have evolved from ascomycetes, which are found in decaying tree branches in streams (Webster and Descals, 1979; Webster, 1992).

Many helicosporous fungi (brown asexual spores with a twist of a minimum of 180°) have been reported from tree canopies (see Table 5.2), and morphological studies have suggested that these are related to multiple families of different classes of ascomycetes (Goos, 1987; Zhao *et al.*, 2007). Tsui and Berbee (2006) recently recognized six convergent lineages of helicosporous fungi in the ascomycetes, and many of these were placed in the Tubeufiaceae (Dothideomycetes). Tsui and Berbee (2006) also confirmed the polyphyly of helicosporous fungi and speculated that the spore forms were a convergent adaptation for dispersal in aquatic habitats. These observations suggest that water-borne fungi can shuttle between aquatic and terrestrial (e.g. canopy, soil) phases as saprophytes and/or endophytes in order to complete their life cycles.

Canopy fungi and nutrient cycling

Throughfall and stemflow commonly contain several potential nutrients (e.g. nitrogen, phosphorus, potassium, calcium, magnesium) (Schroth *et al.*, 2001). Thus, canopy fungi have the advantage of access to additional nutrients in addition to those contained in their immediate substrates (e.g. leaves, bark, twigs, dead litter). Fungi in canopies are of special interest as they can be involved in nutrient cycling and symbiosis, and can interact with arthropods and microfauna (Stone *et al.*, 1996). As seen in epiphytes, fungi are also known to concentrate organic nutrients in canopies (Stone *et al.*, 1996).

A long-term investigation of dead branches in the canopy of the most prominent tree species in Germany (*Acer pseudoplatanus, Fraxinus excelsior, Quercus robur* and *Tilia cordata*) in 2002–2003 by Unterseher *et al.* (2005) identified 118 different fungal taxa. The species richness and composition differed markedly between the four tree species, and many fungi grew on bark or slightly decayed wood with distinct host and substratum specificity.

Conclusions and Future Outlook

Canopy science is interdisciplinary, and its study could lead to a better understanding of the structural and functional role of the canopy in relation to fungal biodiversity. As the information on the role of fungi in canopies remains fragmentary, there are many opportunities for new and significant future discoveries. A basic understanding of the distribution of plants, animals and microbes in forest canopies is vital for estimating energy flow, carbon cycling, resource use and transfer of materials within and across the canopy ecosystem.

Several specific ecological niches serve as habitats for fungi and for their interaction and survival in tree canopies. One example is the tree hole that provides a continuous flow ecosystem for fungi. Interestingly, re-examination of leaf litter in tree holes of south-western Hungary studied in the mid-1980s has confirmed that tree holes are the permanent habitats of water-borne conidial fungi (Gönczöl, 1976; Gönczöl and Révay, 2003). Several alkaline-tolerant fungi have been isolated from tree holes in Thailand (Kladwang *et al.*, 2003), which may be a good source of alkaline-tolerant enzymes for industrial applications.

Floral honey or honeydew excreted by aphids can also serve as ecological niches for mycoflora (Magyar *et al.*, 2005). The trophic interactions between honeydew excreted by aphids and communities of microbes (bacteria, yeasts and filamentous fungi) significantly decreased the concentrations of inorganic nitrogen (NH_4N, NO_3N) and elevated the dissolved organic carbon, as well as dissolved organic nitrogen concentrations, in canopy throughfall (Stadler *et al.*, 2005).

Therefore there is spatial and temporal variability of such trophic links between phytophagous insects and microorganisms, which may influence the energy and nutrient fluxes within the tree canopies. Similarly, gum exudates of tree trunks may also serve as repositories of conidial fungi. Fungal assemblages and diversity in canopy lichens and fungal symbionts of canopy galls are of special interest.

It is surprising to note that an estimate of microfungal biomass in twigs and needle surfaces of old-growth Douglas fir forest canopies was up to 450 kg/ha/year (Carroll *et al.*, 1980). Thus, the fungal biomass may be an important source of food and nutrition to several canopy-dependent fauna. Tree canopies in biodiversity hotspots in the tropics may be important environments for determining overall fungal diversity. In addition to the traditional methods, molecular techniques can be employed to study fungal diversity in minute samples obtained from different canopy habitats (see Baerlocher, 2007). Further investigations are necessary to establish a broader outlook on the role of fungi in canopies and to answer the following questions:

- Do fungi in canopies have any functional significance? Is the occurrence of substantial fungal biomass in the forest canopies a widespread phenomenon?
- What role does fungal biomass play in the ecosystem function of the canopy as a whole?
- Does fungal biomass contribute to the nutritional requirements of leaf-shredding canopy invertebrates?
- What function do fungi have in carbon and nutrient cycling, in decomposition and in the maintenance of biodiversity in the canopy?

References

Ananda, K. and Sridhar, K.R. (2002) Diversity of endophytic fungi in the roots of mangrove species on west coast of India. *Canadian Journal of Microbiology* 48, 871–878.

Ando, K. and Kawamoto, I. (1986) *Arborispora*, a new genus of hyphomycetes. *Transactions of the Mycological Society of Japan* 27, 119–128.

Ando, K. and Tubaki, K. (1984a) Some undescribed hyphomycetes in the raindrops from intact leaf surface. *Transactions of the Mycological Society of Japan* 25, 21–37.

Ando, K. and Tubaki, K. (1984b) Some undescribed hyphomycetes in the rain water draining from intact trees. *Transactions of the Mycological Society of Japan* 25, 39–47.

Azevedo, J.L. Maccheroni, W., Pereira, J.O. and de Araujo, W.L. (2000) Endophytic microorganisms: a review on insect control and recent advances on tropical plants, *EJB Electronic Journal of Biotechnology* 3, 40–65.

Baerlocher, F. (1992) *The Ecology of Aquatic Hyphomycetes.* Springer-Verlag, Berlin.

Baerlocher, F. (2007) Molecular approaches applied to aquatic hyphomycetes. *Fungal Biology Reviews* 21, 19–24.

Bandoni, R.J. (1974) Mycological observations on the aqueous films covering decaying leaves and other litter. *Transactions of the Mycological Society of Japan* 15, 309–315.

Bandoni, R.J. (1981) Aquatic hyphomycetes from terrestrial litter. In: Wicklow, D.T. and Carroll, G.C. (eds) *The Fungal Community – its Organization and Role in the Ecosystem.* Marcel Dekker, New York, pp. 693–708.

Belliveau, M.J.-R. and Baerlocher, F. (2005) Molecular evidence confirms multiple origins of aquatic hyphomycetes. *Mycological Research* 109, 1407–1417.

Carroll, G.C. (1981) Mycological inputs to ecosystem analysis. In: Wicklow, D.T. and Carroll, G.C. (eds) *The Fungal Community – its Organization and Role in the Ecosystem.* Marcel Dekker, New York, pp. 25–35.

Carroll, G.C. (1991) Fungal associates of woody plants and insect antagonists in leaves and stems. In: Barbosa, P., Krischik, V.A. and Jones, C.G. (eds) *Microbial Mediation of Plant–Herbivore Interactions.* Wiley, New York, pp. 253–271.

Carroll, G.C., Carroll, F.E., Pike, L.H., Perkins, J.R. and Sherwood, M. (1980) Biomass and distribution patterns of conifer twig microepiphytes in a Douglas fir forest. *Canadian Journal of Botany* 58, 624–630.

Counts, J.B., Henley, L., Skrabal, M. and Snell, K. (2000) Tree canopy biodiversity in the Great Smoky Mountains National Park. *The Inoculum: Newsletter of the Mycological Society of America,* 51.

Czeczuga, B. and Orłowska, M. (1994) Some aquatic fungi of hyphomycetes on tree leaves. *Roczniki Akademii Medycznej w Bialymstoku* 39, 86–92.

Czeczuga, B. and Orłowska, M. (1997) Hyphomycetes fungi in rainwater falling from building roofs. *Mycoscience* 38, 447–450.

Czeczuga, B. and Orłowska, M. (1998a) Hyphomycetes in rain water draining from intact trees. *Roczniki Akademii Medycznej w Bialymstoku* 43, 66–84.

Czeczuga, B. and Orłowska, M. (1998b) Hyphomycetes in the snow from gymnosperm trees. *Roczniki Akademii Medycznej w Bialymstoku* 43, 85–94.

Czeczuga, B. and Orłowska, M. (1999) Hyphomycetes in rainwater, melting snow and ice. *Acta Mycologica* 34, 181–200.

Dennis, R.W.G. (1970) Fungus flora of Venezuela and adjacent countries. *Kew Bulletin* Additional Series 3, 1–531.

Dreyfuss, M. and Petrini, O. (1984) Further investigations on the occurrence and distribution of endophytic fungi in tropical plants. *Botanica Helvetica* 94, 33–40.

Ellwood, M.D.F. and Foster, W.A. (2004) Doubling the estimate of invertebrate biomass in a rainforest canopy. *Nature* 429, 549–551.

Gilbert, G.S. and Reynolds, D.R. (2005) Nocturnal fungi: airborne spores in the canopy and understorey of a tropical rain forest. *Biotropica* 37, 462–464.

Gönczöl, J. (1976) Ecological observations on the aquatic hyphomycetes of Hungary II. *Acta Botanica Academiae Scientiarum Hungaricae* 22, 51–60.

Gönczöl, J. and Révay, Á. (2003) Treehole fungal communities: aquatic, aero-aquatic and dematiaceous hyphomycetes. *Fungal Diversity* 12, 19–34.

Gönczöl, J. and Révay, Á. (2004) Fungal spores in rainwater: stemflow, throughfall and gutter conidial assemblages. *Fungal Diversity* 16, 67–86.

Gönczöl, J. and Révay Á. (2006) Species diversity in rainborne hyphomycete conidia from living trees. *Fungal Diversity* 22, 37–54.

Goos, R.D. (1987) Fungi with a twist: The helicosporous hyphomycetes. *Mycologia* 79, 1022.
Hammond, P.M. (1992) Species inventory. In: Groombridge, G. (ed.) *Global Biodiversity: Status of the Earth's Living Resources*. Chapman and Hall, London, pp. 17–39.
Hammond, P.M. and Stork, N.E., Brendell, M.J. (1977) Tree crown beetles in context: a comparison of canopy and other ecotone assemblages in a lowland tropical forest in Sulawesi. In: Stork, N.E., Adis, J. and Didham, R. (eds) *Canopy Arthropods*. Chapman and Hall, London, pp. 184–223.
Hawksworth, D.L. (1991) The fungal dimension of biodiversity: magnitude, significance, and conservation. *Mycological Research* 95, 641–655.
Hedger, J. (1985) Tropical agarics, resource relations and fruiting periodicity. In: Moore, D., Casselton, L.A., Wood, D.A. and Frankland, J.C. (eds) *Developmental Biology of Higher Plants*. Cambridge University Press, Cambridge, UK, pp. 41–86.
Hedger, J. (1990) Fungi in tropical forest canopy. *Mycologist* 4, 200–202.
Hedger, J.N., Pickering, V. and Aragundi, J. (1987) Variability of populations of the Witches Broom disease of cocoa (*Crinipellis perniciosa*). *Transactions of the British Mycological Society* 88, 533–546.
Ingold, C.T. (1942) Aquatic hyphomycetes of decaying alder leaves. *Transactions of the British Mycological Society* 25, 339–417.
Karamchand, K.S. and Sridhar, K.R. (2008) Water-borne conidial fungi inhabiting tree holes of the west coast and Western Ghats of India. *Czech Mycology* 60, 63–74.
Kaufman, M.G., Chen, S. and Walker, E.D. (2008) Leaf-associated bacterial and fungal taxa shifts in response to larvae of the tree hole mosquito, *Ochlerotatus triseriatus*. *Microbial Ecology* 55, 673–684.
Kearns, S.G. and Baerlocher, F. (2008) Leaf surface roughness influence colonization success of aquatic hyphomycete conidia. *Fungal Ecology* 1, 13–18.
Kitching, R.L. (1971) An ecological study of water-filled treeholes and their position in the woodland ecosystem. *Journal of Animal Ecology* 40, 281–302.
Kladwang, W., Bhumirattana, A. and Hywel-Jones, N. (2003) Alkaline-tolerant fungi from Thailand. *Fungal Diversity* 13, 69–84.
Kong, R.Y., Chan, J.Y., Mitchell, J.I., Vrijmoid, L.L. and Jones, E.B.G. (2000) Relationships of *Halosarpheia, Lignicola* and *Nais* inferred from partial 18S rDNA. *Mycological Research* 103, 1399-1403.
Kress, W.J. (1986) The systematic distribution of vascular epiphytes: an update. *Selbyana* 9, 2–22.
Laessøe, T. and Lodge, D.J. (1994) Three host-specific Xylaria species. Mycologia 86, 436–446.
Leary, J.J.K., Singleton, P.W. and Borthakur, D. (2004) Canopy nodulation of the endemic tree legume *Acacia koa* in the mesic forests of Hawaii. *Ecology* 85, 3151–3157.
Liew, E.C.Y., Aptroot, A. and Hyde, K.D. (2002) An evolution of the monophyly of *Massarina* based on ribosomal DNA sequences. *Mycologia* 94, 803–813.
Lodge, D. and Cantrell, S. (1995) Fungal communities in wet tropical variation in time and space. *Canadian Journal of Botany* 73(Suppl. 1), S1391–S1398.
Mackinnon, J.A. (1982) Stemflow and throughfall mycobiota of a trembling aspen–red alder forest. MSc thesis, University of British Columbia, Vancouver, Canada.
Magyar, D., Gönczöl, J., Révay, Á., Grillenzoni, F. and Seijo-Coello, M.D.C. (2005) Stauro- and scolecoconidia in floral and honeydew honeys. *Fungal Diversity* 20, 103–120.
Maria, G.L. and Sridhar, K.R. (2003) Endophytic fungal assemblage of two halophytes from west coast mangrove habitats, India. *Czech Mycology* 55, 241–251.
McCutcheon, T.L., Carroll, G.C. and Schwab, S. (1993) Genotypic diversity in populations of a fungal endophyte from Douglas fir. *Mycologia* 85, 180–186.
Moffett, M.W. (2000) What's 'Up'? A critical look at the basic terms of canopy biology. *Biotropica* 32, 569–596.

Murali, T.S., Suryanarayanan, T.S. and Geeta, R. (2006) Endophytic *Phomopsis* species: host range and implications for diversity estimates. *Canadian Journal of Botany* 52, 673–680.
Nadkarni, N.M., Mewin, M.C. and Niedert, J. (2001) Forest canopies, plant diversity. In: *Encyclopedia of Biodiversity*, Vol. 3. Academic Press, New York, pp. 27–40.
Nuñez, M.P. and Ryvarden, L. (1993) Basidiomycetes on twigs at ground level and in the canopy: a comparison. In: Issac, S., Frankland, J.C., Watling, R. and Whalley, A.J.S. (eds) *Aspects of Tropical Mycology*. Cambridge University Press, Cambridge, UK, p. 307.
Osono, T. and Mori, A. (2004) Distribution of phyllosphere fungi within the canopy of giant dogwood. *Mycoscience* 45, 161–168.
Ozanne, C.M.P., Anhuf, D., Boulter, S.L., Keller, M., Kitching, R.L., Körner, C., Meinzer, F.C., Mitchell, A.W., Nakashizuka, T., Silva Dias P.L., Stork, N.E., Wright S.J. and Yoshimura M. (2003) Biodiversity meets the atmosphere: a global view of forest canopies. *Science* 301, 183–186.
Parker, G.G. (1995) Structure and microclimate of forest canopies. In: Lowman, M.D. and Nadkarni, N.M. (eds) *Forest Canopies*. Academic Press, San Diego, California, pp. 73–106.
Pegler, D.N. (1983) *Agaric Flora of the Lesser Antilles*. Kew Bulletin Additional Series 9, 668 pp.
Raviraja, N.S., Maria, G.L. and Sridhar, K.R. (2006) Antimicrobial evaluation of endophytic fungi inhabiting medicinal plants of the Western Ghats of India. *Engineering in Life Sciences* 6, 515–520.
Ryvarden, L. and Nuñez, M. (1992) Basidiomycetes in the canopy of an African rain forest. In: Hallé, F. and Pascal, O. (eds) *Biologie d'une Canopéde Forêt Équatoriale II*. Communication, Lyon, France, pp. 116–118.
Sanders, P.F. and Webster, J. (1978) Survival of aquatic hyphomycetes in terrestrial situations. *Transactions of the British Mycological Society* 71, 231–237.
Schroth, G., Elias, M.E.A., Uguen, K., Seixas, R. and Zech, W. (2001) Nutrient fluxes in rainfall, throughfall and stemflow in tree-based land use systems and spontaneous tree vegetation of central Amazonia. *Agriculture, Ecosystems and Environment* 87, 37–49.
Selosse, M., Vohník, M. and Chauvet, E. (2008) Out of rivers: are some aquatic hypohomycetes plant endophytes? *New Phytologist* 198, 3–7.
Sokolski, S., Piché, Y., Chauvet, E. and Bérubé, J.A. (2006) A fungal endophyte of black spruce (*Picea mariana*) needles is also an aquatic hyphomycete. *Molecular Ecology* 15, 1955–1962.
Sridhar, K.R. and Baerlocher, F. (1993) Aquatic hyphomycetes on leaf litter in and near a stream in Nova Scotia, Canada. *Mycological Research* 97, 1530–1535.
Sridhar, K.R. and Kaveriappa, K.M. (1987) Occurrence and survival of aquatic hyphomycetes in terrestrial conditions. *Transactions of the British Mycological Society* 89, 606–609.
Sridhar, K. R. and Kaveriappa, K. M. (1988) Survival of water-borne fungi imperfecti under non-aquatic conditions. *Proceedings of the Indian National Science Academy* B54, 295–297.
Sridhar, K.R., Karamchand, K.S. and Bhat, R. (2006) Arboreal water-borne hyphomycetes on oak-leaf basket fern *Drynaria quercifolia*. *Sydowia* 58, 309–320.
Stadler, B., Müller, T., Orwig, D. and Cobb, R. (2005) Hemlock woolly adelgid in New England forests: canopy impacts transforming ecosystem processes and landscapes. *Ecosystems* 8, 233–247.
Stevenson, J.A. (1975) *The Fungi of Puerto Rico and the American Virgin Islands*. Reed Herbarium 23, Texas Tech University, Lubbock, Texas, 743 pp.
Stierle, A., Strobel, G. and Stierle, D. (1993) Taxol and taxane production by *Taxomyces andraenae*, an endophytic fungus of the Pacific yew. *Science* 260, 214–216.
Stone, J.F., Sherwood, M.A. and Carroll, G.C. (1996) Canopy microfungi: function and diversity. *Northwest Science* 70, 37–45.
Strobel, G., Stierle, A., Stierle, D. and Hess, W.M. (1993) *Taxomyces andreanae*, a proposed new taxon for a bulbilliferous hyphomycete associated with Pacific yew (*Taxus bervifolia*) *Mycotaxon* 47, 71–80.

Suryanarayanan, T.S., Venkatesan, G. and Murali, T.S. (2003) Endophytic fungal communities in leaves of tropical forest trees: diversity and distribution patterns. *Current Science* 85, 489–493.

Swift, M.J., Healey, I.N., Hibberd, J.K., Sykes, J.M., Bampoe, V. and Nesbitt, M.E. (1976) The decomposition of branch-wood in the canopy and floor of a mixed deciduous woodland. *Oecologia* 26, 139–149.

Tsui, C.K.M. and Berbee, M.L. (2006) Phylogenetic relationships and convergence of helicosporous fungi inferred from ribosomal DNA sequences. *Molecular Phylogenetics and Evolution* 39, 587–597.

Unterseher, M., Otto, P. and Morawetz, W. (2005) Species richness and substrate specificity of lignicolous fungi in the canopy of a temperate, mixed deciduous forest. *Mycological Progress* 4, 117–132.

Unterseher, M., Reiher, A., Finstermeier, K., Otto, P. and Morawetz, W. (2007) Species richness and distribution patterns of leaf-inhabiting endophytic fungi in a temperate forest canopy. *Mycological Progress* 6, 201–212.

Weber, G.F. (1973) *Bacterial and Fungal Diseases of Plants in the Tropics*. University of Florida Press, Gainesville, Florida, 673 pp.

Webster, J. (1992) Anamorph–teleomorph relationships. In: Baerlocher, F. (ed.) *The Ecology of Aquatic Hyphomycetes*. Springer-Verlag, Berlin, pp. 99–117.

Webster, J. and Descals (1979) The teleomorphs of water-borne hyphomycetes from freshwater. In: Kendrick, B. (ed.) *The Whole Fungus*, Vol. 2. National Museums of Canada, Ottawa, Canada, pp. 419–447.

Zhao, G.Z., Liu, X.Z. and Wu, W.P. (2007) Helicosporous hyphomycetes from China. *Fungal Diversity* 26, 313–524.

6 Ecology of Endophytic Fungi Associated with Leaf Litter Decomposition

TAKASHI OSONO[1] AND DAI HIROSE[2]

[1]*Center for Ecological Research, Kyoto University, Japan;*
[2]*Faculty of Pharmacy, Nihon University, Japan*

Introduction

Fungi have diverse and intimate relationships with plants as endophytes, saprobes, pathogens and mutualistic symbionts (Rodriguez and Redman, 1997). Endophytic fungi are defined as those forming symptomless infections, for part or all of their life cycle, within healthy plant tissues. For practical purposes, a fungus is typically considered as an endophyte if it is cultured from surface-disinfected plant material (Wilson, 2000). Endophytic infection by fungi can be found within leaves, petioles, sheaths and roots of plants, as well as on twigs and barks of woody plants. The two major groups of endophytic fungi that have been studied intensively in terms of taxonomy and systematics, biology, ecology, physiology and application are those associated with leaves of trees and grasses (Clay, 1986; Petrini, 1986, 1991; Petrini *et al.*, 1992; Carroll, 1995; Redin and Carris, 1996; Wilson 2000; Clay and Schardl, 2002).

One of the ecological roles of endophytic fungi is as mutualistic symbionts of grasses. Grass endophytes are widespread, and frequently enhance growth and resistance to herbivory and abiotic stresses of infected graminoids (Clay, 1991). They are systemic, seed-borne endophytes belonging to *Epichloë* and *Balansia* in the Clavicipitaceae, with anamorphs known as *Neotyphodium* and *Ehelis,* respectively. Similar effects as insect antagonists have been demonstrated for fungal endophytes of tree leaves (Carroll, 1991; Wilson, 2000), but the ecological roles of many endophytic fungi are yet to be clarified. Some endophytic fungi are believed to be latent pathogens, simply because they are taxonomically related to their pathogenic counterparts. A study of phylogenetic affinities of parasites and endophytes of *Pinus monticola,* however, did not find distinct parasites as endophytes (Ganley *et al.*, 2004), suggesting that endophytic fungi are not necessarily latent pathogens even though they may be closely related to pathogenic forms.

It has been suggested that endophytic fungi have a role in the decomposition of leaf litter. Hudson (1968) reviewed fungal succession on plant remains above the soil and reported that a group of parasites from live leaves occurred in the early stages of fungal succession on dead leaves for grass and tree species. Since then, several authors have confirmed the occurrence of endophytic fungi in the early stages of decomposition of various leaf litters. Osono (2006a) recently summarized the roles of endophytic and epiphytic phyllosphere fungi of forest trees in the development of decomposer fungal communities and the decomposition processes of leaf litter. In this chapter, we review the ecology of endophytic fungi associated with the decomposition of leaf litter of tree species and grasses.

Occurrence of Endophytic Fungi on Leaf Litter

Endophytic fungi are divided into two groups: first, those that occur on both live leaves and leaf litter, and secondly, those that occur exclusively on live leaves and not on leaf litter. There are more reports available on the occurrence of tree endophytes on leaf litter than there are of grass endophytes. Table 6.1 summarizes the numbers of genera of endophytic fungi reported from the leaf litter of various trees. The mean number of endophytic species that occurred on tree leaf litter is 2.5 ($n = 17$), and these species account for some 67% of the total number of endophytic species reported from live leaves (see Table 6.1).

Therefore around two-thirds of the endophytic fungi on live leaves can also occur on leaf litter. The relative proportion of endophytic fungi in decomposer fungal assemblages varies between tree species and ranges from 3 to 92%, with a mean of 31% ($n = 22$) (Table 6.1). The relative proportion of endophytes in decomposer fungal assemblages may decrease (Heredia, 1993; Osono *et al.*, 2004), stay constant (Watson *et al.*, 1974) or increase during decomposition (Osono, 2002). The relative proportion of endophytes may also vary seasonally. Cabral (1985) reported that 33–35% of the fungi on dead leaves of *Eucalyptus viminalis* in summer and winter were endophytes, but this decreased to 2% in autumn and spring. In a similar study, Ruscoe (1971) found that 28–44% of the fungi on dead leaves of *Nothofagus truncate* in autumn, winter and spring were endophytes but that none were present in summer.

Ascomycetes and their related anamorphic genera are commonly well represented in the endophytic fungal assemblages that occur on leaf litter, while zygomycetes and basidiomycetes are generally absent. This distribution pattern is also similar to that seen for endophytic fungal associations on live leaves (Boddy and Griffith, 1989).

Common endophytic fungi reported from tree leaf litter include species of genera such as *Pestalotiopsis* and *Alternaria*. Fungi in these genera are primarily known as pathogenic fungi, but they have recently been found in internal tissues without causing any apparent harm to their hosts, as well as in fallen leaves after leaf death. *Phomopsis, Colletotrichum, Pestalotiopsis,*

Table 6.1. Species number, relative proportion in decomposer fungal assemblages and genera of endophytic phyllosphere fungi occurring on leaf litter of tree species.

Litter type	Location	Species number[a]	Source	Relative proportion[b]	Genera[c]	Reference[d]
Nothofagus truncata (Col.) Ckn.	New Zealand	2/2 (100)	Dead leaves on tree	29	St, Sp	1
Pinus taeda L.	USA	4/4 (100)	Needle from L layer	11	Al, Ni, Pe	2
			Needle from F1 layer	13		
			Needle from F2 layer	15		
Populus tremuloides Michx.	Canada	1/8 (13)	Freshly fallen leaves	92	Ds	3
Eucalyptus viminalis Labill.	Argentina	3/6 (50)	Dead leaves on tree	18	Al, Ma, Co	4
Quercus germana Cham. et Schlecht	Mexico	1/4 (25)	Freshly fallen leaves (day 0)	39	Pe	5
			Decomposing leaves (days 45–157)	6		
Liquidambar styraciflua L.	Mexico	2/2 (100)	Freshly fallen leaves (day 0)	55	Cl, Pe	
			Decomposing leaves (days 45–157)	3		
Abies alba Mill.	Switzerland	3/4 (75)	Decomposing leaves	17	Cr, Gl, Le	6
Rhododendron reticulatum D. Don	Japan	4/4 (100)	Yellow fallen leaves	63	Ph, Cm, Al, Pe	7
R. pulchrum var. speciosum Hara	Japan	4/4 (100)	Yellow fallen leaves	60	Gu, Ph, Cm, Pe	
R. obtusum Planch.	Japan	5/6 (83)	Yellow fallen leaves	51	Ph, Cm, Al, Pe, Di	
R. macrosepalum Maxim.	Japan	5/5 (100)	Yellow fallen leaves	33	Gu, Ph, Cm, Al, Pe	
Pieris japonica D. Don	Japan	1/2 (50)	Yellow fallen leaves	36	Gu	
Fagus crenata Bl.	Japan	3/5 (60)	Freshly fallen leaves	33	Ge, Xy, As	8
			Decomposing leaves	66		
Swida controversa (Hemsl.) Sojak	Japan	1/5 (20)	Decomposing leaves	7	Xy	9
Camellia japonica L.	Japan	2/3 (67)	Decomposing leaves	21	Rh, Ge	10
Castanopsis sieboldii (Makino) Hatusima	Japan	1/2 (50)	Decomposing leaves	8	Ge	11
Shorea obtusa Wall.	Thailand	1/2 (50)	Decomposing leaves	17	Ni	11
Mean		2.5/4 (67)		31		

[a] Species number of endophytic fungi that occurred frequently on leaf litter as compared with the total species number of endophytic fungi on live leaves.
[b] Relative proportion of endophytic fungi in decomposer fungal assemblages in terms of frequency, given as: (the sum of isolation frequency of endophytic fungi)/(the sum of isolation frequency of all fungi on leaf litter) x 100.
[c] Genera of endophytic fungi that occur on leaf litter. St, *Stachylidium*; Sp, Sphaeriaceous; Al, *Alternaria*; Ni, *Nigrospora*; Pe, *Pestalotiopsis*; Ds, dark sterile mycelium; Ma, *Macrophoma*; Co, *Coniothyrium*; Cl, *Cladosporium*; Cr, *Cryptocline*; Gl, *Gloeosporidiella*; Le, *Leptostroma*; Ph, *Phomopsis*; Cm, *Colletotrichum*; Gu, *Guignardia*; Di, *Discostroma*; Ge, *Geniculsporium*; Xy, *Xylaria*; As, *Ascochyta*; Rh, Rhytismataceous.
[d] References : 1, Ruscoe (1971) ; 2, Watson *et al.* (1974); 3, Wildman and Parkinson (1979); 4, Cabral (1985); 5, Heredia (1993); 6, Sieber-Canavesi and Sieber (1993); 7, Okane *et al.* (1998); 8, Osono (2002); 9, Osono *et al.* (2004); 10, Koide *et al.* (2005b); 11, Osono *et al.* (2009).

Alternaria and *Guignardia* have repeatedly been encountered on ericaceous hosts in Japan (Okane *et al.*, 1998). Xylariaceous fungi and their anamorphs, such as *Xylaria* and *Geniculosporium*, are common endophytes of tree leaves (Petrini and Petrini, 1985; Petrini *et al.*, 1995) and have also been found on leaf litter in temperate and tropical forests in Asia (Osono, 2002; Osono *et al.*, 2004; Osono *et al.*, 2008, 2009).

Lophodermium piceae, an endophyte of *Picea abies*, has been shown to form ascomata and ascospores on fallen needles, and also to infect live needles in order to start a new endophytic stage (Osorio and Stephen, 1991). A similar life cycle of colonization via fallen leaves has also been suggested for *Rhabdocline parkeri* on evergreen needles of *Pseudotsuga menziesii* (Stone, 1987), and *Mycosphaerella buna* on leaves of *Fagus crenata* (Kaneko and Kakishima, 2001; Kaneko *et al.*, 2003).

Molecular studies have provided further insights into the phylogenetic affinities of endophytes and saprobes from the same tree host. Müller *et al.* (2001) used DNA-based analyses to identify fungal isolates from needles of *Picea abies*. They found that *Tiarosporella parca*, a low-frequency endophyte, occurred frequently on decomposing brown needles, whereas the most frequent endophyte of the needles, *Lophodermium piceae*, did not appear to be an important needle decomposer. Deckert *et al.* (2002) reported that the DNA sequences of *Lophodermium nitens* isolated from healthy live needles of *Pinus strobus* and those of single ascospores from hysterothecia on needles in the litter showed high levels of genetic similarity. Promputtha *et al.* (2007) also compared sequence data and determined phylogenetic relationships between endophytes and saprobes isolated from leaves and twigs of *Magnolia liliifera*. Their analyses suggested that endophytic isolates of *Colletotrichum*, *Fusarium*, *Guignardia* and *Phomopsis* had high sequence similarities to their saprobic counterparts, and that they were phylogenetically closely related.

Grass endophytes have in general been regarded as mutualistic or parasitic symbionts with their host plants, and consequently little attention has been given to their possible role in the decomposition of leaf litter. As grass endophytes are transmitted either vertically through seeds or horizontally via spores produced on live host tissues, it is believed that they have no need to utilize dead tissues to complete their life cycle. Recently, however, a few studies have shown the occurrence of grass endophytes on dead tissues. Christensen and Voisey (2007) used a genetically modified *Epichloë festucae* that produced a green fluorescent protein to identify hyphae of the endophyte in decaying tissue of perennial ryegrass (*Lolium perenne*), even though many other fungi had begun to colonize it. The hyphae appeared as they had been when the leaf was alive, and had not started to grow saprobically. Therefore, it seems unlikely that the occurrence of the endophyte on the dead tissue indicated active hyphal colonization. In contrast, Tadych and White (2007) demonstrated that an endophytic *Neotyphodium* sp. was capable of growing and sporulating on dead leaves of big bluegrass (*Poa ampla*). *Acremonium chilense*, an endophyte of *Dactylis glomerata*, can proliferate in senescent tissues of the host, which is typical of saprobic fungi (Morgan-Jones *et al.*, 1990). Similar *Acremonium* species have regularly been isolated from grass species (Naffaa *et al.*, 1998). Future studies will probably give further examples of grass endophytes, particularly non-systemic ones that are saprobic on dead host tissues.

Origin and Development of Early Colonizers

Two pathways have been suggested for the origin and development of endophytic fungi as early colonizers of tree leaf litter: persistence from the phyllosphere and direct infection (Osono, 2006a). The persistence of endophytic fungi from living to dead leaves has been considered important in their colonization of leaf litter (Hudson, 1968). This probably holds true for those grass endophytes that infect dead grass tissues. Recently, however, Osono (2002, 2005) isolated xylariaceous endophytes from leaf litter that had been sterilized to exclude previously established fungi, suggesting that some endophytic fungi can colonize fallen leaves directly after litter fall as either hyphae or spores from surrounding litter or the air.

In contrast, rhytismataceous endophytes seem unable to infect leaf litter directly after litter fall, and their persistence in the phyllosphere appears crucial for the colonization of litters. *Coccomyces* and *Lophodermium* endophytes were not found on leaves of *Camellia japonica* and *Pinus thunbergii* when the leaves were sterilized before incubation on litter layers (Koide *et al.*, 2005b; Hirose and Osono, unpublished data). *Ascochyta* sp., a major endophytic fungus of *Fagus crenata*, can directly infect the surface but not the internal tissues of freshly fallen leaves, indicating that persistence in the phyllosphere may be necessary for internal colonization of leaf litter in this case (Osono, 2002).

Stone (1987) described in detail the behaviour of an endophytic fungus *Rhabdocline parkeri* on needles of *Pseudostuga menziesii* during the course of latent colonization in healthy needles, active hyphal growth at the onset of needle senescence and sporulation after needle abscission. This illustrated how endophytic fungi that are latent in host tissues can initiate the active colonization at leaf senescence and persist in fallen leaves from the phyllosphere. Infections by *R. parkeri* are intracellular and confined to a single epidermal cell of a healthy needle. This lasts for a period of 2–5 years until the onset of needle senescence, when active colonization of the needles resumes and haustoria are produced in cells adjacent to the original infection sites. The haustoria gradually enlarge, but the colonization of the needle is confined to a few adjacent cells until the needle is shed. Rapid colonization and sporulation occur immediately after needle abscission, before substantial colonization of the needles by other saprobic fungi.

Prior colonization by endophytic fungi on leaf litter can influence subsequent colonization by other saprobic fungi. For example, on *C. japonica* leaf litter the frequency of a unidentified species of the Dematiaceae (previously denoted as Coelomycete sp.1), which colonized the leaf rapidly after litter fall, was lower in the area previously colonized by rhytismataceous endophytes than in the non-colonized area (Koide *et al.*, 2005b). On *P. thunbergii* needle litter that had been sterilized to exclude previously established fungi and then placed on litter layers, the frequency of not only *Lophodermium* endophytes, but also that of rhizomorphs of *Marasmius* sp., decreased (Hirose and Osono, unpublished data). A causal relationship between these two decreases remains unclear. Conversely, with *F. crenata* and *S. controversa*, an initial leaf sterilization and exclusion of previously established fungi had no effect on other saprobic fungi colonizing these leaves after litter fall (Osono, 2002, 2005). The presence of endophytic fungi can also enhance colonization by mycoparasites, such as *Gonatobotrys simplex* and *Lecanicillium lecanii* that can utilize earlier fungal colonizers as hosts (McKenzie and Hudson, 1976; Mitchell and Millar, 1978).

Succession and Persistence in Decomposition

Most endophytic fungi are considered to dominate soon after leaf senescence and then to disappear in the early stages of leaf litter decomposition. This has been demonstrated in studies of fungal succession on decomposing litter (Hudson, 1968). The relative proportion of endophytic fungi as a group in decomposer fungal communities decreases as the decomposition progresses (see Table 6.1), and only a few endophytic species can persist until the late stages. Osono (2006a) suggested that factors such as life cycle, the ability to utilize substrates, micro-environmental conditions and competitive and antagonistic mycelial interactions that mediate the persistence of both endophytic and epiphytic phyllosphere fungi were important. Of these, life cycle, the ability to utilize substrates and antagonistic interactions are of particular importance for endophytes, and are discussed here.

Some endophytes utilize fallen leaves as a habitat for sexual reproduction and sporulation in order to infect live leaves and complete their life cycle. *Lophodermium piceae* is latent in green needles for several years, and conidiomata and ascomata start to form on dead needles (Osorio and Stephan, 1991). The drying of the needle tissue at needle senescence seems to trigger the development of fruiting bodies, and ascomata develop relatively quickly on dead needles. Endophytes such as *R. parkeri* in *P. menziesii* needles (Stone, 1987), *Mycosphaerella buna* in *F. crenata* leaves (Kaneko *et al.*, 2003) and *Coccomyces* sp. in *C. japonica* leaves (Koide *et al.*, 2005b) have similar life cycles. In these cases their disappearance from litter is due to their life cycle rather than to the exhaustion of available substrates in litter, and this has been shown for a *Coccomyces* sp. that caused further decomposition when re-inoculated on to leaf pieces that had been sterilized following previous colonization (Osono and Hirose, 2009).

Fungi that are endophytic in tree leaves can utilize various organic compounds as carbon sources, including starch, cellulose, xylan, mannan, lignin and derivatives, pectin, lipids and proteins (Carroll and Petrini, 1983; Petrini *et al.*, 1991; Sieber-Canavesi *et al.*, 1991; Kumaresan and Suryanarayanan, 2002; Lumyong *et al.*, 2002; Urairuj *et al.*, 2003; Kudanga and Mwenje, 2005). Carroll and Petrini (1983) tested the patterns of substrate utilization by endophytes from coniferous foliages and divided these fungi into two groups. The first group, such as *Xylaria* and *Phomopsis* species, showed a broad range of substrate utilization capabilities, including cellulolytic activity. The second group, such as *Phyllosticta* species, showed restricted substrate utilization capabilities and were generally unable to utilize cellulose and hemicelluloses. These enzymatic abilities can be expected to play an important role, not only in infection and substrate acquisition in live leaves, but also in saprobic growth on dead leaves.

The disappearance of some endophytic fungi from leaf litter may be associated with the exhaustion of available substrates in litter, and the persistence of a fungus until late stages of decomposition can depend on the ability of the fungus to utilize residual substrates of less availability and increased chemical complexity (Osono, 2007b). Generally, components in litter that are readily available to fungi, such as sugars and non-lignified holocellulose, are decomposed faster than less available or recalcitrant components, such as lignin and lignified holocellulose (Osono and Takeda, 2005a). The persistence of xylariaceous endophytes until the late stages of decomposition, when the readily available substrates have been exhausted and the remaining main components are lignin and lignified holocellulose, is thus partly attributable to their ligninolytic activity (Osono and Takeda, 2001b, 2002b).

Grass endophytes have also been shown to be able to utilize a wide range of simple carbon sources such as glucose, fructose, sucrose and xylose (Davis *et al.*, 1986; White *et al.*, 1991). As these simple components are abundant in recently-dead tissues, endophytes have an advantage in being able to utilize them prior to fungi that colonize after tissue death, and some grass endophytes can grow on recently dead tissues as saprobes. It however seems unlikely that grass endophytes, particularly systemic ones, are capable of utilizing more recalcitrant polymers such as cellulose and lignin, although there are no data available to support this theory. The disappearance of grass endophytes from decomposing leaf litter might be associated with the exhaustion of available substrates in litter.

Some endophytic fungi produce antibiotic metabolites (Fisher *et al.*, 1984a,b; Miller, 1986; Noble *et al.*, 1991; Bills *et al.*, 1992; Petrini *et al.*, 1992; Polishook *et al.*, 1993; Peláez *et al.*,1998; Schulz *et al.*, 2002; Weber *et al.*, 2004; Wicklow *et al.*, 2005). These metabolites may potentially be necessary for the defence of their colonies on fallen leaves against competitors. For example, *Lophodermium pinastri*, an endophyte of *Pinus* species, produces antibiotic substances that suppress the growth of saprobic fungi *in vitro* (Hata, 1998).

The Role of Endophytic Fungi in Decomposition Processes

Chemical decomposition of leaf litter follows a sequential pattern, with different classes of organic compounds dominating the process as it proceeds (Berg and McClaugherty, 2003; Osono and Takeda, 2005a,b). In general, loss of soluble components occurs during the early stage, followed by holocellulose decomposition during the late stages. Finally, lignin becomes a dominant component in the third stage, when the rate of mass loss of litter is slowed down and the litter approaches humus. Studies of fungal succession have emphasized the occurrence of endophytic fungi on freshly fallen leaves and in the early stages of decomposition. Only a few studies, using tree leaves, have examined chemical changes and succession of endophytic fungi simultaneously during decomposition in order to estimate the possible role of colonization by endophytes in the chemical changes in litter (Osono, 2007b).

Many endophytic fungi occur on leaves in the early stage of decomposition, which is characterized by the rapid loss of soluble components. The loss of soluble components includes not only leaching through the soil but also microbial respiration. Tietema and Wessel (1994) estimated the relative importance of leaching and respiration in leaf litter of *Quercus robur.* They found that leaching and respiration accounted for 21 and 64%, respectively, of total mass loss of oak leaves during the first 6 months of decomposition. In a similar experiment with leaves of *Acer saccharum*, McClaugherty (1983) found that after 1 year of decomposition 21% of the original carbohydrates in litter remained, 15% were leached and 64% were respired by microbes or transformed into insoluble compounds. The 'respiration' in these studies included not only that by endophytic fungi but also that by other phylloplane and saprobic fungi and bacteria. The relative contribution of these microbes to respiration remains unclear, but a part of the loss would be attributable to the respiration of endophytic fungi. However, Osono (2005), Koide *et al.* (2005b) and Hirose and Osono (unpublished) reported that the exclusion of phyllosphere fungi in initially sterilized leaves did not affect the mass loss of leaves within the first few months of decomposition, compared with the mass loss of non-sterilized leaves. These studies suggest that the role of endophytic fungi in the initial stage of decomposition is relatively small in spite of their frequent colonization of leaf litter.

Recently, the roles of endophytic fungi in the decomposition of structural components, such as holocellulose and lignin, have been examined in a series of experiments in Japan and other regions. In *F. crenata* leaves, for example, the ligninolytic activity of xylariaceous endophytes is seen as bleached areas on the surface of brown fallen leaves. There are higher frequencies of endophytes and lower lignin concentrations in these areas than in the surrounding, non-bleached areas (Osono and Takeda, 2001a). Similarly, xylariaceous endophytes have been implicated in lignin decomposition in *S. controversa* leaf litter (Osono *et al.*, 2004; Osono, 2005). *Coccomyces* sp., a ligninolytic endophyte of *C. japonica* leaves, caused selective delignification as early as 3 weeks after litter fall (Koide and Osono, 2003; Koide *et al.*, 2005b) and *Lophodermium pinastri*, a major endophyte of *Pinus* needles, has cellulolytic activity (Sieber-Canavesi *et al.*, 1991) and is responsible for cellulose hydrolysis during the first year of decomposition (Mitchell and Millar, 1978; Ponge, 1991). Our personal observations also suggest that some *Lophodermium* species on trees leaves are involved in lignin decomposition in leaf litter (see Fig. 6.1a).

Leaf litter is a major source of essential macronutrients, such as nitrogen and phosphorus, for primary production in forest trees (Likens and Bormann,

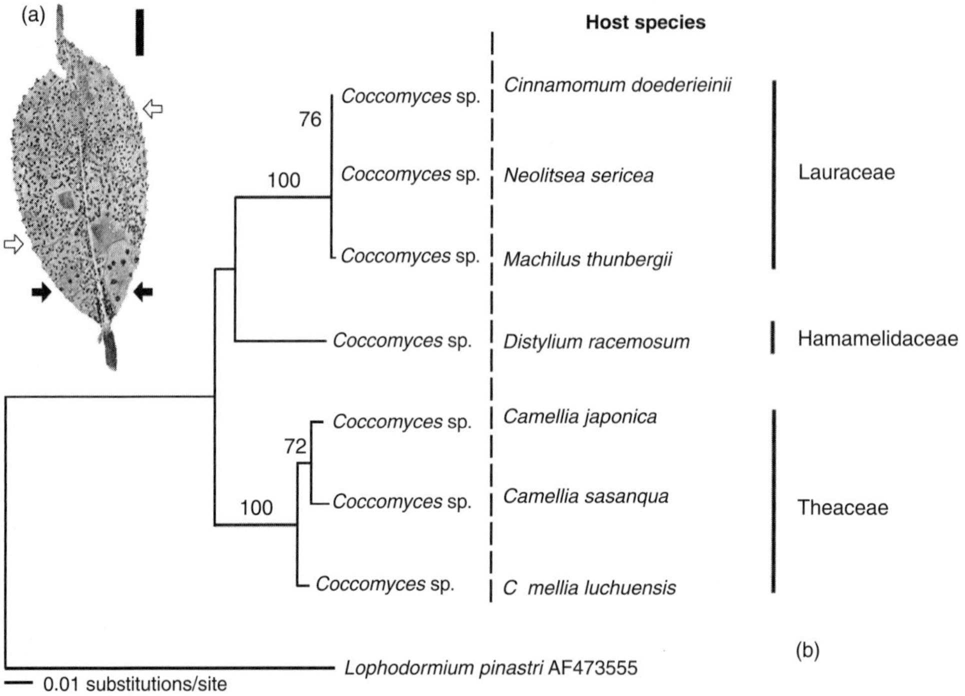

Fig. 6.1. (a) A bleached leaf litter of *Camellia japonica*. Filled arrows, colonies of *Coccomyces* sp.; open arrows, colonies of *Lophodermium* sp. Bar = 1 cm. (b) Neighbour-joining tree based on sequences of the rDNA-ITS region from *Coccomyces* spp. associated with seven different host tree species in a subtropical forest in southern Japan. Numbers above the branches are bootstrap values. *Lophodermium pinastri* sequence was used as outgroup.

1995). The dynamics of nitrogen and phosphorus show leaching, accumulation and release phases during decomposition, and the accumulation and release of nitrogen and phosphorus are associated with microbial decomposition of lignin (Berg and McClaugherty, 1989; Osono and Takeda, 2004). Endophytic fungi associated with the decomposition of structural components therefore have the potential to influence nitrogen and phosphorus dynamics. The selective decomposition of holocellulose in preference to lignin by xylariaceous endophytes during the first 2 years of beech leaf litter decomposition has been associated with the incorporation of nitrogen and phosphorus in decomposing litter from surrounding environments (Osono and Takeda, 2001b), and xylariaceous fungi have been shown to accumulate phosphorus in litter in a pure culture decomposition test (Osono and Takeda, 2002a). Similarly, selective holocellulose decomposition by *Lophodermium* endophytes in the early stages of pine needle litter decomposition has been associated with the accumulation of nitrogen and phosphorus in litter (Bhatta, 2003). In contrast, the selective delignification of *C. japonica* leaves by rhytismataceous endophytes has been reported to enhance the release of nitrogen (Koide *et al.*, 2005a). Chemical pathways related to nitrogen and phosphorus accumulation in decomposing litter are not fully understood, but some authors have suggested that the formation of lignin-like substances complexed with nitrogen and phosphorus is responsible for such accumulations (Aber and Mellilo, 1982; Berg, 1986; Osono and Takeda, 2004; Osono *et al.*, 2006).

Prior decomposition by endophytic fungi can influence the decomposition of lignin and holocellulose in residual leaf litter by succeeding fungi. Pre-treatment of *F. crenata* leaf litter by two endophytic fungi led to a decrease in litter decomposition by ascomycetes and zygomycetes, but enhanced lignin decomposition by two *Mycena* species (Osono, 2003). These basidiomycetes appear to be physiologically predisposed to the selective removal of lignin in residual litter that has previously been partly decomposed by endophytic fungi. Koide *et al.* (2005a) also reported that the prior colonization of *C. japonica* leaf litter by the ligninolytic endophyte *Coccomyces* sp. enhanced the decomposition of lignin and holocellulose by succeeding fungal communities. This was subsequently verified in pure culture, where pre-treatment of *C. japonica* leaf litter by *Coccomyces* sp. enhanced litter decomposition by two succeeding ascomycetes (Osono and Hirose, 2009). These results indicate that the effects of pre-decomposition by endophytes may depend not only on the ligninolytic activity of the endophytes but also on the succeeding fungi. Similar effects of pre-decomposition have been reported in a laboratory study using wood blocks (Tanaka *et al.*, 1988) and in a field study using *Pinus sylvestris* needle litter (Cox *et al.*, 2001).

The colonization of tree leaf litter by endophytic fungi can influence the accumulation of soil organic matter on the forest floor. Osono (2006a) estimated that the ligninolytic endophytes of *C. japonica* leaves had the potential to reduce approximately 60% of the amount of soil organic matter formed as decomposition products compared with when the litter was not colonized by endophytes. In contrast, it is generally expected that cellulose decomposition by endophytic fungi may retard litter mass loss and contribute to the accumulation of soil organic matter (Bhatta, 2003). Interestingly, Osono (2003) showed that cellulolytic endophytes can also contribute to the decrease of soil organic matter when the residue is colonized by ligninolytic basidiomycetes. This was seen in pure culture where the prior decomposition of *F. crenata* leaf litter by endophytic fungi stimulated lignin decomposition in the residue by *Mycena* spp. Ligninolytic basidiomycetes cause bleaching and enhance the decomposition of soil organic matter (Harris, 1945; Saito, 1957; Hintikka, 1970; Miyamoto *et al.*, 2000). Therefore, the ligninolytic activity of endophytes, the succeeding fungi, and the effect of endophytes on the ligninolytic activity of succeeding fungi all need to be considered in evaluating the role of endophytes in the accumulation and decomposition of soil organic matter.

A few studies have shown slightly lower nitrogen contents for litter infected by grass endophytes compared with non-infected litter, but the differences reported were not statistically significant. Omacini *et al.* (2004) found no significant difference in total nitrogen content between endophyte-infected (E+) and non-infected (E–) *Lolium* litters (4.1–4.9 versus 5.1–5.5 mg/g^{-1}). Lemons *et al.* (2005) also reported no significant difference in total nitrogen content between E+ and E– litters (10.4 versus 11.2 mg/g). In contrast,

Lyons *et al.* (1990) reported an increase of total soluble amino acids and NH_4^+ in live *Festuca* leaves from infected plants, but this was apparent only when nitrogen fertilizer was applied at a high rate. Malinowski *et al.* (2000) also reported an increase in some minerals such as phosphorus, calcium and zinc in live roots of infected *Festuca*. However, in the latter two studies, it is not known whether the litter quality reflected the qualitative differences between the live tissue from the infected and non-infected plants.

Recent studies have shown a negative effect on litter decomposition by a *Neotyphodium* endophyte. Omacini *et al.* (2004) found decomposition rates in garden microcosm and greenhouse experiments to be 18% slower on average in E+ litter (k = 1.28/year) than in litter previously non-infected by endophytes (k = 1.57/year). Lemons *et al.* (2005) also found that decomposition rates were 6% slower in E+ litter (k = 1.08/year) than in E– litter (k = 1.19/year) in an agricultural field experiment. However, in another greenhouse pot experiment, these authors found decomposition rates were 5% slower in E+ than in E– litter, but the difference was not statistically significant. The greater accumulation of soil carbon in highly endophyte-infected *Festuca* pastures than in low-infected ones (Franzluebbers *et al.*, 1999; Schomberg *et al.*, 2000) is possibly attributable to a reduction in decomposition rates of E+ litter.

Omacini *et al.* (2004) and Lemons *et al.* (2005) suggested that litter deposited by endophyte-infected grasses may have chemical or physical properties that reduced its suitability as a substrate for decomposers, and that this might be due to the production of toxic alkaloids. Negative effects on decomposition, such as the production of alkaloids by endophytes, are likely but have yet to be demonstrated. Some alkaloids are water soluble (Koulman *et al.*, 2007) and susceptible to photodegradation, and so may not persist in the environment, and Hume *et al.* (2007) have shown that endophyte alkaloid concentrations in dried grass herbage rapidly decrease after 2 weeks in the field to between 0 and 40% of the original concentration. Interestingly, Perumbakkam *et al.* (2007) reported that anaerobic incubation of earthworm (*Eisenea fetida*) homogenate with pure ergovaline resulted in a 60% reduction in ergovaline concentration, suggesting that diverse anaerobic bacteria may have roles in alkaloid detoxification. More studies are necessary on alkaloid contents in decomposing infected litter. Patterns of changes in nitrogen and lignin contents should be investigated in infected and non-infected litters to evaluate the effect of alkaloids on the long-term decomposition processes. In addition, bioassays of the toxicity of alkaloids to decomposer fungi and animals can provide insights into the mechanisms underlying the effects of alkaloids on decomposition (Osono, 2007a).

Conclusion and Future Perspectives

This chapter has shown that several of the common endophytic fungi of tree leaves are primarily saprobes, and that they are specifically adapted to colonize and utilize dead host tissue. Endophytic fungi of trees and grasses have the potential to influence short- and long-term decomposition processes, although it is clear that relatively little research has been undertaken on their ecology and roles during decomposition. More comparative studies are needed on leaf-associated endophytes and litter fungi, and more detailed analyses are required into the phylogenetic affinities of endophytes and their litter counterparts (Promputtha *et al.*, 2007). Data from tropical regions and for grasses associated with non-systemic endophytes are particularly scarce (Arnold, 2005; Murali *et al.*, 2007). The use of environmental PCR to examine the diversity of endophytic fungi in not only asymptomatic foliage, but also in dead leaves, has potential as a future research direction (Arnold *et al.*, 2007), and environmental PCR with taxon-specific primers may be useful for detecting and monitoring endophytic fungi during litter decomposition. Some current studies on the functional fungal diversity in litter and soil have considered the diversity and expression of laccase genes that encode for enzymes used in lignin oxidation (Luis *et al.*, 2004, 2005a,b; Blackwood *et al.*, 2007). This approach could be used to obtain direct evidence for the involvement of endophytic fungi in the decomposition processes.

Alternatively, Osono (2006b) has developed an approach to examine the abundance and activity of ligninolytic fungi in tropical and temperate forests based on the occurrence of bleached areas on the surface of leaf litter (see Fig. 6.1a). The presence of bleached areas was associated with fungal colonization of leaf tissues and decomposition of lignin (Osono and Takeda, 2001a), and endophytic fungi were identified as ligninolytic when they fruited on the bleached areas on leaf litter (Koide *et al.*, 2005a,b). A preliminary survey in a subtropical forest in southern Japan has found that endophytic *Lophodermium* species occurred on the bleached areas of leaf litter from trees of 17 species from 15 genera in 11 families, and endophytic *Coccomyces* species were found on trees of 11 species from 9 genera in 6 families (Osono and Hirose, unpublished data). The bleached colonies of these rhytismataceous endophytes occupied only a few per cent of the total leaf area in leaves of some tree species, but almost 100% of the total leaf area in other tree species. A preliminary phylogenetic analysis suggested that different *Coccomyces* species were associated with host trees at family level (see Fig. 6.1b). Recent results suggest that the ligninolytic activities of these endophytic fungi contribute to the rapid disappearance of leaf litter on the forest floor and to the low accumulation of soil organic matter in the forest stand. Studies on the ecology and diversity of endophytic fungi associated with leaf litter decomposition will provide insights into the ecological consequence of endophytic infection on ecosystem processes.

References

Aber, J.D. and Mellilo, J.M. (1982) Nitrogen immobilization in decaying hardwood leaf litter as a function of initial nitrogen and lignin content. *Canadian Journal of Botany* 60, 2263–2269.

Arnold, A.E. (2005) Diversity and ecology of fungal endophytes in tropical forests. In: Deshmukh S.K. and Rai. M.K. (eds) *Biodiversity of Fungi, their Role in Human Life*. Science Publishers, Enfield, New Hampshire, USA, pp. 49–68.

Arnold, A.E., Henk, D.I., Eells, R.L., Lutzoni, F. and Vilgalys, R. (2007) Diversity and phylogenetic affinities of foliar fungal endophytes in loblolly pine inferred by culturing and environmental PCR. *Mycologia* 99, 185–206.

Berg, B. (1986) Nutrient release from litter and humus in coniferous forest soils – a mini review. *Scandinavian Journal of Forest Research* 1, 359–369.

Berg, B. and McClaugherty, C. (1989) Nitrogen and phosphorus release from decomposing litter in relation to the disappearance of lignin. *Canadian Journal of Botany* 67, 1148–1156.

Berg, B. and McClaugherty, C. (2003) *Plant Litter, Decomposition, Humus Formation, Carbon Sequestration*. Springer Verlag, Berlin, Germany, 286 pp.

Bhatta, B.K. (2003) Effects of soil condition on the decomposition process of Japanese black pine needle in a pine plantation forest. PhD thesis, Kyoto University, Kyoto, Japan, 77 pp.

Bills, G.F., Gaicobbe, R.A., Lee, S.H., Peláez, F. and Kacz, J.S. (1992) Tremorgenic mycotoxins, paspalitrem A and C, from a tropical *Phomopsis*. *Mycological Research* 96, 977–983.

Blackwood, C.B., Waldrop, M.P., Zak, D.R. and Sinsabaugh, R.L. (2007) Molecular analysis of fungal communities and laccase genes in decomposing litter reveals differences among forest types but no impact of nitrogen deposition. *Environmental Microbiology* 9, 1306–1316.

Boddy, L. and Griffith, G.S. (1989) Role of endophytes and latent invasion in the development of decay communities in sapwood of angiospermous trees. *Sydowia* 41, 41–73.

Cabral, D. (1985) Phyllosphere of *Eucalyptus viminalis*: dynamics of fungal populations. *Transactions of the British Mycological Society* 85, 501–511.

Carroll, G.C. (1991) Fungal associates of woody plants as insect antagonists in leaves and stems. In: Barbosa, P. Krischik V.A. and Jones C.G. (eds) *Microbial Mediation of Plant–Herbivore Interactions.* John Wiley & Sons, New York, USA, pp. 253–271.

Carroll, G.C. (1995) Forest endophytes: pattern and process. *Canadian Journal of Botany* 73(Suppl. 1), S1316–S1324.

Carroll, G.C. and Petrini, O. (1983) Patterns of substrate utilization by some fungal endophytes from coniferous foliage. *Mycologia* 75, 53–63.

Christensen, M.J. and Voisey, C.R. (2007) The biology of the endophyte/grass partnership. In: Popay, A.J. and Thom, E.R. (eds) *Proceedings of the 6th International Symposium on Fungal Endophytes of Grasses.* Grassland Research and Practice Series No. 13, New Zealand Grassland Association, Dunedin, New Zealand, pp. 123–134.

Clay, K. (1986) Grass endophytes. In: Fokkema, N.J. and van den Heuvel, J. (eds) *Microbiology of the Phyllosphere.* Cambridge University Press, Cambridge, UK, pp. 188–204.

Clay, K. (1991) Fungal endophytes, grasses, and herbivores. In: Barbosa, P., Krischik, V.A. and Jones, C.G. (eds) *Microbial Mediation of Plant–Herbivore Interactions.* John Wiley & Sons, New York, USA, pp. 199–226.

Clay, K. and Schardl, C. (2002) Evolutionary origins and ecological consequences of endophyte symbiosis with grasses. *American Naturalist* 160, S99–S127.

Cox, P., Wilkinson, S.P. and Anderson, J.M. (2001) Effects of fungal inocula on the decomposition of lignin and structural polysaccharides in *Pinus sylvestris* litter. *Biology and Fertility of Soils* 33, 246–251.

Davis, N.D., Clark, E.M., Schrey, K.A. and Diener, U.L. (1986) *In vitro* growth of *Acremonium coenophialum,* an endophyte of toxic tall fescue grass. *Applied and Environmental Microbiology* 52, 888–891.

Deckert, R.J., Hsiang, T. and Peterson, R.L. (2002) Genetic relationships of endophytic *Lophodermium nitens* isolates from needles of *Pinus strobus*. *Mycological Research* 106, 305–313.

Fisher, P.J., Anson, A.E. and Petrini, O. (1984a) Novel antibiotic activity of an endophytic *Cryptosporiopsis* sp. isolated from *Vaccinium myrtillus*. *Transactions of the British Mycological Society* 83, 145–148.

Fisher, P.J., Anson, A.E. and Petrini, O. (1984b) Antibiotic activity of some endophytic fungi from ericaceous plants. *Botanica Helvatica* 94, 249–253.

Franzluebbers, A.J., Nazih, N., Stuedemann, J.A., Fuhrmann, J.J., Schomberg, H.H. and Hartel, P.G. (1999) Soil carbon and nitrogen pools under low- and high-endophyte-infected tall fescue. *Soil Science Society of America Journal* 63, 1687–1694.

Ganley, R.J., Brunsfeld, S.J. and Newcombe, G. (2004) A community of unknown, endophytic fungi in western white pine. *Proceedings of the National Academy of Sciences* 101, 10107–10112.

Harris, G.C.M. (1945) Chemical changes in beech litter due to infection by *Marasmius peronatus* (Bolt.) Fr. *Annals of Applied Biology* 32, 38–39.

Hata, K. (1998) Endophytic mycobiota and their dynamics in *Pinus* needles. PhD thesis, Kyoto University, Kyoto, Japan, 102 pp.

Heredia, G. (1993) Mycoflora associated with green leaves and leaf litter of *Quercus germana, Quercus sartorii* and *Liquidambar styraciflua* in a Mexican cloud forest. *Cryptogamie Mycologie* 14, 171–183.

Hintikka, V. (1970) Studies on white-rot humus formed by higher fungi in forest soils. *Communicationes Instituti Forestalis Fenniae* 69, 1–68.

Hudson, H.J. (1968) The ecology of fungi on plant remains above the soil. *New Phytologist* 67, 837–874.

Hume, D.E., Hickey, M.J. and Tapper, B.A. (2007) Degradation of endophyte alkaloids in field-dried cut ryegrass herbage. In: Popay, A.J. and Thom, E.R. (eds) *Proceedings of the 6th International Symposium on Fungal Endophytes of Grasses.* Grassland Research and Practice Series No. 13, New Zealand Grassland Association, Dunedin, New Zealand, pp. 167–170.

Kaneko, R. and Kakishima, M. (2001) *Mycosphaerella buna* sp. nov. with a *Pseudocercospora* anamorph isolated from the leaves of Japanese beech. *Mycoscience* 42, 55–69.

Kaneko, R., Kakishima, M. and Tokumasu, S. (2003) The seasonal occurrence of endophytic fungus, *Mycosphaerella buna,* in Japanese beech, *Fagus crenata. Mycoscience* 44, 277–281.

Koide, K. and Osono, T. (2003) Chemical composition and mycobiota of bleached portion of *Camellia japonica* leaf litter at two stands with the different nitrogen status. *Journal of the Japanese Forestry Society* 85, 359–363.

Koide, K., Osono, T. and Takeda, H. (2005a) Fungal succession and decomposition of *Camellia japonica* leaf litter. *Ecological Research* 20, 599–609.

Koide, K., Osono, T. and Takeda, H. (2005b) Colonization and lignin decomposition of *Camellia japonica* leaf litter by endophytic fungi. *Mycoscience* 46, 280–286.

Koulman, A., Lane, G.A., Christensen, M.J., Fraser, K. and Tapper, B.A. (2007) Peramine and other fungal alkaloids are exuded in the guttation fluid of endophyte-infected grasses. *Phytochemistry* 68, 355–360.

Kudanga, T. and Mwenje, E. (2005) Extracellular cellulase production by tropical isolates of *Aureobasidium pullulans. Canadian Journal of Microbiology* 51, 773–776.

Kumaresan, V. and Suryanarayanan, T.S. (2002) Endophyte assemblages in young, mature and senescent leaves of *Rhizophora apiculata*: evidence for the role of endophytes in mangrove litter degradation. *Fungal Diversity* 9, 81–91.

Lemons, A., Clay K. and Rudgers, J.A. (2005) Connecting plant–microbial interactions above and below ground: a fungal endophyte affects decomposition. *Oecologia* 145, 595–604.

Likens, G.E. and Bormann, F.H. (1995) *Biogeochemistry of a Forested Ecosystem.* Springer, Heidelberg, Berlin, 159 pp.

Luis, P., Walther, G., Kellner, H., Martin, F. and Buscot, F. (2004) Diversity of laccase genes from basidiomycetes in a forest soil. *Soil Biology and Biochemistry* 36, 1025–1036.

Luis, P., Kellner, H., Martin, F. and Biscot, F. (2005a) A molecular method to evaluate basidiomycete laccase gene expression in forest soils. *Geoderma* 128, 18–27.

Luis, P., Kellner, H., Zimdars, B., Langer, U., Martin, F. and Biscot, F. (2005b) Patchiness and spatial distribution of laccase genes of ectomycorrhizal, saprotrophic, and unknown basidiomycetes in the upper horizons of a mixed forest cambisol. *Microbial Ecology* 50, 570–579.

Lumyong, S., Lumyong, P., McKenzie, E.H.C. and Hyde, K.D. (2002) Enzymatic activity of endophytic fungi of six native seedling species from Doi Suthep-Pui National Park, Thailand. *Canadian Journal of Microbiology* 48, 1109–1112.

Lyons, P.C., Evans, J.J. and Bacon, C.W. (1990) Effects of the fungal endophyte *Acremonium coenophialum* on nitrogen accumulation and metabolism in tall fescue. *Plant Physiology* 92, 726–732.

Malinowski, D.P., Alloush, G.A. and Belesky, D.P. (2000) Leaf endophyte *Neotyphodium coenophialum* modified mineral uptake in tall fescue. *Plant and Soil* 227, 115–126.

McClaugherty, C.A. (1983) Soluble polyphenols and carbohydrates in throughfall and leaf litter decomposition. *Acta Oecologica* 4, 375–385.

McKenzie, E.H.C. and Hudson, H.J. (1976) Mycoflora of rust-infected and non-infected plant material during decay. *Transactions of the British Mycological Society* 66, 223–238.

Miller, J.D. (1986) Toxic metabolites of epiphytic and endophytic fungi of conifer needles. In: Fokkema, N.J. and van der Heuvel, J. (eds) *Microbiology of the Phyllosphere*. Cambridge University Press, Cambridge, UK, pp. 221–231.

Mitchell, C.P. and Millar, C.S. (1978) Mycofloral succession on corsican pine needles colonized on the tree by three different fungi. *Transactions of the British Mycological Society* 71, 303–317.

Miyamoto, T., Igarashi, T. and Takahashi, K. (2000) Lignin-degrading ability of litter-decomposing basidiomycetes from *Picea* forests of Hokkaido. *Mycoscience* 41, 105–110.

Morgan-Jones, G., White, J.F. Jr. and Pointelli, E.L. (1990) Endophyte–host associations in foliage grasses. 8. *Acremonium chilense,* an undescribed endophyte occurring on *Dactylis glomerata* in Chile. *Mycotaxon* 39, 441–454.

Müller, M.M., Valjakka, R., Suokko, A. and Hantula, J. (2001) Diversity of endophytic fungi of single Norway spruce needles and their role as pioneer decomposers. *Molecular Ecology* 10, 1801–1810.

Murali, T.S., Suryanarayanan, T.S. and Venkatesan, G. (2007) Fungal endophyte communities in two tropical forests of southern India: diversity and host affiliation. *Mycological Progress* 6, 191–199.

Naffaa, W., Ravel, C. and Guillaumin, J.J. (1998) A new group of endophytes in European grasses. *Annals of Applied Biology* 132, 211–226.

Noble, H.M., Langley, D., Sidebottom, P.J., Lane, S.J. and Fisher, P.J. (1991) An echinocandin from an endophytic *Cryptosporiopsis* sp. and *Pezicula* sp. in *Pinus sylvestris* and *Fagus sylvatica*. *Mycological Research* 95, 1439–1440.

Okane, I., Nakagiri, A. and Ito, T. (1998) Endophytic fungi in leaves of ericaceous plants. *Canadian Journal of Botany* 76, 657–663.

Omacini, M., Chaneton, E.J., Ghersa, C.M. and Otero, P. (2004) Do foliar endophytes affect grass litter decomposition? A microcosm approach using *Lolium multiflorum*. *Oikos* 104, 581–590.

Osono, T. (2002) Phyllosphere fungi on leaf litter of *Fagus crenata*: occurrence, colonization, and succession. *Canadian Journal of Botany* 80, 460–469.

Osono, T. (2003) Effects of prior decomposition of beech leaf litter by phyllosphere fungi on substrate utilization by fungal decomposers. *Mycoscience* 44, 41–45.

Osono, T. (2005) Colonization and succession of fungi during decomposition of *Swida controversa* leaf litter. *Mycologia* 97, 589–597.

Osono, T. (2006a) Role of phyllosphere fungi of forest trees in the development of decomposer fungal communities and decomposition processes of leaf litter. *Canadian Journal of Microbiology* 52, 701–716.

Osono, T. (2006b) Fungal decomposition of lignin in leaf litter: comparison between tropical and temperate forests. In: Meyer, W. and Pearce, C. (eds) *Proceeding for the 8th International Mycological Congress*, Cairns, Australia. Medimond, Bologna, Italy, pp. 111–117.

Osono, T. (2007a) Role of endophytic fungi in grass litter decomposition. In: Popay, A.J. and Thom, E.R. (eds.) *Proceedings of the 6th International Symposium on Fungal Endophytes of Grasses*. Grassland Research and Practice Series No. 13,. New Zealand Grassland Association, Dunedin, New Zealand, pp. 103–105.

Osono, T. (2007b) Ecology of ligninolytic fungi associated with leaf litter decomposition. *Ecological Research* 22, 955–974.

Osono, T. and Hirose, D. (2009) Effects of delignification of *Camellia japonica* leaf litter by an endophytic fungus on the subsequent decomposition by fungal colonizers. *Mycoscience* 50, 52–55.

Osono, T. and Takeda, H. (2001a) Effects of organic chemical quality and mineral nitrogen addition on lignin and holocellulose decomposition of beech leaf litter by Xylaria sp. *European Journal of Soil Biology* 37, 17–23.

Osono, T. and Takeda, H. (2001b) Organic chemical and nutrient dynamics in decomposing beech leaf litter in relation to fungal ingrowth and succession during 3-year decomposition processes in a cool temperate deciduous forest in Japan. *Ecological Research* 16, 649–670.

Osono, T. and Takeda, H. (2002a) Nutrient contents of beech leaf litter decomposed by fungi in Basidiomycota and Ascomycota. *Applied Forest Science Kansai* 11, 7–11.

Osono, T. and Takeda, H. (2002b) Comparison of litter decomposing ability among diverse fungi in a cool temperate deciduous forest in Japan. *Mycologia* 94, 421–427.

Osono, T. and Takeda, H. (2004) Accumulation and release of nitrogen and phosphorus in relation to lignin decomposition in leaf litter of 14 tree species in a cool temperate forest. *Ecological Research* 19, 593–602.

Osono, T. and Takeda, H. (2005a) Decomposition of organic chemical components in relation to nitrogen dynamics in leaf litter of 14 tree species in a cool temperate forest. *Ecological Research* 20, 41–49.

Osono, T. and Takeda, H. (2005b) Limit values for decomposition and convergence process of lignocellulose fraction in decomposing leaf litter of 14 tree species in a cool temperate forest. *Ecological Research* 20, 51–58.

Osono, T., Bhatta, B.K. and Takeda, H. (2004) Phyllosphere fungi on living and decomposing leaves of giant dogwood. *Mycoscience* 45, 35–41.

Osono, T., Hobara, S., Koba, K., Kameda, K. and Takeda, H. (2006) Immobilization of avian excreta-derived nutrients and reduced lignin decomposition in needle and twig litter in a temperate coniferous forest. *Soil Biology and Biochemistry* 38, 517–525.

Osono, T., Ishii, Y. and Hirose, D. (2008) Fungal colonization and decomposition of *Castanopsis sieboldii* leaf litter in a subtropical forest. *Ecological Research* 23, 909–917.

Osono, T., Ishii, Y., Takeda, H., Seramethakun, T., Khamyong, S., To-Anun, C., Hirose, D., Tokumasu, S. and Kakishima, M. (2009) Decomposition and fungal succession on *Shorea obtusa* leaf litter in a tropical seasonal forest in northern Thailand. *Fungal Diversity* 36, in press.

Osorio, M. and Stephan, B.R. (1991) Life cycle of *Lophodermium piceae* in Norway spruce needles. *European Journal of Forest Pathology* 21, 152–163.

Peláez, F., Collado, J., Arenal, F., Basilio, A., Cabello, A., Díez Matas, M.T., García, J.B., González, del Val A., González, V., Gorrochategui, J., Hernández, P., Martín, I., Platas, G. and Vicente, F. (1998) Endophytic fungi from plants living on gypsum soils as a source of secondary metabolites with antimicrobial activity. *Mycological Research* 102, 755–761.

Perumbakkam, S., Rattray, R.M., Delorme, M.J.M., Duringer, J.M. and Craig, A.M. (2007) Discovery of novel microorganisms involved in ergot detoxification: an approach. In:

Popay, A.J. and Thom, E.R. (eds) *Proceedings of the 6th International Symposium on Fungal Endophytes of Grasses.* Grassland Research and Practice Series No. 13, New Zealand Grassland Association, Dunedin, New Zealand, pp. 395–398.

Petrini, L.E. and Petrini,O. (1985) Xylariaceous fungi as endophytes. *Sydowia* 38, 216–234.

Petrini, L.E., Petrini, O., Leuchtmann, A. and Carroll, G.C. (1991) Conifer-inhabiting species of *Phyllosticta. Sydowia* 43, 148–169.

Petrini, O. (1986) Taxonomy of endophytic fungi of aerial plant tissues. In: Fokkema, N.J. and van der Heuvel, J. (eds.) *Microbiology of the Phyllosphere.* Cambridge University Press, Cambridge, UK, pp. 175–187.

Petrini, O. (1991) Fungal endophytes of tree leaves. In: Andrews J.H. and Hirano S.S. (eds) *Microbial Ecology of Leaves.* Springer Verlag, Berlin, Germany, pp.179–197.

Petrini, O., Sieber, T.N., Toti, L. and Viret, O. (1992) Ecology, metabolite production, and substrate utilization in endophytic fungi. *Natural Toxins* 1, 185–196.

Petrini, O., Petrini, L.E. and Rodriguez, K.F. (1995) Xylariaceous endophytes: an exercise in biodiversity. *Fitopatologia Brasiliera* 20, 531–539.

Polishook, J.D., Dombrowski, A.W., Tsou, N.N., Salituro, G.M. and Curotto, J.E. (1993) Preussomerin D from the endophyte *Hormonema dematioides. Mycologia* 85, 62–64.

Ponge, J.F. (1991) Succession of fungi and fauna during decomposition of needles in a small area of Scots pine litter. *Plant and Soil* 138, 99–113.

Promputtha, I., Lumyong,S., Dhanasekara,n V., McKenzie, E.H.C., Hyde, K.D. and Jeewon, R. (2007) A phylogenetic evaluation of whether endophytes become saprotrophs at host senescence. *Microbial Ecology* 53, 579–590.

Redin, S.C. and Carris, L.M. (1996) *Endophytic Fungi in Grasses and Woody Plants. Systematics, Ecology, and Evolution.* APS Press, Eagan, Minnesota, USA.

Rodriguez, R.J. and Redman, R.S. (1997) Fungal life-style and ecosystem dynamics: biological aspects of plant pathogens, plant endophytes and saprophytes. *Advances in Botanical Research* 24, 169–193.

Ruscoe, Q.W. (1971) Mycoflora of living and dead leaves of *Nothofagus truncata. Transactions of the British Mycological Society* 56, 463–474.

Saito, T. (1957) Chemical changes in beech litter under microbiological decomposition. *Ecological Review* 14, 209–216.

Schomberg, H.H., Stuedemann, J.A., Franzluebbers, A.J. and Wilkinson, S.R. (2000) Spatial distribution of extractable phosphorus, potassium, and magnesium as influenced by fertilizer and tall fescue endophyte status. *Agronomy Journal* 92, 981–986.

Schulz, B., Boyle, C., Draeger, S., Römmert, A.K. and Krohn, K. (2002) Endophytic fungi: a source of novel biologically active secondary metabolites. *Mycological Research* 106, 996–1004.

Sieber-Canavesi, F. and Sieber, T.N. (1993) Successional patterns of fungal communities in needles of European silver fir (*Abies alba* Mill.) *New Phytologist* 125, 149–161.

Sieber-Canavesi, F., Petrini, O., and Sieber, T.N. (1991) Endophytic *Leptostroma* species on *Picea abies, Abies alba,* and *Abies balsamea*: a cultural, biochemical, and numerical study. *Mycologia* 83, 89–96.

Stone, J.K. (1987) Initiation and development of latent infections by *Rhabdocline parkeri* on Douglas fir. *Canadian Journal of Botany* 65, 2614–2621.

Tadych, M. and White, J.F. Jr. (2007) Ecology of epiphyllous stages of endophytes and implications for horizontal dissemination. In: Popay, A.J. and Thom, E.R. (eds) *Proceedings of the 6th International Symposium on Fungal Endophytes of Grasses.* Grassland Research and Practice Series No. 13, New Zealand Grassland Association, Dunedin, New Zealand, pp. 157–161.

Tanaka, H., Enoki, A., Fuse, G. and Nishimoto, K. (1988) Interactions in successive exposure of wood to varying wood-inhabiting fungi. *Holzforschung* 42, 29–35.

Tietema, A. and Wessel, W.W. (1994) Microbial activity and leaching during initial oak leaf litter decomposition. *Biology and Fertility of Soils* 18, 49–54.
Urairuj, C., Khanongnuch, C. and Lumyong, S. (2003) Ligninolytic enzymes from tropical endophytic Xylariaceae. *Fungal Diversity* 13, 209–219.
Watson, E.S., McClurkin, D.C. and Huneycutt, M.B. (1974) Fungal succession on loblolly pine and upland hardwood foliage and litter in north Mississippi. *Ecology* 55, 1128–1134.
Weber, R.W.S., Stenger, E., Meffert, A. and Hahn, M. (2004) Brefeldin A production by *Phoma medicaginis* in dead pre-colonized plant tissue: a strategy for habitat conquest? *Mycological Research* 108, 662–671.
White, J.F. Jr, Breen, J.P. and Morgan-Jones, G. (1991) Substrate utilization in selected *Acremonium*, *Atkinsonella* and *Balansia* species. *Mycologia* 83, 601–610.
Wicklow, D.T., Roth, S., Deyrup, S.T. and Gloer, J.B. (2005) A protective endophyte of maize: *Acremonium zeae* antibiotics inhibitory to *Aspergillus flavus* and *Fusarium verticillioides*. *Mycological Research* 109, 610–618.
Wildman, H.G. and Parkinson, D. (1979) Microfungal succession on living leaves of *Populus tremuloides*. *Canadian Journal of Botany* 57, 2800–2811.
Wilson, D. (2000) Ecology of woody plant endophytes. In: Beacon, C.W. and White, J.F. (eds) *Microbial Endophytes*. Marcel Dekker, New York, pp. 389–420.

7 Brewing Yeast in Action: Beer Fermentation

Pieter J. Verbelen and Freddy R. Delvaux

Centre for Malting and Brewing Science, Faculty of Bioscience Engineering, Katholieke Universiteit Leuven, Heverlee, Belgium

Introduction

Production of beer by fermentation of a liquid malt extract is one of the oldest biotechnologies known. The first records of the use of yeast by man are concerned with the production of a type of acid beer in Egypt, dating from about 6000 BC. Although yeasts were as essential for those ancient processes as they are today, their role was elucidated only relatively recently. In 1876, Louis Pasteur demonstrated in his book *Etudes sur la Bière* that alcoholic fermentation was a process caused by living organisms. Years passed before Emil Christian Hansen first isolated brewing yeasts and propagated them in pure culture. From then on, much has been learnt about the characteristics of brewing yeasts and the behaviour of brewery fermentations. As a consequence, the brewing process has changed a lot since then.

Nowadays, alcoholic beverage production represents a significant contribution to the economies of many countries. *Circa* 250 million hectolitres (hl) of beer and 10,000 beer brands are available around the world. In many countries, beer is the most important alcoholic beverage, although its consumption has declined in the last 40 years in competition with other alcohol-containing drinks, predominantly wine.

The principal raw materials of beer are water, malted barley, hops and yeast. The brewing process involves extracting and breaking down the carbohydrates from the malting barley in the mashing process, followed by a filtration. Then, the liquid extract (the so-called wort) is hopped and boiled. After boiling, the hot 'trub' is removed and the clarified wort is cooled down and aerated. In the fermentation process, the sugar-rich wort is converted into an alcoholic flavour-rich beer. This is brought about by the activity of brewer's yeast, which metabolizes the wort sugars and converts them primarily into ethanol and carbon dioxide. As a rule of thumb, it can be stated that 100 g/l sugar is converted in 50 g/l CO_2 and 50 g/l ethanol. In addition, many secondary metabolites are produced, which are of major importance to the flavour of beer. It is generally accepted that the yeast strain used in a brewery is a fundamental component in defining the unique beer flavour characteristics. Also, these flavour compounds are related to yeast growth and consequently the physiological state of the yeast.

Brewer's yeast and the fermentation process, which are both highly critical to the quality and the characteristics of the final beer, will be discussed in detail in this chapter.

Characteristics of Brewing Yeast

Taxonomy and genetic constitution of brewing yeast

Ninety per cent of the worldwide beer production is of lager beers, 5% is of ale beers (produced using ale yeasts) and the rest is produced using mixed or spontaneous fermentations by yeasts and bacteria. Traditionally, brewers have distinguished two types of brewer's yeasts, ale (top-fermenting) yeasts and lager (bottom-fermenting) yeasts, namely *Saccharomyces cerevisiae* and *S. carlsbergensis*, respectively. Brewing yeast strains have been subject to various taxonomic revisions over the years. Currently, the *Saccharomyces* genus includes two groups of species, the *S. sensu stricto* group (species closely related to *S. cerevisiae*, which is composed of four species: *S. cerevisiae*, *S. bayanus*, *S. pastorianus* and *S. paradoxus*) and *S. sensu lato* (species more distantly related to *S. cerevisiae*) (Rainieri *et al.*, 2003).

Most of the industrial yeast strains (including brewing, wine, bread, cider and sake cultures) belong to the *sensu stricto* group. This classification is not yet considered accurate and is very ambiguous, especially in relation to *S. bayanus* (which includes wine and cider strains) and *S. pastorianus* (which includes lager brewing strains). *S. bayanus* and *S. pastorianus* contain diverse strains, with different genetic and metabolic characteristics, that may have a hybrid origin. On the basis of physiological and molecular tests, *S. bayanus* has recently been divided into two groups: *S. bayanus* var. *bayanus* and *S. bayanus* var. *uvarum*. *S. pastorianus* also contains hybrid strains with different characteristics, including lager-brewing strains. Lager strains are thought to originate from a natural hybridization that occurred between a *S. cerevisiae* strain and a non-*S. cerevisiae* strain. In a recent study, strains of *S. bayanus* and *S. pastorianus* were identified as either 'pure' strains, containing a single type of genome or 'hybrid' strains, that contained portions of the genomes from the 'pure' lines, as well as alleles termed 'Lager' that represent a third genome commonly associated with lager brewing strains (Rainieri *et al.*, 2006). In contrast, ale yeasts can be regarded as 'pure' *S. cerevisiae* strains.

Laboratory strains of *S. cerevisiae*, used for scientific research, are mainly haploid (n) or diploid (2n), whereas most industrial brewing strains are polyploid. The number of copies of the chromosomes is, however, not the same for all the chromosomes, making the lager brewing strains aneuploid.

A direct consequence of this genomic character is that brewing strains do not have a mating type, sporulate poorly and, when sporulation occurs, the spores are almost always non-viable. Traditional genetic analyses, such as hybridization, mutation and selection, can not be applied. The polyploid nature of brewer's yeasts makes them genetically stable: the more copies of a gene that are present, the less likely that mutation of the phenotype will occur. However, genetic and physiological instability is a major concern for brewers, because it can affect the fermentation consistency and performance, and thus the quality of the resulting beer. Therefore, yeast management is an important issue in a brewery and it is essential that the original production strains are stored at very low temperatures, preventing genetic alterations.

Distinguishing between yeast strains

Notwithstanding technologically important factors of lager yeasts such as production of sulfury flavours, their ability to ferment at lower temperatures (10–15°C) than ale yeasts (20–28°C) and their property to sediment to the bottom instead of rising to the surface of the fermentation broth, the major taxonomic distinction between the two groups of yeasts is the inability of ale yeasts to ferment the disaccharide melibiose (α-D-galactose-(1-6)-α-D-glucose). Despite the controversy about their taxonomic position they can therefore be readily distinguished by the lager yeast's ability to ferment melibiose.

From the brewer's point of view, distinguishing between yeast strains is of much more importance, especially when different strains are used in one brewery. Many tests have been formulated for separating different yeast strains (Hammond, 1993). The most comprehensive method is mimicking the behaviour of yeast in an industrial fermentation by assessing the fermentation performance in 1.5–2.0 l-tall tube fermentors (Walkey and Kirsop, 1969). Several parameters can be followed up: growth, deposit formation, the decrease in density, alcohol formation and flavour compound analysis. These tests can provide valuable information, but they are too laborious for routine analyses. A number of simple differentiating protocols have been developed, such as yeast flocculation tests (see later), giant-colony tests, carbon assimilation tests, immunological tests and oxygen requirement tests (Boulton and Quain, 2001).

Recently, genetic tests have been developed to make DNA fingerprints, e.g. restriction fragment length polymorphisms (RFLP), polymerase chain reaction (PCR) or karyotyping (Laidlaw *et al.*, 1996). However, none of these tests can distinguish absolutely between closely related brewery strains, and a combination of tests remains necessary.

The yeast cell cycle during fermentation

Polyploid brewing strains reproduce asexually via bud formation. Sexual reproduction (mating or conjugation) and sporulation, which apply to haploid and diploid cells, are not considered here. During budding, one cell gives rise to a 'daughter' cell, which is genetically identical to the original mother cell. These virgin daughter cells themselves divide, becoming mother cells and so on. However, cell division is not a never-ending process; the lifespan of brewing strains is limited, eventually reaching a senescent phase, which culminates in cell death.

The brewing yeast lifespan is not chronological, but relates to the number of divisions an individual cell has undertaken. The number of daughters produced by a mother cell, termed the divisional age, indicates the relative age of the cell, while the maximum lifespan potential of a cell is referred to as the Hayflick limit. For brewing yeast, the mean lifespan is strain specific within the range of 10 to 30 divisions (Powell *et al.*, 2000). The number of divisions an individual cell has undertaken is the most widely accepted indicator of cell age, and can be directly calculated by analysis of the cell wall for bud scars. However, in the brewing industry, yeast age is often measured chronologically in terms of the number of times a yeast population is serially repitched (recycled for reuse in successive fermentations).

The yeast cell cycle involves a progression through a series of events, which enable the cell to complete a round of vegetative reproduction known as division. Normally, an actively growing population will contain cells at all stages of this cycle. The cell cycle can be divided into five stages: G_0, G_1, S, G_2 and M (see Fig. 7.1; Wheals, 1987).

G_1 and G_2 represent two 'gap' periods of variable length, during which organelle production and normal cell processes such as growth and development take place. G_1 represents the time prior to starting a new round of cell division. At the end of G_1 a decision point, termed START, represents a collection of cellular events necessary for division. One important issue before the start of the cell cycle is whether the environmental conditions are positive for replication. Also, a minimal size has to be reached. DNA synthesis occurs during the S phase. The post-synthetic G_2 phase coincides with

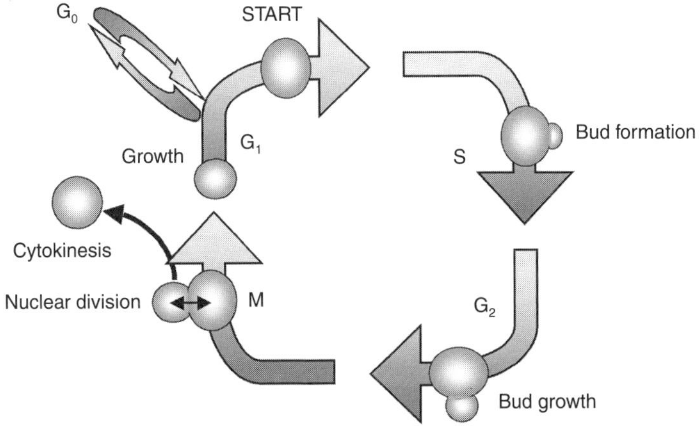

Fig. 7.1. The cell cycle of *Saccharomyces cerevisiae*.

significant bud growth and organelle formation, which then culminates in mitosis (the M phase), where segregation of chromosomes takes place.

The final steps in the cell cycle are: (i) cytokinesis, where daughter and mother cells are separated by septa; and (ii) cell separation, although complete separation is not a prerequisite for continued reproduction, which can result in the formation of chains. G_0 is an `off-cycle' state during which no net increase in cell number occurs. Entry into G_0 is triggered by adverse environmental factors, one of which is nutrient depletion (Werner-Washburne *et al.*, 1993). During the later stages of fermentation the yeast enters the stationary phase and remains in this state during storage and re-pitching. A characteristic of organisms in G_0 phase is an increased resistance to stresses for an extended period of time, causing cells to retain viability and vitality during cold storage.

Normally, brewing yeast is harvested from a previous fermentation and then the yeast slurry finds itself in the stationary phase. Upon pitching, the starved yeast slurry (G_0) comes into contact with fresh wort, senses the presence of nutrients and shifts into the G_1 phase and START. During the first hours of fermentation there is a lag phase during which very little appears to happen. However, within the yeast slurry many changes are taking place. Initially there is an increase in cell volume, associated with a decrease in cell mass. This is probably due to the breakdown of glycogen, a reserve carbohydrate present in yeast, needed for the formation of sterols and unsaturated fatty acids. These lipid compounds are essential for a healthy membrane and are a critical requirement for cell division. Entry into the S phase is apparent after a few hours, with the onset of buds (Boulton and Quain, 2001). After 6 hours, 90% of the population is budding, the cell number starts to increase and the density of the wort starts to decline.

During active fermentation there are about three to four divisions per cell, as cell numbers can increase from 10 to 100 million cells/ml. In the now heterogeneous population, there are typically 50% virgin daughter cells (without bud scars), 25% single budded mothers and 25% multi-budded mother cells. Interestingly, cell division does not continue throughout fermentation, but ends in the middle of fermentation. It is generally accepted that division is not arrested by nutrient deficiency in the fermenting wort, but through dilution by cell division of essential lipids over mother and daughter cells. When the yeast cells stop growing, they go into the quiescent stationary phase (G_0).

The yeast cell wall and flocculation

The brewing yeast cell wall accounts for 15–25% of the cell dry weight. The cell wall is composed of four major fractions, namely:

- A glucan layer; β-glucans account for 30–60% of the cell wall, although they are restricted to the inner layer. They have a structural role and interconnect to all the different fractions.

- Mannoproteins; these have a complex structure, account for 25–50% of the cell wall and can be found at the outer cell wall layer. These glyco proteins are receptors in the flocculation process.
- Chitin; this polymer is found, almost exclusively, in scars left on the mother cell surface after the cell has undergone vegetative reproduction through budding.
- Proteins; these can be found in the cell wall where they have structural, enzymatic or antigenic roles.

Brewer's yeast strains have the natural ability to adhere to inert surfaces as well as to other yeast cells, the latter process being called 'flocculation'. Yeast flocculation is a reversible, asexual and calcium-dependent process in which cells adhere to form flocs consisting of thousands of cells. Upon formation, these flocs rapidly separate from the bulk medium by sedimentation (lager yeasts) or by rising to the surface (ale yeasts). Ale yeasts tend to be more hydrophobic and, as a consequence, are more able to adhere to CO_2 bubbles, forming a yeast head.

The ability of yeast cells to flocculate is of considerable importance for the brewing industry, as it provides an effective and simple way to separate most of the yeast cells from the green beer at the end of fermentation. Therefore, strong and complete flocculation is a desirable property for every brewer's yeast. However, the yeast cells should not flocculate before the wort is completely attenuated. Such premature flocculation causes sluggish, so-called 'hanging' fermentations, which can result in beers with severe off-flavours.

Flocculation involves lectin-like proteins, called 'flocculines', which stick out of the yeast cell wall and selectively bind mannose residues present on the cell walls of adjacent yeast cells (see Fig. 7.2). Calcium ions are needed in order to activate the flocculines.

Yeast flocculation is a complex process that depends on the expression of several specific genes, called the *FLO* genes. Two flocculation phenotypes can be distinguished by their sugar inhibition pattern: (i) the Flo1 phenotype, caused by expression of *FLO1* and its homologues and inhibited only by mannose; and (ii) the NewFlo type, which is inhibited by mannose, glucose, maltose and sucrose (Stratford, 1989). Yeast strains of the Flo1 type flocculate constitutively in wort, because mannose is never present in wort. Brewing yeasts are all members of the NewFlo group, so that flocculation occurs at the end of fermentation, by which time glucose and maltose are depleted in the fermenting wort.

Flocculation is affected by numerous parameters, such as nutrient conditions, agitation, Ca^{2+} concentration, pH, fermentation temperature, yeast age, ethanol content and yeast handling (pitching, cropping, storing and acid washing) (Sampermans *et al.*, 2005; Fig. 7.2). However, these parameters have not been systematically studied as yet and it is hard to predict the impact of the medium on cell adhesion. In addition, flocculation is a strain-specific phenomenon and the flocculation phenotype is extremely variable (Verstrepen *et al.*, 2005).

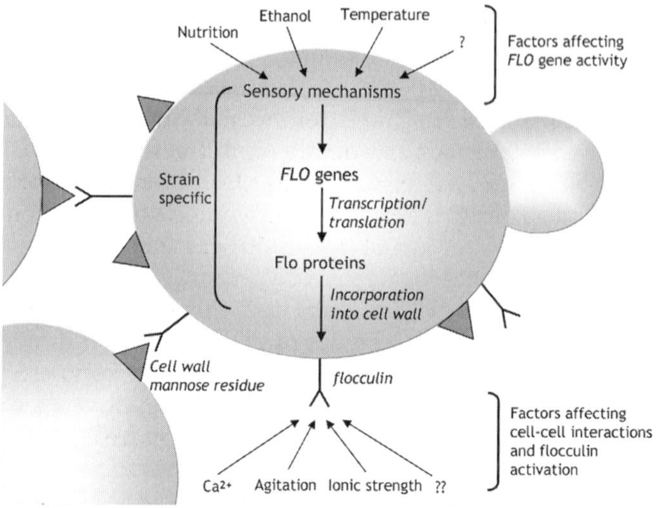

Fig. 7.2. A schematic overview of the lectin-binding model of flocculation and the factors affecting flocculation according to their mode of operation (from Verstrepen *et al.*, 2003b).

The yeast flocculation process is not yet fully understood, and the complexity and the high variability between and within yeast strains make it difficult to control the process. Genetic engineering could provide a solution to steer the process towards the brewer's needs (Verstrepen *et al.*, 1999). However, public opinion is currently somewhat negative about the use of genetically modified organisms.

Yeast metabolism during fermentation

The biochemistry of brewery fermentations is complex, and many of its aspects remain to be elucidated. Wort is a complex medium and provides a complete growth medium for yeast. As a result of yeast growth, and consequently yeast metabolism, the by-product beer is produced. In this process, the choice of the variable yeast strain is often restricted, whilst fermentation management (including fermentation parameters such as wort composition, temperature, pitching rate and aeration) has to regulate the conditions so that by-products of yeast growth and metabolism are formed in desired quantities within acceptable time limits.

Carbohydrate metabolism

Uptake and metabolism of wort sugars

Carbohydrates in wort make up 90–92% of wort solids. The carbohydrate composition of wort can differ significantly, but a typical all-malt wort has more or less 25% as an un-fermentable fraction, mainly dextrin. Dextrins are not taken up by yeast, but contribute to beer quality by means of imparting body or fullness. From the other 75%, maltose is quantitatively the most dominant sugar in wort (60–65%). The second most abundant sugar is maltotriose, accounting for 20–25% of the total fermentable extract. Glucose accounts for 10–15%, while sucrose and fructose are present in low concentrations. Adding adjuncts, such as glucose or maltose syrups, sucrose, (un-)malted wheat, flaked maize and other carbohydrate-rich sources can reduce or increase the fermentability of the wort.

Brewer's yeast consumes wort sugars in a sequential manner and glucose is consumed first, together with sucrose. Completion of glucose assimilation is followed by uptake of maltose. Finally, maltotriose is utilized slowly after maltose. This sequential uptake pattern is a consequence of the glucose repression pathway, or the catabolite repression pathway (Gancedo, 1998). This pathway down-regulates several genes involved in the uptake and metabolism of other carbohydrates, as well as genes involved in gluconeogenesis and respiration. The repression of respiration by glucose and sucrose, which is known as the 'Crabtree effect', means that irrespective of the presence of oxygen at the beginning of fermentation, the metabolism is always fermentative (Lagunas, 1986). Even when glucose is taken up, all the oxygen has disappeared from the wort so that fermentation occurs, with production of ethanol and CO_2.

The yeast species *S. cerevisiae* transports monosaccharides across the cell membrane using several hexose transporters (Dickinson, 1999). Sucrose is assimilated via the action of an invertase, which is located in the cell periplasm (between the cell membrane and cell wall). This enzyme is encoded by *SUC2* and is under transcriptional repression of glucose. Once hydrolysed, the fructose and glucose released are taken up by hexose transporters. Maltose utilization is accomplished by any of five *MAL* loci: *MAL1* to *MAL4* and *MAL6* (Bisson *et al.*, 1993). Each locus consists of three genes: gene 1 encodes a maltose transporter; gene 2 encodes a maltase (α-glucosidase); and gene 3 encodes a post-transcriptional activator of the other two genes. In addition, three other maltose transporters have been reported in some strains of *S. cerevisiae*, encoded by *AGT1*, *MPH2* and *MPH3*.

The maltose uptake system is an active process, requiring cellular energy and is inactivated by glucose. The second most abundant fermentable sugar in wort, maltotriose, is often poorly utilized, leading to incomplete fermentation. Moreover, ale yeasts are frequently less effective in maltotriose uptake then lager yeasts. Maltotriose can be assimilated by the aid of Agt1p, Mph2p, Mph3p and Malx1p (x = 1–4 or 6), but with a lower uptake rate (Day *et al.*, 2002). Moreover, lager yeast strains contain the gene *MTT1*, which encodes a maltotriose transporter (Dietvorst *et al.*, 2005). After transport in the yeast, maltotriose is hydrolysed by α-glucosidases.

All yeasts predominantly utilize the Embden–Meyerhof–Parnas glycolytic pathway for the generation of ATP via substrate-level phosphorylation. The product of glycolysis, pyruvate, is a major branching point in metabolism. It can be metabolized directly into acetyl-CoA and subsequently be oxidized via the tricarboxylic acid (TCA) cycle or into ethanol via acetaldehyde, releasing a molecule of CO_2. However, due to the catabolite repression of glucose, the yeast cannot develop functional mitochondria and, consequently, ethanol is produced under aerobic conditions in the wort. A small proportion of the sugars is catabolized via the hexose monophosphate shunt for generation of NADPH for use in lipid synthesis.

Storage carbohydrates

Saccharomyces cerevisiae accumulates two classes of storage carbohydrates, glycogen and trehalose, both having important roles in brewery fermentations.

GLYCOGEN Glycogen apparently serves as a true reserve carbohydrate, which may be mobilized during periods of starvation. It is a high-molecular mass, branched polysaccharide of linear α-(1,4)-glycosyl chains with α-(1,6)-linkages. It is synthesized from glucose, via glucose-6-phosphate and glucose-1-phosphate. Glycogen synthase catalyses chain elongation by successive transfer of glucosyl units to the growing α-(1,4)-linked glucose polymer. The branching enzyme forms the α-(1,6)-glucosidic bonds. Dissimilation of glycogen is brought about by glycogen phosphorylase and a debranching enzyme (François and Parrou, 2001).

At the beginning of fermentation an immediate glycogen mobilization is observed, when yeast cells are pitched in aerated wort. Since exogenous sugars cannot be taken up due to lack of membrane function in sterol-depleted yeast, the endogenous glycogen reserves have to be mobilized for the synthesis of sterols in the presence of oxygen (Boulton, 2000). After the disappearance of oxygen from the wort, glycogen accumulates during the exponential growth phase. Maximum glycogen levels are reached towards the end of primary fermentation, when yeast growth is finished, where it can represent up to 40% of the cell dry weight. In the final stationary phase, when fermentation is complete, glycogen levels decline slowly.

Glycogen thus fulfils two vital roles in brewery fermentations. It provides the energy for the synthesis of sterols and unsaturated fatty acids during the aerobic phase of fermentation, and energy for the cellular maintenance functions during the stationary phase of fermentation and in the storage phase between cropping and pitching (Quain and Tubb, 1982). During prolonged fermentation and storage, the glycogen levels can decline to such levels that insufficient growth can take place in the following fermentation, suggesting that the glycogen content of yeast is directly related to subsequent fermentation performance. As a result, yeast should be stored at low temperature, without agitation and under a nitrogen or carbon dioxide atmosphere, in order to minimize glycogen breakdown.

TREHALOSE Trehalose is a disaccharide that contains two molecules of glucose (α-glucopyranosyl-(1,1)-α-glucopyranoside). Trehalose biosynthesis is catalysed by the trehalose synthase complex, composed of four subunits, which forms trehalose-6-phosphate from UDP-glucose and glucose-6-phosphate and, subsequently, de-phosphorylates it to trehalose. Trehalose is degraded by the neutral (Nth1p) or the acid trehalase (Ath1p) (François and Parrou, 2001).

In brewing yeast harvested from the fermentor, trehalose levels are usually less then 5% of cell dry weight. Under stress conditions, this molecule can accumulate to form up to 15% of the cell dry weight (Majara *et al.*, 1996). Due to activation of trehalase by glucose, trehalose is degraded at the beginning of fermentation. After the glucose concentration drops in the medium, trehalose accumulates. Studies have illustrated the extraordinary property of this molecule to protect membranes and proteins. However, this protective role is not well understood at the molecular level. It has been suggested that trehalose stabilizes membranes under adverse conditions by binding the hydroxyl groups of the disaccharide to the polar head groups of phospholipids otherwise occupied by water molecules (Iwahashi *et al.*, 1995).

During the stationary phase a portion of the carbon, originating from glycogen, is utilized for trehalose synthesis, providing protection for cells during this resting phase (Boulton, 2000). It has been observed that trehalose can also rapidly accumulate in response to environmental changes, such as dehydratation, heating and osmotic stress (high-gravity brewing) (Majara *et al.*, 1996). It is believed that the *TPS2* gene is triggered by a general stress responsive promoter (STRE), suggesting that trehalose is a stress indicator (François and Parrou, 2001).

Nitrogen metabolism

Wort nitrogenous components are heterogeneous in nature. All malt worts contain about 65–100 mg/100 ml nitrogen of which about 20% are proteins, 30–40% are polypeptides, 30–40% are amino acids and 10% nucleotides and other nitrogenous compounds (Ingledew, 1975). Although the nitrogen requirements for protein synthesis in yeast can be met by ammonium ions, amino acids are the preferred source. Proteolysis does not take place in fermenting wort, unless autolysis of yeast occurs. Moreover, small peptides can also be assimilated.

The uptake of wort amino acids uses a number of permeases, some specific for individual amino acids and a general amino acid permease (GAP) with a broad specificity. Under conditions of high nitrogen levels, e.g. at the beginning of fermentation, GAP is repressed, and amino acids are absorbed selectively (Grenson, 1992). The uptake of amino acids is an active process, requiring energy. In general, brewery worts have a relatively constant spectrum of amino acids. Jones and Pierce (1964) classified wort amino acids on the basis of assimilation during fermentation (see Table 7.1). Those in the first group are assimilated immediately after the yeast cells come into contact with wort. Those of group B are taken up more slowly, whereas the amino acids of group C are not utilized until class A amino acids have disappeared from the wort. Proline is the sole member of group D, and this is not taken up from the wort as its dissimilation requires the presence of a mitochondrial oxidase not active under the repressed and anaerobic conditions of fermentation.

Once inside the yeast cell, the amino acids and peptides are not assimilated intact, but instead pass through a transaminase system, which removes the amino group, leaving the carbon skeleton to be metabolized. The amino groups released can be donated to other carbon skeletons, formed by anabolic and catabolic pathways. In this way, yeast is not reliant on a sufficient supply of all amino acids, but can synthesize amino acids from other amino acids or sugars. Moreover, these reactions are of great interest to the brewer since keto acids, produced by the action of transaminases, are precursors of the aldehydes, higher alcohols and vicinal diketones (see below), which play an important role in beer flavour.

Yeast growth increases in an almost linear fashion up to α-amino nitrogen levels of 100 ppm. Therefore, it is believed that minimal amounts of free amino nitrogen (FAN) of 140–150 ppm are recommended for adequate fermentations of normal gravity (10–12°P) (O'Connor-Cox and Ingledew, 1991). High-gravity worts, prepared with nitrogen-deficient adjuncts, can exhibit more pronounced nitrogen limitation effects due to the increased ratio of sugar over nitrogen. The adequacy of the nitrogen supply is more critical with ale yeast, where higher rates of yeast growth are achieved. Parameters that affect growth, such as the extent of aeration, pitching rate, temperature and agitation, also influence the nitrogen needs of the yeast. Moreover, beers with excessive amounts of FAN are more vulnerable to infections by spoilage organisms.

Lipid metabolism and the necessity for oxygen

Although it has long been known that oxygen is required for satisfactory brewery fermentations, it was not until relatively recently that the role of oxygen was elucidated. As discussed before, under aerobic conditions and in

Table 7.1. Wort amino acid classification in order of assimilation during fermentation (from Jones and Pierce, 1964).

Group A	Group B	Group C	Group D
Arginine	Histidine	Alanine	Proline
Asparagine	Isoleucine	Ammonia	
Aspartate	Leucine	Glycine	
Glutamate	Methionine	Phenylalanine	
Glutamine	Valine	Tryptophan	
Lysine		Tyrosine	
Serine			
Threonine			

the presence of repressing concentrations of sugar, the metabolism is fermentative. In traditional brewing, yeast is harvested after fermentation and stored for reuse during subsequent fermentations. Yeast cells then contain insufficient sterols and unsaturated fatty acids, and will only grow when appropriate levels of these essential membrane compounds are synthesized, which requires oxygen. In normal wort, the quantities of these essential lipids are too low to satisfy the yeast's needs for normal growth. However, supplementing these compounds can eliminate the necessity for wort aeration. Next to lipid synthesis, oxygen is also used in other cellular processes, such as mitochondrial development (O'Connor-Cox *et al.*, 1993). Finally, exposure to excessive oxygen is considered to cause oxidative stress, due to the toxic effect of reactive oxygen species (Gibson *et al.*, 2007).

Sterols

Sterols contribute less than 1% (w/w) to the cell dry mass. Aerobic yeast cells contain mainly ergosterol (75–90%), lanosterol (5–15%) and squalene (2–15%) (Daum *et al.*, 1999). If oxygen is limited, the sterol content decreases during yeast growth, because the sterols are distributed over mother and daughter cells, and below a threshold of 0.1% (w/w) no further growth is possible (Aries and Kirsop, 1977). A build-up of sterols during brewery fermentations to 1–10 mg/g cell dry weight has been considered normal. Sterol contents strongly depend on the yeast strain used, the lipid composition of the culture medium and the oxygen availability. Sterols have structural and functional roles. They are important structural lipids, and they play a role in regulating membrane fluidity. They also regulate membrane permeability and influence the activity of membrane-bound enzymes (Lees *et al.*, 1995). Sterols are synthesized using carbon, which originates from glycogen and is derived from glycolysis.

The first part of the synthesis is an anaerobic process, which involves the conversion of acetyl-CoA to squalene, a branched C_{30} isoprenoid. In the second part, oxygen is used to form 2,3-epoxysqualene via squalene epoxidase, followed by cyclization to the first sterol, lanosterol (Rosenfeld and Beauvoit, 2003). Other sterols, such as ergosterol, are formed from lanosterol and involve several oxygen-dependent desaturation reactions. Only free sterols play an active role in the membrane function. Sterols can be esterified to form inactive steryl esters, which represent a reserve pool of sterols, located in intracellular lipid particles.

Fatty acids

Under aerobic conditions, the major fatty acids in yeast are palmitoleate (C16:1) and oleate (C18:1) (representing 60–80% of the total fatty acids), palmitate (C16:0) (10%) and stearate (C18:0) (5%). Under anaerobic conditions the fatty acid profile of the yeast shifts to a more saturated pattern (Ahvenaïnen, 1982). The unsaturated fatty acids contribute to membrane fluidity, exert effects upon the formation of esters and play a role in mitochondrial development (O'Connor-Cox *et al.*, 1993). Synthesis of sterols is probably the more important since unsaturated fatty acids are present in wort and can be taken up and used by yeast. Regarding critical unsaturated fatty acid levels, Casey *et al.* (1984) found a growth limiting level of 5 mg/g. As with sterols, the biosynthesis of unsaturated fatty acids proceeds from acetyl-CoA and involves the formation of saturated fatty acids, followed by an oxygen-dependent desaturation reaction. Three key enzymes are involved: acetyl-CoA carboxylase, the multifunctional fatty acid synthase complex and a desaturase (Rosenfeld and Beauvoit, 2003).

Formation of flavour compounds

A multitude of compounds contribute to beer flavour. Many of these compounds are directly derived from the raw materials (hops and malt) or formed during wort production. However, fermentation has the most significant impact on flavour development. Yeast secondary metabolites can make a positive or a negative contribution, depending on their flavour characteristics and concentration in beer. The important flavour compounds produced by yeast can be classified into five categories: alcohols, esters, organic acids, carbonyl compounds (aldehydes and vicinal diketones) and sulfur-containing compounds. In addition, some speciality beers can contain important concentrations of volatile phenol compounds (Vanbeneden *et al.*, 2007). The interrelation between yeast metabolism and the production of bio-flavouring byproducts is illustrated in Fig. 7.3 and in Table 7.2; the thresholds of important flavour compounds in lager beer are depicted.

Higher alcohols

During beer fermentation, higher alcohols (also called 'fusel alcohols') are produced by yeast cells as by-products and represent the major fraction of volatile compounds. More than 35 higher alcohols have been identified, and the most important are given in Table 7.2. Isoamyl alcohol (3-methyl-1-butanol) and 2-phenylethanol can be found around their threshold concentrations in lager beer and can thus contribute significantly to the flavour of lagers. However, other aliphatic alcohols can also contribute to the alcoholic perception of the beer and can produce a 'warm' taste. In addition, higher alcohols have an indirect flavour importance as they are precursors for the production of esters (see above). The aromatic alcohol 2-phenylethanol has a sweet, rose-like aroma and is believed to mask any dimethyl sulfide (DMS) present (Hegarty *et al.*, 1995). Tyrosol and tryptophol are undesirable in beer, but are only present above their thresholds in some top-fermented beers.

Higher alcohols are synthesized by yeast during fermentation via the catabolic pathway (Ehrlich) and the anabolic pathway (Genevois) (Chen, 1978). In the catabolic pathway, the yeast uses amino acids from the wort to produce the corresponding α-keto acids via a transamination reaction. An outsider in this pathway is propanol, which is derived from threonine via an

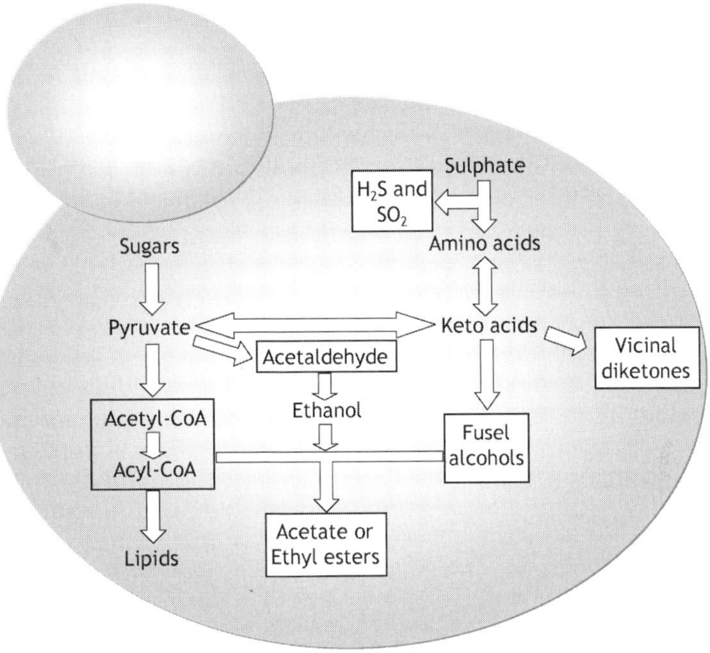

Fig. 7.3. Interrelation between yeast metabolism and the production of flavour compounds.

Table 7.2. Important flavour compounds produced by yeasts, their concentration range in lager beer and their flavour threshold (from Engan, 1972; Meilgaard, 1975).

Compound	Concentration range (mg/l)	Threshold (mg/l)	Flavour description
Higher alcohols			
Propanol	7–19	600–800	Alcoholic
Isobutanol	4–20	100–200	Alcoholic
Amyl alcohol	9–25	50–70	Banana, solvent-like
Isoamyl alcohol	25–75	50–65	Alcoholic, banana
2-Phenylethanol	11–51	40–125	Roses, perfume
Tyrosol	5–32	20–200	Bitter, chemical
Tryptophol	0.2–3.5	200	Almonds, solvent
Esters		30	
Ethyl acetate	8–32	1.2	Solvent, fruity
Isoamyl acetate	0.3–3.8	3.8	Banana, solvent
2-Phenylethyl acetate	0.1–0.73	0.21	Roses, honey
Ethyl caproate	0.05–0.30	0.9	Apple, aniseed
Ethyl caprylate	0.04–0.53		Apple, fruity
Organic acids		8	
Caproic acid	0.7–2.9	14	Goaty, fatty acid
Caprylic acid	2.1–7.4	10	Goaty, fatty acid
Capric acid	0.1–2.4		Waxy, rancid

oxidative deamination. The excess oxo-acids are subsequently decarboxylated to aldehydes and further reduced (by alcohol dehydrogenase) to higher alcohols. In the anabolic pathway, the higher alcohols are synthesized from α-keto acids during the synthesis of amino acids from the carbohydrate source. The importance of the anabolic pathway decreases as the number of carbon atoms in the alcohol increases, and increases in the later stage of a conventional batch fermentation as wort amino acids are depleted (Chen, 1978).

Higher alcohols are mainly produced during the active growth phase of the fermentation and are therefore influenced by each factor affecting the yeast growth, such as high levels of nutrients (amino acids, oxygen, lipids and zinc), increased temperature and agitation. Higher pressures can reduce the extent of yeast growth and therefore the production of higher alcohols (Landaud *et al.*, 2001). As a result, the yeast strain, fermentation conditions and wort composition all have significant effects on the pattern and concentrations of the higher alcohols that are formed.

Esters

Esters are the most important flavour compounds in beer, since their low flavour thresholds can be exceeded in lager beers (see Table 7.2). The major esters in beer are ethyl acetate, isoamyl acetate, ethyl caproate, ethyl caprylate and phenylethyl acetate. They are desirable components of beer when present in appropriate quantities and contribute to the fruity, flowery character of the beer. The esters found in beer can be divided into acetate esters and C_6–C_{10} medium-chain fatty acid ethyl esters.

Esters are produced by yeast during both the growth phase (60%) and stationary phase (40%). They are formed by the condensation reaction between acetyl/acyl-CoA and ethanol/higher alcohols catalysed by the alcohol acyltransferases of the yeast. Several different enzymes are involved, the best-characterized ones being AATase I and II. Other enzymes involved in ester production are Lg-Atflp, an AATase found in lager yeast that is homologous to Atflp, Ehtlp (ethanol hexanoyl transferase) and Eeb1p, which are two enzymes responsible for the formation of ethyl esters (Saerens *et al.*, 2006). Furthermore, it has been shown that the balance between ester-synthesizing enzymes and esterases, which hydrolyse esters, might be important for the net rate of ester accumulation (Fukuda *et al.*, 1998). The evolution of acetate ester synthesis and yeast growth can be viewed as a bell-shaped curve. So, between the two extremes of yeast growth, the level of esters should reach a maximum.

Fundamentally, two factors determine the rate of ester formation: the availability of the two substrates (acetyl/acyl-CoA and fusel alcohols) and the activity of the enzymes. Supplementation of the wort with medium-chain fatty acids increases the production of ethyl esters, while higher concentrations of higher alcohols, such as isoamyl alcohol, increase the production rate of the corresponding acetate ester (Calderbank and Hammond, 1994). Ester formation is also influenced by temperature, top pressure, stirring, wort zinc concentration, growth rate, carbon source and nitrogen concentration (Peddie, 1990; Landaud *et al.*, 2001; Verstrepen *et al.*, 2003a). High-gravity brewing significantly increases the concentration of esters, although the type of sugar added to the wort plays an important role.

Dissolved oxygen affects ester synthesis through reduced availability of acetyl-CoA (used for growth and lipid synthesis) and inhibition of AATases by a direct inhibition of *ATF1* gene transcription. Unsaturated fatty acids also exercise an inhibiting effect at the level of gene expression. In addition, ester production is highly dependent on the yeast strain used. The different parameters that influence ester production provide several opportunities for brewers to exert a certain influence on ester concentrations in beer. However, as so many factors are involved in both the regulation of AATase activity (including multiple regulation mechanisms at the level of *ATF* expression) and in the regulation of substrate availability (including carbon, nitrogen and fatty acid metabolism), the control over ester formation is extremely complex and difficult to predict.

Organic acids

Over 100 different organic acids have been reported in beer. These include pyruvate, acetate, lactate, succinate, malate, citrate, α-ketoglutarate and the medium-chain fatty acids, caproic (C_6), caprylic (C_8) and capric (C_{10}) acid (Coote and Kirsop, 1976). Although most of these have relatively high thresholds, they are important to beer in lowering the pH. The majority of organic acids are overflow products of the incomplete TCA cycle during beer fermentation. Their excretion can be explained by the lack of any mechanism for further oxidation, the need to maintain a neutral intracellular pH and the fact that they are not required for anabolic reactions. Their accumulation in beer depends on fermentation vigour and the yeast strain used.

The medium-chain fatty acids account for 85–90% of the fatty acids in beer and impart an undesirable goaty, sweaty or 'yeasty' flavour (see Table 7.2). These fatty acids are produced *de novo* by yeast during anaerobic fermentation and are not the result of β-oxidation of wort or yeast long-chain fatty acids. Because of the close relationship between ester formation and fatty acid biosynthesis (both pathways use acyl-CoA as a substrate), it is likely that the control mechanisms of esters will also regulate medium-chain fatty acid synthesis. The presence of these medium-chain fatty acids in beer is also related to yeast autolysis, the process where cell constituents are broken down and intracellular compounds are released (Masschelein, 1981). Yeast autolytic off-flavours are stimulated at high temperatures, high yeast concentrations and prolonged contact times at the end of primary fermentation and during secondary fermentation.

Vicinal diketones and other carbonyl compounds

VICINAL DIKETONES One of the objectives of the maturation of green beer is the removal of vicinal diketones (VDK), especially diacetyl (2,3-butanedione), which are unwanted aroma compounds. Diacetyl has a very low threshold (0.08–0.15 ppm) and imparts a buttery aroma to the beer. 2,3-Pentanedione has a threshold which is approximately ten times higher than that of diacetyl and is thus quantitatively of minor importance. During the primary fermentation, diacetyl is produced as a by-product in the synthesis of isoleucine, leucine and valine (ILV pathway) (see Fig. 7.4; Wainwright, 1973).

Because the formation of α-acetolactate is related to the amino acid metabolism, more α-acetolactate will be produced with increasing yeast growth. In addition, the production of α-acetolactate is very dependent on the yeast strain used. For a classical fermentation, *circa* 0.6 ppm α-acetolactate is formed. Higher wort aeration, temperature and agitation can increase this value to 1.2–1.5 ppm. Because brewer's yeast does not possess α-acetolactate decarboxylase activity, α-acetolactate is excreted from the cell and non-enzymatically converted to diacetyl by an oxidative decarboxylation. This step is the rate-limiting step and proceeds faster at high temperature and lower pH (Inoue, 1992).

Subsequently, diacetyl is reassimilated in the yeast cell, which possesses the necessary enzymes (reductases) to reduce diacetyl to the flavour-inactive acetoin and further to 2,3-butanediol. The reduction of diacetyl occurs at the end of primary fermentation and during maturation, provided that sufficient yeast is present; e.g. early flocculation lowers the diacetyl reduction efficiency. Although α-acetolactate is not flavour-active, the component must be converted to diacetyl completely, because residual α-acetolactate can be converted to diacetyl during pasteurization or during storage, which can not then be further reduced.

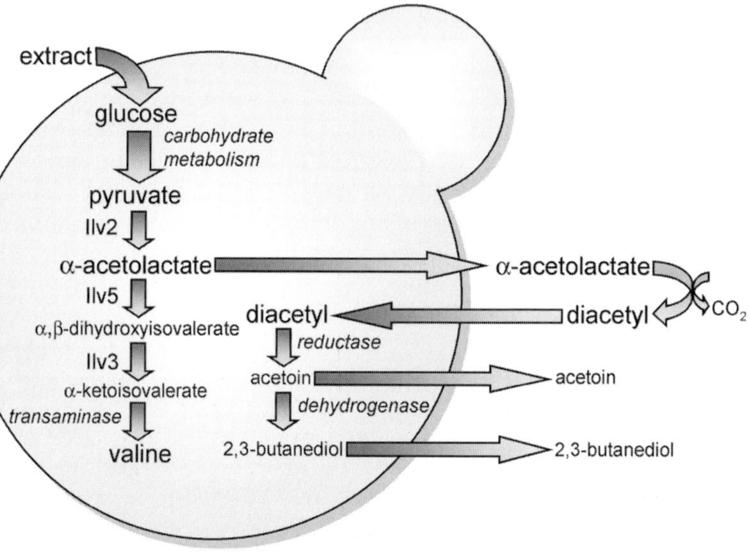

Fig. 7.4. Schematic presentation of the formation, reassimilation and reduction of diacetyl.

Several strategies can be adopted to lower the diacetyl concentration in beer, such as: (i) a warm rest before maturation; (ii) optimalization of the FAN content of the wort; (iii) the use of a continuous maturation system with immobilized yeast; (iv) addition of the enzyme α-acetolactate decarboxylase, which converts α-acetolactate directly into acetoin; and (v) the use of genetically modified yeast strains (encoding the α-acetolactate decarboxylase or overexpressing *ILV5*) (Willaert, 2007).

ALDEHYDES Aldehydes have flavour threshold concentrations which are significantly lower than the corresponding alcohols. Almost without exception, they have unpleasant flavours, variously described as grassy, green leaves and cardboard. A 'carbonylic' note is characteristic of the wort aroma. During fermentation, the yeast will reduce these carbonylic compounds. In low- or zero-alcohol beers, made by limited fermentation, the 'worty' flavour is however retained due to limited reduction and is considered undesirable in these beers.

The most abundant aldehyde in beer, acetaldehyde (green apple, grassy; threshold: 10–20 ppm), is important to beer flavour and is an intermediate in the formation of ethanol (see Fig. 7.4). Its synthesis is linked to yeast growth. The acetaldehyde concentration is maximal at the end of the growth phase, and is reduced at the end of primary fermentation and during maturation (Geiger and Piendl, 1976). As with diacetyl, the acetaldehyde levels increase with stimulated yeast metabolism, and reduction requires enough yeast at the end of fermentation: premature flocculation results in unacceptably high acetaldehyde concentrations.

Finally, carbonyl compounds are considered to be the main reason for the typical aged flavour of beer, mostly described as cardboard. They increase during beer storage due to a myriad of chemical reactions including lipid oxidation, Strecker degradation and Maillard reactions (Vanderhaegen *et al.*, 2006).

Sulfur compounds

Many sulfur-containing compounds found in beer derive directly from the raw materials, malt and hops, but some are produced through yeast metabolism. The most important ones are sulfite (pungent, threshold: 10–25 ppm), hydrogen sulfide (rotten egg, threshold: 10 ppb) and dimethyl sulfide (DMS) (cooked cabbage, threshold: 30 ppb). In addition to these, other sulfur-containing compounds have been reported in beer at much lower levels, such as thiols (e.g. methanethiol) and thioesters (e.g. methyl and ethyl thioacetate) (Walker and Simpson, 1993). These compounds are very potent and can contribute in trace amounts to flavour characteristics of lager beer.

Dimethyl sulfide is an important compound in lager beers when present at concentrations between 30 and 60 ppb. The two main routes leading to the formation of DMS are the thermal degradation of S-methylmethionine (SMM) into DMS and homoserine, and the reduction of dimethyl sulfoxide (DMSO) by yeast during fermentation (Anness and Bamforth, 1982).

The general precursor of DMS is SMM, present in green malt (unkilned malt). SMM is not present in barley, but is formed during germination. During kilning and boiling, DMS and DMSO are formed in the thermal degradation of SMM. Because DMS has a very low boiling point, most of the DMS will evaporate during boiling. The final concentration of DMS in beer depends therefore on the DMS and DMSO amounts in the pitching wort, the DMS formed by yeast during fermentation minus the DMS removed by CO_2 stripping. When DMS formation during brewing is restricted, the DMS level in beer is determined by DMSO reduction. This reduction is achieved by the yeast enzyme methionine sulfoxide reductase. Typically, yeasts can reduce 25% of the wort DMSO, but the conversion rate depends on the yeast strain, fermentation temperature, pH, the wort extract and the wort composition (Anness and Bamforth, 1982).

Hydrogen sulfide (H_2S) and sulfite (SO_2) play an important role in the flavour and stability of lager beers. SO_2 has a high threshold, but is nevertheless important to beer flavour through an indirect effect. In the first place SO_2 is an antioxidant, and it can also bind reversibly with carbonyl compounds to form non-volatile adducts. This implies that SO_2 can bind aldehydes, formed during beer ageing, which results in masking off-flavours during staling. However, the production of sulfite must be kept under control during fermentation, because yeast is not capable of reducing sulfite-bound aldehydes (Dufour, 1991). Later in the processing, e.g. during storage, the aldehydes can be released and become flavour-active. More interesting, SO_2 could be added after fermentation, but these additions have a legal limit in some countries.

Sulfite and hydrogen sulfide are intermediates in the biosynthesis of the sulfur-containing amino acids methionine and cysteine. An organic sulfate is taken up by the yeast cells with the aid of a permease. Subsequently, reduction takes place to sulfide via the intermediates adenylyl sulfate (APS), phosphoadenylyl sulfate (PAPS) and sulfite. H_2S and SO_2, which are not incorporated in S-containing amino acids, are excreted in the fermenting wort (Thomas and Surdin-Kerjan, 1997). Moreover, H_2S and SO_2 can also be formed from the catabolism of the sulfur-containing amino acids. The production of H_2S and SO_2 depends on the yeast strain, sulfate content of the wort and growth conditions. It has been shown that a deficiency of the vitamin pantothenate, which is required in the metabolism of methionine, causes overproduction of H_2S (Wainwright, 1970). During secondary fermentation, H_2S is re-utilized by the yeast, lowering the concentration in the matured beer.

Fermentation Technology

Batch fermentation

Before the 1970s, beer fermentations were performed in open, and later in closed, fermenting vessels, which were cleaned manually and had a low capacity (50–500 hl). Because fermentation is the most important process step for flavour formation, brewers were always averse to the introduction of new technologies. With the introduction of bottom-fermenting yeast, the need for easy separation of yeast and medium became necessary. The first major installations of cylindroconical tanks were built in the 1960s, with much larger capacities (1200–4000 hl) than the former vessels (Maule, 1986). The advantages of cylindroconical tanks are as a consequence of their geometry: their tall and slender geometry occupies little ground area, resulting in faster fermentation through more vigorous agitation, efficient cooling, simplified automation, more efficient utilization of hop, ease of cleaning and reduced losses of beer. The use of very large cylindroconical tanks was limited, because of the lack of flexibility, long filling and emptying times and flavour deteriorations (e.g. lower ester concentrations, due to high pressures and high diacetyl concentrations, in turn due to increased agitation).

Because yeast, which is reused in a consecutive fermentation, is depleted in sterols and unsaturated fatty acids, wort has to be aerated to ensure adequate levels of these lipids. Only the first generation of yeast slurry, which is freshly propagated, has a good membrane function and does not need to be aerated. Normally the wort is supplied with oxygen at air saturation (6–8 ppm). The concentration of oxygen needed depends on the yeast strain being used and its history, which makes knowledge of the oxygen control and requirements of the yeast very important to a brewer. To overcome over- and under-aeration problems, the yeast slurry can be preoxygenated before fermentation. Studies have shown that preoxygenation ensures the synthesis of sterols and unsaturated fatty acids, without growth, which results in an improved physiological condition of the yeast, a better fermentation performance and more consistent fermentations (Depraetere *et al.*, 2003).

A constant pitching rate is crucial for consistent fermentations, because it is related to yeast growth. Generally, brewers pitch lager yeast strains at a rate of 15–20 x 10^6 viable cells/ml and ale yeasts at a rate of 5–10 x 10^6 viable cells/ml. The pitching rate has to be increased when using higher-gravity worts, to ensure a complete attenuation (Suihko *et al.*, 1993).

Not only must the pitching yeast have good viability (e.g. 90–95%), but the physiological conditions and the age of the yeast are also important parameters for fermentation performance. Many researchers have therefore suggested varying the pitching rate as a function of the yeast glycogen content, or the concentration of essential lipids in the yeast. Because these analyses are not routinely available in a brewery, there is a need for a simple method to determine the yeast's 'vitality'. Practical procedures to measure the physiological condition of yeast are still being investigated (Mochaba *et al.*, 1998). Older yeast populations have increased flocculation properties, an increased mortality and a lower division rate. Therefore, brewers have to be aware of age stratification in the cone of the tank and select the right pitching yeast population for the following fermentation (Powell *et al.*, 2004).

At the end of primary fermentation, yeast flocculation should occur consistently. After sedimentation of lager yeast, the yeast is harvested, washed with acid and stored for the next fermentation. In the case of ale yeast, the harvesting can take place at the head of the fermentation or after sedimentation at the bottom of the cone. Delayed removal of yeast from the green beer has the risk of yeast autolysis releasing unwanted flavour compounds (e.g. caprylic flavours) or hydrolytic enzymes (such as esterases) and increasing the pH.

Maturation can take place in the same tank (single-vessel processing) or in other tanks (called 'lager' tanks). During the secondary fermentation or maturation of beer, several objectives are realized: (i) further sedimentation of yeast; (ii) improvement of colloidal stability by sedimentation of polyphenol–protein complexes; (iii) beer saturation with CO_2; (iv) removal of unwanted aroma compounds (e.g. diacetyl, H_2S and acetaldehyde); (v) fermentation of the remaining extract; (vi) adjustment of the beer colour (if necessary) by addition of colouring substances; and (vii) addition of hop products to add hoppy flavours.

After maturation, the beer will be filtered, pasteurized, stabilized and bottled. In the case of some speciality beers, yeast and sugar will be added to initiate re-fermentation in the bottle.

Continuous fermentation

Continuous fermentation systems offer important economic advantages in comparison with traditional systems, including faster fermentation rates, improved product consistency, reduced product losses and environmental advantages. One important aspect of continuous fermentation is the high volumetric efficiency, which is usually obtained by increased yeast cell concentrations in the reactor compared with traditional batch systems. In the 1960s, the interest in continuous beer fermentation rose intensively, giving birth to a series of systems. However, these continuous beer fermentation processes never became commercially successful due to many practical problems, such as increased risk of contamination (mainly because of the necessity of storing wort in supplementary holding tanks), variations in beer flavour, complex system designs and a lack of flexibility.

Only the continuous beer production process of Dominion Breweries in New Zealand by Morton Coutts has been implemented successfully (see Fig. 7.5). This system uses flocculent yeast and consists of a buffer tank, where the wort is kept microbiologically stable by lowering the pH. Two continuous, stirred reactors keep the flocculent yeast in suspension and consequently ferment the wort, after which a yeast sedimentation tank separates the yeast from the beer. Finally, the yeast is recycled to control the fermentation rate. Beer with an alcohol percentage of 5.5% is produced in barely 2 days (Dunbar *et al.*, 1988).

In the 1970s there was a revival in the development of new continuous beer fermentation systems, owing to the development of immobilized cell technology. The main advantages of using immobilized cells for the production of beer are enhanced volumetric productivities, improved cell stability, easier implementation of continuous operation, improved operational control,

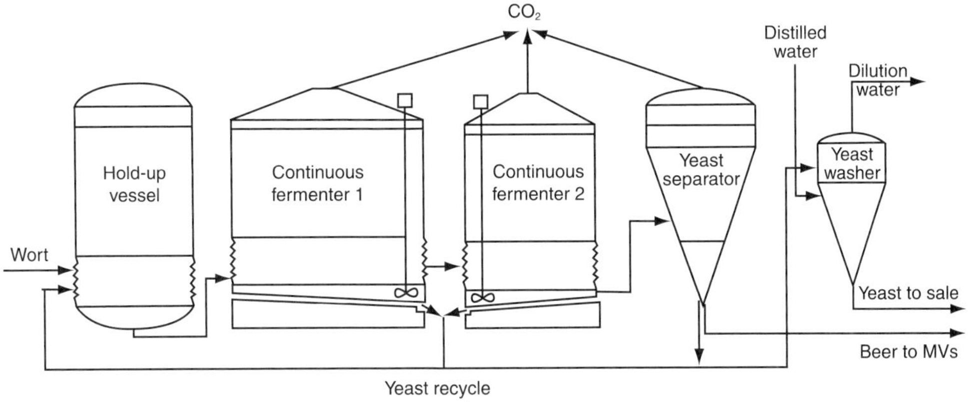

Fig 7.5. The continuous fermentation plant of Dominion Breweries. MVs, maturation vessels.

facilitated cell recovery and re-use, and simplified downstream processing (Verbelen *et al.*, 2006). Different immobilization types can be used, including inclusion in hydrogels, adsorption on porous and non-porous carriers, self-aggregation (flocculation) and containment behind membranes. In continuous immobilized yeast fermentation systems, different types of bioreactors are now being used, such as packed-bed reactors, fluidized bed reactors, airlift reactors and stirred reactors. The choice of bioreactor is related to the type of immobilization, to the metabolism of cells and to the mass and heat transfer requirements (Obradovic *et al.*, 2004).

At present, only beer maturation and alcohol-free beer production are obtained by means of commercial-scale immobilized yeast reactors. In these processes, no real yeast growth and flavour formation is required.

Recent Advances

The beer fermentation process has a long tradition, but recent scientific and technological advances make it a modern biotechnological discipline. Nowadays, the biochemical reactions and metabolic pathways related to the conversion of wort to beer are well studied. Recent technological developments, such as high-gravity brewing, continuous fermentation with immobilized yeast, pre-oxygenation and process automatization, have increased the productivity of the process and consistency of the beer produced.

However, many questions about the physiology of yeast during fermentation remain to be solved, and many processes are not yet fully understood (e.g. ester synthesis, yeast flocculation). Genetic engineering may be the tool to develop yeast strains with ideal properties for a given brewing process and product. Manipulated brewing strains have been constructed to accelerate the maturation of beer, for increased attenuation, improved beer filterability, control of sulfite production and control of yeast flocculation. However, the constraining factors to the successful exploitation of genetically engineered yeast strains include undesirable side effects, the stability of the introduced property and regulatory approval and consumer acceptance. It is to be expected that genetic manipulation of brewing yeast will not be adopted directly in beer production yet, but instead the selection of an ideal yeast strain seems the best way to fit the desired yeast properties in a given brewery process.

Acknowledgements

The first author acknowledges the financial support from the Institute for the Promotion of Innovation through Science and Technology in Flanders (IWT-Vlaanderen).

References

Ahvenaïnen, J. (1982) Lipid composition of aerobically and anaerobically propagated brewer's bottom yeast. *Journal of the Institute of Brewing* 88, 367–370.

Anness, B.J. and Bamforth, C.W. (1982) Dimethyl sulphide – a review. *Journal of the Institute of Brewing* 88, 244–252.

Aries, V. and Kirsop, B.H. (1977) Sterol synthesis in relation to growth and fermentation by brewing inoculated at different concentrations. *Journal of the Institute of Brewing* 83, 220–223.

Bisson, L.F., Coons, D.M., Kruckeberg, A.L. and Lewis, D.A. (1993) Yeast sugar transporters. *Critical Reviews in Biochemistry and Molecular Biology* 28, 259–308.

Boulton, C. (2000) Trehalose, glycogen and sterol. In: Smart K.A. (ed.) *Brewing Yeast Fermentation Performance*. Blackwell Science Ltd., Oxford, UK, pp. 10–19.

Boulton, C. and Quain, D. (2001) *Brewing Yeast and Fermentation*, 1st edn. Blackwell Science Ltd., Oxford, UK.

Calderbank, J. and Hammond, J.R.M. (1994) Influence of higher alcohol availability on ester formation by yeast. *Journal of the American Society of Brewing Chemists* 52(2), 84–90.

Casey, G.P., Magnus, C.A. and Ingledew, W.M. (1984) High-gravity brewing: effects on nutrition on yeast composition, fermentative ability and alcohol production. *Applied and Environmental Microbiology* 48(3), 639–646.

Chen, E.C.-H. (1978) The relative contribution of Ehrlich and biosynthetic pathways to the formation of fusel alcohols. *Journal of the American Society of Brewing Chemists* 36(1), 39–43.

Coote, N. and Kirsop, B.H. (1976) The content of some organic acids in beer and in other fermented media. *Journal of the Institute of Brewing* 80, 474–483.

Daum, G., Tuller, G., Nemec, T., Hrastnic, C. and Balliano, G. (1999) Systematic analysis of yeast strains with possible defects in lipid metabolism. *Yeast* 15, 601–614.

Day, R.E., Rogers, P.J., Dawes, I.W. and Higgins, V.J. (2002) Molecular analysis of maltotriose transport and utilization by *Saccharomyces cerevisiae*. *Applied and Environmental Microbiology* 68(11), 5326–5335.

Depraetere, S.A., Winderickx, J. and Delvaux, F.R. (2003) Evaluation of the oxygen requirement of lager and ale yeast strains by preoxygenation. *Master Brewers Association Technical Quarterly* 40(4), 283–289.

Dickinson, J.R. (1999) Carbon metabolism. In: Dickinson J.R. and Schweizer M. (eds) *The Metabolism and Molecular Physiology of* Saccharomyces cerevisiae. Taylor and Francis Ltd., London, pp. 23–55.

Dietvorst, J., Londesborough, J. and Steensma, H.Y. (2005) Maltotriose utilization in lager yeast strains: MTT1 encodes a maltotriose transporter. *Yeast* 22, 775–788.

Dufour, J.P. (1991) Influence of industrial brewing and fermentation working conditions on beer sulphur dioxide level and flavour stability. In: *Proceedings of the 23rd European Brewery Convention,* Lisbon, Portugal, pp. 209–216.

Dunbar, J., Campbell, S.I., Banks, D.J. and Warren, D.R. (1988) Metabolic aspects of a commercial continuous fermentation system. In: *Proceedings of the 20th Convention of the Institute of Brewing (Asia and New Zealand Section),* Brisbane, Australia, pp. 151–158.

Engan, S. (1972) Organoleptic threshold values of some alcohols and esters in beer. *Journal of the Institute of Brewing* 78, 33–36.

François, J. and Parrou, J.L. (2001) Reserve carbohydrates metabolism in the yeast *Saccharomyces cerevisiae. FEMS Microbiology Reviews* 25(1), 125–145.

Fukuda, K., Yamamoto, N., Kiyokawa, Y., Yanagiuchi, T., Wakai, Y., Kitamoto, K., Inoue, Y. and Kimura, A. (1998) Balance of activities of alcohol acetyltransferase and esterase in *Saccharomyces cerevisiae* is important for production of isoamyl acetate. *Applied and Environmental Microbiology* 64(10), 4076–4078.

Gancedo, J.M. (1998) Yeast carbon catabolite repression. *Microbiology and Molecular Biology Reviews* 62(2), 334–361.

Geiger, E. and Piendl, A. (1976) Technological factors in the formation of acetaldehyde during fermentation. *Master Brewers Association Technical Quarterly* 13(1), 51–61.

Gibson, B.R., Lawrence, S.J., Leclaire, J.P.R., Powell, C.D. and Smart, K.A. (2007) Yeast responses to stresses associated with industrial brewery handling. *FEMS Microbiology Reviews* 31(5), 535–569.

Grenson, M. (1992) Amino acid transporters in yeast: structure, function and regulation. *New Comprehensive Biochemistry* 21, 219–245.

Hammond, J.R.M. (1993) Brewer's yeasts. In: Rose A.H. and Harrison J.S. (eds) *The Yeasts,* 2nd. edn. Academic Press, London, pp. 8–67.

Hegarty, P.K., Parsons, R., Bamforth, C.W. and Molzahn, S.W. (1995) Phenyl ethanol – a factor determining lager character. In: *Proceedings of the 25th European Brewery Convention,,* Brussels, pp. 515–522.

Ingledew, W.M. (1975) Utilisation of wort carbohydrates and nitrogen by *Saccharomyces cerevisiae. Master Brewers Association Technical Quarterly* 12, 146–150.

Inoue, T. (1992) A review of diacetyl control technology. In: *Proceedings of the 22nd Convention of the Institute of Brewing, Australia and New Zealand Section,* Melbourne, Australia, pp. 76–79.

Iwahashi, H., Obuchi, K., Fuji, S. and Komatsu, Y. (1995). The correlative evidence suggesting that trehalose stabilizes membrane structure in the yeast *Saccharomyces cerevisiae. Cellular and Molecular Biology* 41(6), 763–769.

Jones, M. and Pierce, J.S. (1964) Absorption of amino acids from wort by yeasts. *Journal of the Institute of Brewing* 70, 307–315.

Lagunas, R. (1986) Misconceptions about the energy metabolism of *Saccharomyces cerevisiae. Yeast* 2, 221–228.

Laidlaw, L., Tompkins, T.A., Savard, L. and Dowhanick, T.M. (1996) Identification and differentiation of brewing yeasts using specific and RAPD polymerase chain reaction. *Journal of the American Society of Brewing Chemists* 54(2), 97–102.

Landaud, S., Latrille, E. and Corrieu, G. (2001) Top pressure and temperature control of the fusel alcohol/ester ratio through yeast growth in beer fermentation. *Journal of the Institute of Brewing* 10(2), 107–117.

Lees, N.D., Skaggs, B., Kirsch, D.R. and Bard, M. (1995) Cloning of late genes in the ergosterol biosynthetic pathway of *Saccharomyces cerevisiae*: a review. *Lipids* 30, 221–226.

Majara, N., O'Connor-Cox, E.S.C. and Axcell, B.C. (1996) Trehalose – a stress protectant and stress indicator compound for yeast exposed to adverse conditions. *Journal of the American Society of Brewing Chemists* 54(4), 221–227.

Masschelein, C.A. (1981) Role of fatty acids in beer flavour. In: *EBC Symposium on Beer Flavour, Monograph VII*, Copenhagen, pp. 211–221.

Maule, D.R. (1986) A century of fermenter design. *Journal of the Institute of Brewing* 92, 137–145.

Meilgaard, M.C. (1975) Flavor chemistry of beer: Part II: flavor threshold of 239 aroma volatiles. *Master Brewers Association Technical Quarterly* 12(3), 151–168.

Mochaba, F., O'Connor-Cox, E.S.C. and Axcell, B.C. (1998) Practical procedures to measure yeast viability and vitality prior to pitching. *Journal of the American Society of Brewing Chemists* 56(1), 1–6.

Obradovic, B., Nedovic, V.A., Bugarski, B., Willaert, R.G. and Vunjak-Novakovic, G. (2004) Immobilised cell bioreactors. In: Nedovic V. and Willaert R. (eds) *Fundamentals of Cell Immobilisation Biotechnology*. Kluwer Academic Publishers, Dordrecht, Netherlands, pp. 411–436.

O'Connor-Cox, E.S.C. and Ingledew, W.M. (1991) Alleviation of the effects of nitrogen limitation in high-gravity worts through increased inoculation rates. *Journal of Industrial Microbiology* 7, 89–96.

O'Connor-Cox, E.S.C., Lodolo, E.J. and Axcell, B.C. (1993) Role of oxygen in high-gravity fermentations in the absence of unsaturated lipid biosynthesis. *Journal of the American Society of Brewing Chemists* 51(3), 97–107.

Peddie, H.A.B. (1990) Ester formation in brewery fermentations. *Journal of the Institute of Brewing* 96, 327–331.

Powell, C.D., Van Zandycke, S.M., Quain, D. and Smart, C.A. (2000) Replicative ageing and senescence in *Saccharomyces cerevisiae* and the impact in brewing fermentations. *Microbiology* 146, 1023–1034.

Powell, C.D., Quain, D.E. and Smart, K.A. (2004) The impact of sedimentation on cone yeast heterogeneity. *Journal of the American Society of Brewing Chemists* 62(1), 8–17.

Quain, D.E. and Tubb, R.S. (1982) The importance of glycogen in brewing yeast. *Master Brewers Association Technical Quarterly* 19, 19–23.

Rainieri, S., Zambonelli, C. and Kaneko, Y. (2003) *Saccharomyces sensu stricto*: systematics, genetic diversity and evolution. *Journal of Bioscience and Bioengineering* 96, 1–9.

Rainieri, S., Kodama, Y., Kaneko, Y., Mikata, K., Nakao, Y. and Ashikari, T. (2006) Pure and mixed genetic lines of *Saccharomyces bayanus* and *Saccharomyces pastorianus* and their contribution to the lager brewing strain genome. *Applied and Environmental Microbiology* 72(6), 3968–3974.

Rosenfeld, E. and Beauvoit, B. (2003) Role of the non-respiratory pathways in the utilization of molecular oxygen by *Saccharomyces cerevisiae. Yeast* 20(13), 1115–1144.

Saerens, S.M.G., Verstrepen, K.J., Van Laere, S.D.M., Voet, A.R.D., Van Dijck, P., Delvaux, F.R. and Thevelein, J.M. (2006) The *Saccharomyces cerevisiae* EHT1 and EEB1 genes encode novel enzymes with medium-chain fatty acid ethyl ester synthesis and hydrolysis capacity. *Journal of Biological Chemistry* 281(7), 4446–4456.

Sampermans, S., Mortier, J. and Soares, E.V. (2005) Flocculation onset in *Saccharomyces cerevisiae:* the role of nutrients. *Journal of Applied Microbiology* 98(2), 525–531.

Stratford, M. (1989) Evidence for two mechanisms of flocculation in *Saccharomyces cerevisiae. Yeast* 5, 441–445.

Suihko, M.-L., Vilpola, A. and Linko, M. (1993) Pitching rate in high-gravity brewing. *Journal of the Institute of Brewing* 99, 341–346.

Thomas, D. and Surdin-Kerjan, Y. (1997) Metabolism of sulfur amino acids in *Saccharomyces cerevisiae. Microbiology and Molecular Biology Reviews* 61(4), 503–532.

Vanbeneden, N., Gils, F., Delvaux, F. and Delvaux, F.R. (2007) Formation of 4-vinyl and 4-ethyl derivates from hydroxycinnamic acids: occurrence of volatile phenolic flavour compounds in beer and distribution of Pad1-activity among brewing yeasts. *Food Chemistry* 107(1), 221–230.

Vanderhaegen, B., Neven, H., Verachtert, H. and Derdelinckx, G. (2006) The chemistry of beer aging – a critical review. *Food Chemistry* 95, 357–381.

Verbelen, P.J., De Schutter, D.P., Delvaux, F., Verstrepen, K.J. and Delvaux, F.R. (2006) Immobilized yeast cell systems for continuous fermentation applications. *Biotechnology Letters* 28(19), 1515–1525.

Verstrepen, K., Bauer, F.F., Michiels, C., Derdelinckx, G., Delvaux, F.R. and Pretorius, I.S. (1999) Controlled expression of FLO1 in *Saccharomyces cerevisiae.* In: *EBC Symposium on Yeast Physiology: a New Era of Opportunity*, Nutfield, New Hampshire. Fachverlag Hans Carl, Nümberg, Germany, pp. 30–40.

Verstrepen, K.J., Derdelinckx, G., Dufour, J.-P., Winderickx, J., Thevelein, J.M., Pretorius, I.S. and Delvaux, F.R. (2003a) Flavor-active esters: adding fruitiness to beer. *Journal of Bioscience and Bioengineering* 96(2), 110–118.

Verstrepen, K.J., Derdelinckx, G., Verachtert, H. and Delvaux, F.R. (2003b) Yeast flocculation: what brewers should know. *Journal of Applied Microbiology and Biotechnology* 61, 197–205.

Verstrepen, K.J., Jansen, A. and Fink, G.R. (2005) Intragenic tandem repeats generate functional variability. *Nature Genetics*, 1–5.

Wainwright, T. (1970) Hydrogen sulphide production by yeast under conditions of methionine, pantothenate or vitamin B6 deficiency. *Journal of General Microbiology* 61, 107–119.

Wainwright, T. (1973) Diacetyl – a review. *Journal of the Institute of Brewing* 79, 451–470.

Walker, M.D. and Simpson, W.J. (1993) Production of volatile sulphur compounds by ale and lager brewing strains of *Saccharomyces cerevisiae. Letters in Applied Microbiology* 16, 40–43.

Walkey, R.J. and Kirsop, B.H. (1969) Performance of strains of *Saccharomyces cerevisiae* in batch fermentation. *Journal of the Institute of Brewing* 75, 393–398.

Werner-Washburne, M., Braun, E., Johnston, G.C. and Singer, R.A. (1993) Stationary phase in the yeast *Saccharomyces cerevisiae. Microbiological Reviews* 57(2), 383–401.

Wheals, A.E. (1987) Biology of the cell cycle in yeasts. In: Rose A.H. and Harrison J.S. (eds) *The Yeasts: Vol. 1, Biology of Yeasts.* Academic Press, London, pp. 283–377.

Willaert, R. (2007) The beer brewing process: wort production and beer fermentation. In: Huie Y.H. (ed.) *Handbook of Food Products Manufacturing,* 1st. edn. John Wiley & Sons, New Jersey, pp. 443–506.

8 Genomic Adaptation of *Saccharomyces cerevisiae* to Inhibitors for Lignocellulosic Biomass Conversion to Ethanol

ZONGLIN LEWIS LIU[1] AND MINGZHOU SONG[2]

[1]*US Department of Agriculture, Agricultural Research Service, National Center for Agricultural Utilization Research, Peoria, USA;*
[2]*Department of Computer Science, New Mexico State University, Las Cruces, USA*

Introduction

As interest in alternative energy sources rises, the concept of agriculture as an energy producer has become increasingly attractive (Outlaw *et al.*, 2005). Renewable biomass, including lignocellulosic materials and agricultural residues, provide low-cost materials for bioethanol production (Bothast and Saha, 1997; Wheals *et al.*, 1999; Zaldivar *et al.*, 2001).

For economic reasons, dilute acid hydrolysis, which hydrolyses the hemicellulose fraction and increases fibre porosity to allow enzymatic saccharification and fermentation of the cellulose fraction, is commonly used in biomass degradation (Bothast and Saha, 1997; Saha, 2003). However, one major limitation of this method is the generation of numerous by-products and compounds that inhibit microbial growth and subsequent fermentation.

The stress conditions involved in the distinct lignocellulosic biomass conversion process have been a technical barrier in biomass conversion to ethanol. Biomass pre-treatment generates varied harsh conditions, including high temperatures, extreme pH, high substrate concentration, osmotic shifts and toxic compounds that inhibit yeast growth and fermentation. Furfural and 5-hydroxymethylfurfural (HMF) are major inhibitors commonly recognized from biomass pre-treatment. Genetic mechanisms involved in the tolerance of stresses such as those caused by furfural and HMF are unknown, and few yeast strains are available that are tolerant to these inhibitors.

Development of stress-tolerant ethanologenic yeasts is one of the significant challenges for cost-competitive bioethanol production. Recently, progress has been made in developing more tolerant strains to detoxify furfural and HMF *in situ*. This chapter summarizes current knowledge in this regard and discusses future applications using genomic adaptation for new strain development.

Biomass Conversion Inhibitors

Furfural and HMF derived from biomass pretreatment

More than 100 compounds derived from pre-treatment of biomass substrates have been shown to have potential inhibitory effects on microbial fermentation (Luo *et al.*, 2002). Common inhibitors have been identified and classified into the four groups of aldehydes, ketones, organic acids and phenols (Palmqvist and Hähn-Hägerdal, 2000; Klinke *et al.*, 2004; Liu and Blaschek, 2009). Furfural and 5-hydroxymethylfurfural derived from biomass pre-treatment are among the most significant and potent inhibitors to yeast growth and fermentation (Chung and Lee, 1985; Olsson and Hähn-Hägerdal, 1996; Taherzadeh *et al.*, 2000a).

During biomass degradation by dilute acid treatment, cellulose components are converted to hexoses, and hemicellulose components are converted to pentoses. Furfural and HMF are derived from further dehydration of pentoses and hexoses, respectively (see Fig. 8.1; Dunlop, 1948; Antal *et al.*, 1990, 1991; Larsson *et al.*, 1999; Lewkowski, 2001).

These compounds reduce enzymatic and biological activities, break down DNA, inhibit protein and RNA synthesis and damage yeast cell walls (Sanchez and Bautista, 1988; Khan and Hadi, 1994; Modig *et al.*, 2002; Gorsich and Liu, unpublished data). Most yeasts, including industrial strains, are susceptible to the complexes associated with dilute acid hydrolysis pre-treatment and can be killed by low concentrations of inhibitory complexes (Palmqvist *et al.*, 1999; Taherzadeh *et al.*, 2000a; Martin and Jonsson, 2003; Liu *et al.*, 2004). Additional remediation treatments, including physical, chemical, or biochemical detoxification procedures, are often required to remove these inhibitory compounds and to allow fermentation. However, these additional steps add cost and complexity to the process and generate extra waste products (Martinez *et al.*, 2000; Mussatto and Roberto, 2004).

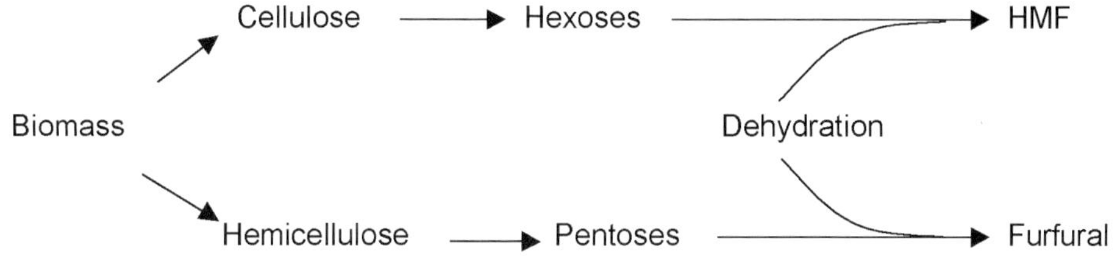

Fig. 8.1. Furfural and HMF derived from biomass pre-treatment. During biomass pre-treatment, cellulose is degraded to hexoses, and hemicellulose to pentoses, from which 5-hydroxymethylfurfural (HMF) and furfural are derived from sugar dehydrations, respectively.

Metabolic conversion pathways of furfural and HMF

Furfural can be converted to furan methanol (FM, furfuryl alcohol) by yeasts (see Fig. 8.2), and it is believed that FM is further reduced to pyromucic acid (Morimota and Murakami, 1967; Nemirovskii and Kostenko, 1991; Villa *et al.*, 1992; Mohsenzadeh *et al.*, 1998; Liu, 2006). Furfural can also be broken down to form formic acid (Palmqvist and Hähn-Hägerdal, 2000). The biotransformation of furfural and HMF by yeasts is due to NADH- and NADPH-coupled enzymes (Palmqvist *et al.*, 1999; Larroy *et al.*, 2002; Liu, 2006; Petersson *et al.*, 2006; Liu *et al.*, 2008b).

In the presence of furfural the intracellular concentration of ATP is low, cell replication is limited and glycerol formation is reduced. Furfural is an electron acceptor and so it can cause a shortage of NADH (Wahlbom and Hähn-Hägerdal, 2002). As furfural reduction competes for NADH, it interferes with cell glycolysis during regeneration of NAD^+. As a result, the presence of furfural can lead to an accumulation of acetaldehyde and a delay in acetate and ethanol production. A reduction in xylitol excretion has also been reported during xylose fermentation when furfural was added to the medium (Wahlbom and Hähn-Hägerdal, 2002). Reduced furfural tolerance was observed with some selective deletion mutants of genes in the pentose phosphate pathway (Gorsich *et al.*, 2005), and these observations can be related to NADPH-dependent reactions involved in pentose phosphate pathways. Enzyme cofactor imbalances also appeared to be affected by furfural.

Unlike furfural, knowledge of HMF conversion has been limited because there has not been a readily available commercial source for an HMF conversion product. Following the furfural conversion route, it has been assumed that HMF is converted to HMF alcohol (Nemirovskii *et al.*, 1989). Recently, an HMF metabolic conversion product was isolated and identified

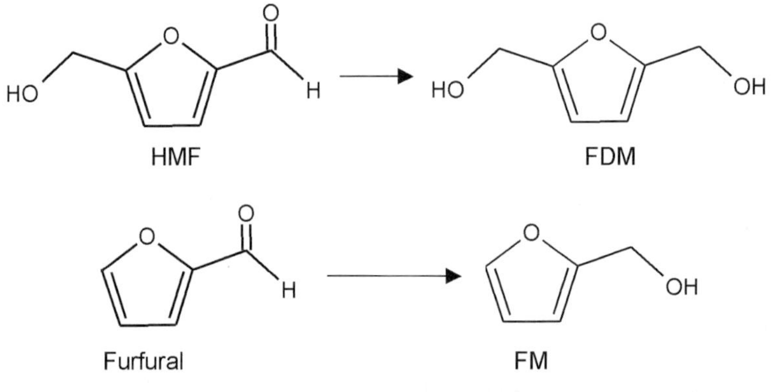

Fig. 8.2. Inhibitor conversion pathways. 5-Hydroxymethylfurfural (HMF) metabolic conversion product is identified as 2,5-bis-hydroxymethylfuran (also called furan-2,5-dimethanol, FDM), with the formula $C_6H_8O_3$ and a molecular mass of 128 by GC-MS and NMR; and furfural is converted to furan methanol (FM) (from Liu *et al.*, 2008a).

as furan-2,5-dimethanol (FDM), also termed 2,5-bis-hydroxymethylfuran (Liu *et al.*, 2004, 2008a; Liu, 2006; Fig. 8.2). FDM has been isolated from cell free culture supernatant, purified and characterized using mass and NMR spectra analysis (Liu *et al.*, 2004). Signals for the aldehyde proton and the asymmetric spectra of HMF were absent when the purified HMF-conversion product was analyzed using NMR, and the NMR spectra were consistent with that of a symmetrical molecule with a furan ring. The metabolite was found to have a composition of $C_6H_8O_3$ and a molecular mass of 128 g/mol (see Fig. 8.2). The characterization of FDM has clarified the existing literature and provides a basis for new studies on mechanisms for the detoxification of the inhibitor.

Yeast Adaptation to Furfural and HMF

Microbial performance is the key for cost-efficient improvement

Fermentation is among the oldest microbial applications in human history. Although a tremendous amount of knowledge has been accumulated through years of experience and development of modern technology, many alternative fermentation processes remain unknown. The economics of fermentation-based bioprocesses rely extensively on the performance of microbial biocatalysts for their industrial application. Development of yeast strains that can efficiently utilize heterogeneous sugars and withstand stress conditions in the bioethanol conversion process is key for sustainable and cost-competitive conversion of lignocellulosic biomass to ethanol. However, many of the industrially interesting microorganisms obtained so far are not sufficiently robust.

Genetically manipulated yeast strains have generally enhanced ethanol fermentation performance, due to improvements in their sugar utilization and enzyme production (Ho *et al.*, 1998; Jeffries and Shi, 1999; Ostergaard *et al.*, 2000; Hähn-Hägerdal *et al.*, 2001). Development of genetically engineered strains with greater tolerance to inhibitors, especially to furfural and HMF, is a promising alternative to the costly traditional inhibitor remediation steps (Liu and Slininger, 2005; Liu *et al.*, 2005). However, development of such strains is hindered due to a lack of understanding of the basic mechanisms underlying stress tolerance in ethanol production by *Saccharomyces cerevisiae*. The fast life cycle and genetic diversity of ethanologenic yeasts are invaluable resources for strain improvement, and an efficient utilization of these characteristics will lead to more cost-effective fermentation and processing in the future.

Dose-dependent response

On a defined medium, yeast strain Y-12632 showed clear dose-dependent cell growth and metabolic conversion activities in response to varied doses of HMF and/or furfural, and cell growth was delayed at concentrations of 10, 30 and 60 mM (Liu *et al.*, 2004). Metabolic transformation of HMF into FDM, furfural into FM and glucose into ethanol was also delayed with the increase in inhibitor doses (Liu *et al.*, 2004). The lag phase lasted from a few hours to several days, depending upon the concentration of the inhibitors and the strains used. Once cell growth had recovered, cultures inoculated with sublethal doses of the inhibitors were able to utilize glucose and produce ethanol. This demonstrated a clear dose-dependent inhibition of the yeast by furfural and HMF. However, this lag phase was not observed at the higher concentration of 120 mM, where cells were completely repressed and no biological activity or HMF transformation was observed (Liu *et al.*, 2008b).

The duration of the lag phase may be interpreted as a measure of the level of tolerance to furfural and HMF. This suggests that some yeast strains have more effective mechanisms than others to withstand these inhibitors. The prolonged lag phase before the recovery of cell growth could reflect a genetic response, and result in a shift in physiology of the cells adapting to the chemical stress. It has been suggested that certain enzymes may be induced during the lag phase (Kang and Okada, 1973; Liu and Slininger, 2005, 2006), and important metabolic enzymes – including alcohol dehydrogenase, aldehyde dehydrogenase and pyruvate dehydrogenase – have been reported to be inhibited by furfural and HMF *in vitro* (Modig *et al.*, 2002). Numerous enzymes have been found to have significantly enhanced expression in the presence of furfural and HMF (Liu *et al.*, 2008a, b), and yeasts are apparently stimulated to undergo an adaptation process in response to HMF during the lag phase (Liu and Slininger, 2006).

Enhanced biotransformation and tolerance

Adaptation of *S. cerevisiae* to furfural and HMF has been observed, and methods used to overcome inhibitory effects have included the use of increased yeast inoculum level, increased biomass or fed-batch mode fermentation (Banerjee *et al.*, 1981; Chung and Lee, 1985; Villa *et al.*, 1992; Taherzadeh *et al.*, 2000b). An adapted *Pichia stipitis* that gave improved ethanol production from hemicellulose hydrolysate has been reported (Nigam, 2001). The dose-dependent yeast response to furfural and HMF has been used to generate strains tolerant to either furfural or HMF by a directed adaptation method (Liu *et al.*, 2005).

Recently, a newly developed strain, Y-50049, has shown tolerance to both furfural and HMF and was able to complete ethanol fermentation in 48 h, while a wild type failed to establish a culture under the same conditions (Liu *et al.*, 2008b). This strain did not require a pre-build biomass but functioned as an initial inoculum to establish a culture. The tolerant strain showed significantly enhanced biotransformation, converting furfural into FM, HMF into FDM and produced a normal yield of ethanol. In contrast, a normal control strain failed to establish a culture in the presence of the inhibitors 48 h after inoculation. This showed that a qualitative change could be derived from evolution by quantitative adaptation, leading to a genetic adaptation in the response to the inhibitors.

Strain improvement by directed evolutionary adaptation

The directed evolution method has been used to obtain numerous strains with improved performance. In addition to the yeast strains tolerant to furfural and HMF (Liu *et al.*, 2004, 2005, 2008b; Liu and Slininger, 2006), a preliminary improvement to inhibitor tolerance was recently obtained for a sugarcane bagasse hydrolysate by adaptation of an engineered xylose utilization strain (Martin *et al.*, 2007). At least two phenotypically different populations have been recovered from a recombinant strain by selection pressures applied during an evolution method (Sonderegger and Sauer, 2003). Under such conditions, multiple mutations appeared necessary to obtain more integrated functional adaptations. Efficient xylose utilization strains of *S. cerevisiae* were obtained through directed evolution with a minimum of recombinant engineering of exogenous xylose isomerase (Kuyper *et al.*, 2005). Strains of *S. cerevisiae* tolerant to HMF and also able to grow on xylose have been obtained in the laboratory through directed evolution methods (Liu *et al.*, unpublished data).

The persistence of specific altered gene expression over time supports the hypothesis that yeasts are stimulated to undergo an adaptation process during the lag phase in response to inhibitors (Liu, 2006; Liu and Slininger, 2006). Directed evolutionary adaptation therefore appears promising for the development of desirable characteristics in ethanologenic yeasts (Kuyper *et al.*, 2005; Liu *et al.*, 2005, 2008a; Liu, 2006). The adaptation approach can be an alternative means of improving microbial strain performance, and such adapted strains may be suitable for further genetic manipulation. Further study in this area can be expected to lead to new strain development.

Adaptation is not a new process in yeast utilization, and success depends upon the genetic potential of the yeast. Earlier studies have shown a link between genetic potential and the ability of a yeast to withstand and transform furfural and HMF. The genetic potential for all of the stress conditions encountered in the bioethanol conversion process has not been experimentally tested. Enrichment of the genetic background of the candidate yeast can be achieved by introducing exogenous gene functions, following proper genetic manipulation and adaptation. Development of tolerant strains with enhanced detoxification using directed enzyme evolution has shown some promising results (Moon and Liu, unpublished data). Enhanced laboratory procedures can significantly speed up evolutionary adaptation to the stress condition and may maintain desirable ethanol production characteristics.

Functional Genomics of Ethanologenic Yeast

Understanding mechanisms of tolerance using functional genomics

Single-gene studies have contributed significantly to our knowledge of gene functions in the past 50 years and will continue to do so in the future. However, the new advances in genomics have revolutionized our understanding and changed our view on yeast-processing events. A biological process often cannot be explained by a single gene function, and is frequently the result of a complex control system. As thousands of genes in a genome are required to maintain a living yeast system, significant gene alteration can have a significant impact on the responses of other genes in the system. Significant gene interactions and genomic regulatory networks need to be considered for efficient genetic manipulation for strain development (Liu and Slininger, 2005), and a few enhanced functional genes are unlikely to address the challenges encountered in bioethanol conversion.

An understanding of the genomic mechanisms involved in the integration and balance of these functions in individual strains through directed evolutionary adaptation is needed. Such knowledge provides a fundamental insight into the integrated alteration of genome architecture, transcriptional profiling and gene regulatory networks that underline heterogeneous sugar utilization and stress tolerance. This new technology will provide great flexibility and power in the design and development of more desirable and robust biocatalysts for cost-effective and highly productive lignocellulosic conversion to ethanol for the next decade. In order to understand mechanisms of stress tolerance to furfural and HMF, it is necessary to identify key gene functions, gene interaction networks and regulatory elements involved at the genome level.

Yeasts live in ever-changing environments and need constantly to adapt to external stimuli for survival. As documented in numerous reports, yeast adaptation to stress conditions is common and accomplished via a variety of molecular mechanisms (Gasch and Werner-Washburne, 2002; Erasmus *et al.*, 2003). Laboratory strains of yeasts have been used extensively as model organisms in studies of genomic expression profiling with varied environmental stimulants (Gasch *et al.*, 2000; Causton *et al.*, 2001; Brejning *et al.*, 2003; Zhang *et al.*, 2003; Lucau-Danila *et al.*, 2005).

Common stress-tolerant genes have been reported, and the transient expression responses to stimuli appear to be common. Genome expression and transcriptome dynamics to environmental stress and other fermentation stress conditions, including HMF, have also been studied for some industrial yeasts (Chen *et al.*, 2003; Erasmus *et al.*, 2003; James *et al.*, 2003; Devantier *et al.*, 2005; Liu and Slininger, 2006). However, systematic information on the inhibitory stress tolerances involved in the bioethanol conversion process at the genome level is not yet available. Due to the heterogeneity of experimental conditions and a lack of common quality control for multiple microarray experiments, it is impossible to make comparisons between existing studies. Currently, an integrated functional genomic approach has been taken to study furfural and HMF stress tolerance involved in bioethanol conversion and to develop more inhibitor-tolerant strains (Liu and Slininger, 2005).

Quality control issues for gene expression analysis

The significance of the proper application of quality controls cannot be overestimated in ensuring the reliability and reproducibility of microarray expression data. Quality control has been a very important issue since the emergence of high-throughput gene expression technology (Schena *et al.*, 1995; Brazma *et al.*, 2001; Badiee *et al.*, 2003), and ever-increasing attention has been drawn to concerns over the application of expression data (Baker *et al.*, 2005; Bammler *et al.*, 2005; Larkin *et al.*, 2005). The need for standard controls across different platforms of gene expression analysis has been recognized (Dallas *et al.*, 2005; Etienne *et al.*, 2005; External RNA Control Consortium, 2005; Irizarry *et al.*, 2005).

Six species of exogenous nucleotides have been used as a set of universal external RNA quality controls, which were developed specifically for microbial gene expression analysis across different platforms of microarray and real-time quantitative RT-PCR (Liu and Slininger, 2007). The DNA sequences of these control genes were compared with those in the microbial gene sequence database (Peterson *et al.*, 2001). The selected control genes had no homology or similarity to the yeast genome and bacterial system, and therefore avoided interference with microbial gene expression signals. The linearity of signal intensity of the control genes allowed them to serve as a quantitative calibration and normalization reference for quantitative measurement of gene expression analysis. Using these quality controls, a coefficient of variance can be calculated and analysis of variance can be applied to results (Kerr and Churchill, 2001; Churchill, 2004).

Unlike housekeeping genes usually affected by environmental conditions, these controls demonstrate consistency in mRNA detection independent of environmental factors. Such quality control measurements provide a consistent reference, allow an estimate of variation for microarray experiments, reduce variability and increase the reliability and reproducibility of microarray data. The application of universal quality controls will allow confirmation and comparison of data obtained from different microarray experiments and platforms. In order to generate high-quality data from microarray experiments, it is strongly recommended that the complete length of cDNA populations of labelled probes be evaluated using a microgel or slide gel electrophoresis system (Lage *et al.*, 2002; Liu and Slininger, 2005, 2007). Such a quality control measurement cannot be substituted by quantitative measurements of the probe-using spectrophotometer.

Genomic expression response to furfural and HMF

Yeast genes have been shown to give an immediate response, at least 10 min after exposure to furfural or HMF (Liu and Slininger, 2006). The expression levels of several hundred genes were significantly different for the yeast under HMF stress conditions compared with those of an untreated normal control. These genes demonstrated significant differential expression patterns during a lag phase under HMF stress, and are more likely to be involved in HMF stress tolerance. Thus stress responses are not single gene-controlled events but are an organized global expression response involving multiple genes. Of the genes that were significantly induced, members of the pleiotropic drug resistance gene family were suggested as playing an important role for cell survival in coping with inhibitor stress (Liu *et al.*, 2006).

Unlike the transient changes reported for laboratory strains, constant functional mRNA expression was observed for ethanologenic yeast in response to the HMF stress during the lag phase. Some genes showed continued enhanced or repressed expression, while others demonstrated significant dynamics of reversed expression (Liu and Slininger, 2006). Genes involved in biological processes, cellular components and molecular function were identified and, of these, some appeared to be HMF- and/or furfural-specific, while others shared functions with those in a core set of common stress genes. Regulatory elements and transcription factors significant for HMF tolerance have been identified (Liu and Sinha, 2006). However, interpretation of some genes was limited by incomplete annotations or lack of known functions. Further studies are needed at the genome level. Many valuable genome resources are available, such as the commonly used *Saccharomyces* Genome Database (Fisk *et al.*, 2006), YEASTRACT (Teixeira *et al.*, 2006), Kyoto Encyclopedia of Genes and Genomics (Kanehisa *et al.*, 2006) and Gene Ontology (The Gene Ontology Consortium, 2000).

Computational modelling of gene regulatory networks for HMF stress

Computational modelling to infer gene regulatory networks (GRN) has recently shown that various potentially interesting gene interactions are involved in inhibitor detoxification (Song and Liu, 2007). Discrete dynamic system models using first-order linear difference equations have been built for the transcriptional interactions among genes in yeast during the earlier exposure to the inhibitor HMF. In a discrete dynamic system model, the expression change rate of a gene is a linear function of the concentrations of potential regulator genes, and one equation is used for each gene. A GRN is derived from a discrete dynamic system model by creating an edge from every potential regulator to each gene it regulates. These models were developed based on mRNA abundance over five time points in the presence or absence of HMF (see Fig. 8.3).

A reconstructed GRN with a subset of 46 gene nodes plus an HMF node showed complex gene interactions under HMF stress (see Fig. 8.4). Forty-six significantly induced expressed genes from the HMF treatment were selected based on ANOVA and cluster analysis, and these were used for the prototype computational model development. This system model captured temporal dependencies among the 46 genes and HMF during the earlier exposure to the inhibitor in the yeast fermentation process.

The system model underlying the GRN is an optimal solution after searching all possible directed graphs with 47 nodes, except that the HMF node is not allowed to have incoming edges and the maximum number of incoming edges for a gene node is, at most, 5. The existence of an edge from *YAP*1 to *DDI1* indicates a temporal dependency of the rate of change in *DDI1* expression on the mRNA level of *YAP1*. The number 1.2e-07, positioned next

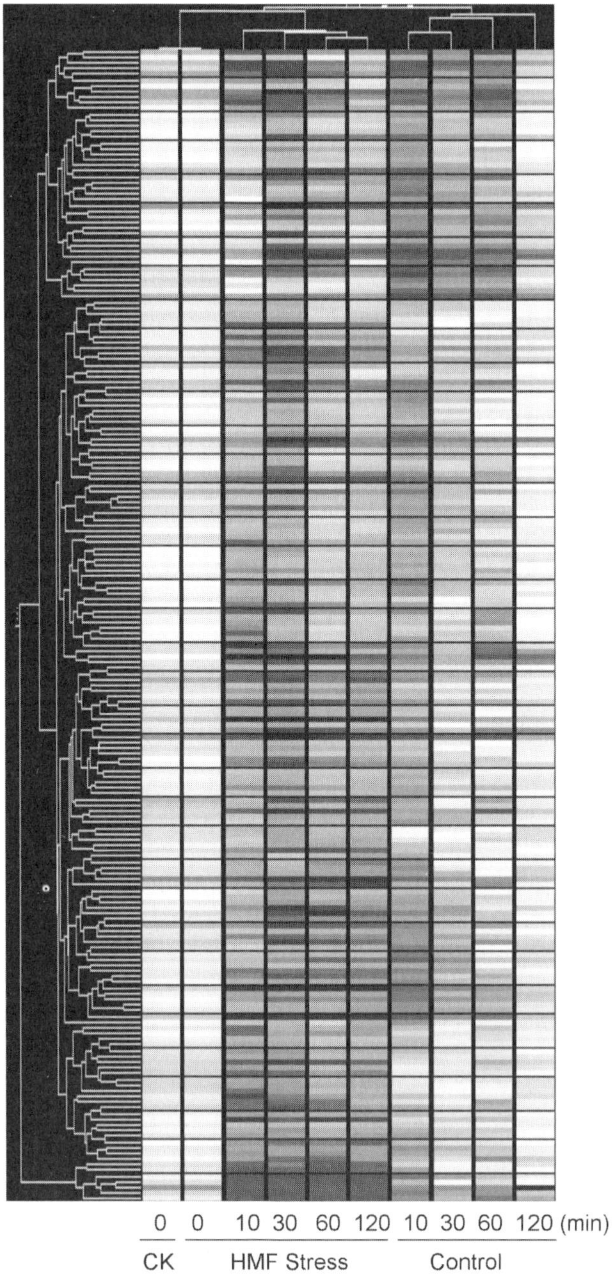

Fig. 8.3. The genomic adaptation to HMF stress. Interactions of significantly expressed genes of ethanologenic yeast *Saccharomyces cerevisiae* under a normal control condition and 5-hydroxymethylfurfural (HMF) stress condition from 0, 10, 30, 60 and 120 min after the treatment, showing significantly induced (blue) and repressed (red) mRNA expression caused by the HMF stress on a defined medium. Yellow colour indicates mRNA equally expressed under different conditions. Varied colours between yellow and red, or yellow and blue, indicate varied relative quantitative measurements of mRNA expression levels in a log scale.

to the edge (see Fig. 8.4), is the *p*-value of this temporal dependency. The original system matrix was stabilized by scaling all eigenvalues by the spectral norm of 3.09. The overall *p*-value, 1.6e-5, of the entire system model indicates that the model is statistically significant. The *p*-value is based on a stringent standard and the resulting model has high levels of consistency with biological observations, because the probability of the model arising by chance is as low as 1.6e-5.

Three known transcription factors, *PDR1*, *PDR3*, and *YAP1*, were examined and, in this subset, *YAP1* was shown by the model to be one of the most influential regulators in the early response to HMF stress (see Fig. 8.4). This is strongly supported by current knowledge and documented experimental observations (Teixeira *et al.*, 2006). For example, the following edges have been reported as transcriptional regulations, including *YAP1* to

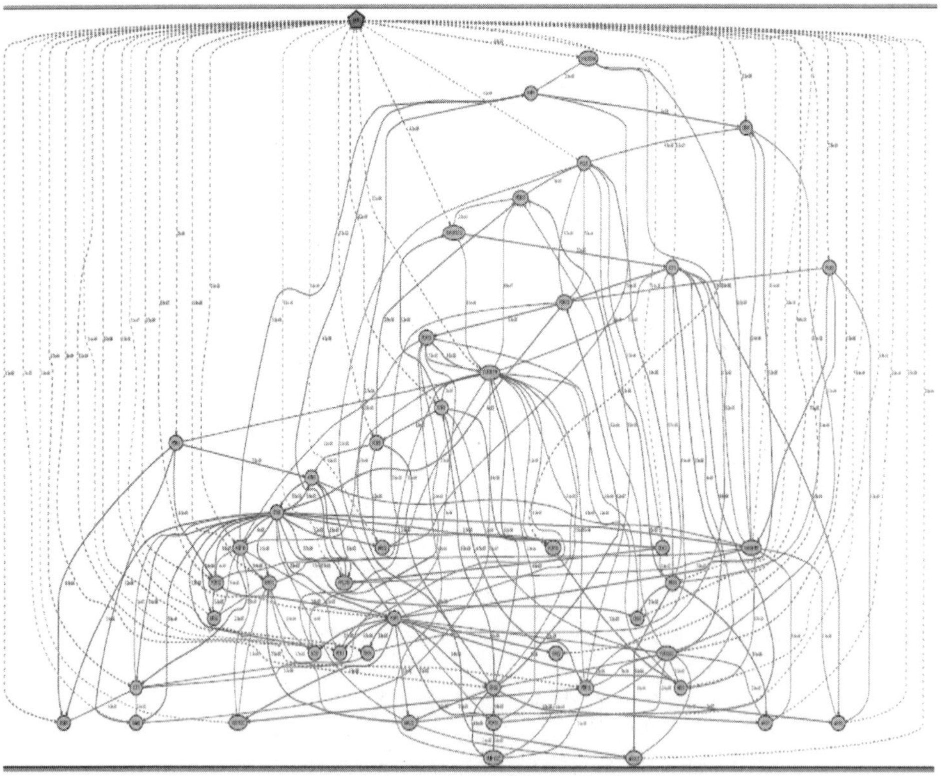

Fig. 8.4. Computation modelling of gene regulatory networks of HMF stress. Temporal interactions for a subset of 46 genes in response to HMF for biomass conversion to ethanol by the ethanologenic yeast. The *p*-values of each edge are displayed. A solid directed edge in green from the first gene node to the second gene node with an arrowhead indicates enhancement of the second gene by the first gene; an edge in red from the first gene node to the second gene node with a solid dot indicates repression of the second gene by the first gene. The dashed edges represent the external influence from HMF to each gene: red for repressing and green for enhancing (from Song and Liu, 2007).

DDI1 (Haugen *et al.*, 2004), *YAP1* to *ATM1* (Haugen *et al.*, 2004), *YAP1* to *GRE2* (Lee *et al.*, 1999), *YAP1* to *SNQ2* (Lee *et al.*, 2002; Lucau-Danila *et al.*, 2005) and *YAP1* to *TPO1* (Lucau-Danila *et al.*, 2005). Four more edges from *YAP1* demonstrated enhancement to *SCS7, PDR1, PDR11* and *HIS3*, suggesting regulatory roles of *YAP1* to these genes (see Fig. 8.4). The genes *SCS7, PDR1, PDR11*, and *HIS3* are considered potential transcriptional regulatees of *YAP1* based on sequence motifs (YEASTRACT, 2006). In addition, the transcription factor PDR3 showed a regulatory role to *RSB1* as demonstrated in this model, which is in agreement with and supported by previous documented observations (Devaux *et al.*, 2002). PDR3 also showed enhancement to *SAM3, ATM1* and *PDR12*. It is very encouraging that the GRN model developed in this study is highly consistent with the current knowledge, including documented experimental observation and sequence motif-based analysis. More importantly, the model was able to demonstrate statistical significance for the temporal dependencies.

This system model also presented numerous interesting interactions among genes with potential significance. For example, *STE6, SNQ2, ARG4* and *YOR1* enhanced directly or indirectly for 15, 8, 5 and 4 other genes, respectively. These genes have been observed to be core stress response genes, and many related genes are reported to be involved in survival under HMF stress. Resolution of such interactions could have a significant impact in elucidating the mechanisms of detoxification and stress tolerance caused by HMF. Although they have not been reported, such statistically significant gene interactions presented by this model could be potentially biologically significant in predicting unknown gene interactions. With the high consistency between the system model on *YAP1* presented in this study and current knowledge, it is reasonable to assume potential relationships for associations with significant *p*-values in this model. Although it is highly homologous with *PDR3*, the common transcription factor *PDR1* did not appear to respond in the same way as *PDR3*, and did not appear to have a significant regulatory role towards the selected subset of genes used in this model.

Another output of the system model is to prescribe the desired system behaviours by applying perturbation to the system. A perturbation can be used to change the concentration level of the inhibitor HMF, silence a subset of genes in the GRN or mutate a subset of genes. In order to increase the tolerance to the inhibitor HMF, one can consider adjusting the influential genes to achieve an effect similar to the transcriptome profile observed in the absence of HMF. This model indicated the following genes to be potentially significant in gene interactions for detoxification and HMF stress tolerance: *STE6* (15/46), *YCR061W* (14/46), *YAP1* (12/46), *YGR035C* (10/46), *SNQ2* (8/46), *HSP10* (7/46) and *YAR066W* (7/46) (see Fig. 8.4). By perturbing these potential regulators, one may exert most control over the expression of other genes, which might be economically desirable. Approaches using linear discrete dynamic system models have shown promising potential to infer complex gene interactions in this computational modelling prototype. More accurate gene regulatory networks will be further defined by continued efforts using additional data and cross-examinations.

Mechanisms of *in situ* detoxification

A recently described furan-2,5-dimethanol preparation procedure can be used as a standard for HPLC metabolic profiling (Liu *et al.*, 2007), and this has in turn allowed studies on mechanisms of *in situ* detoxification of HMF. Studies with the tolerant strain have shown that the furfural and HMF conversion products FM and FDM accumulate in the medium as yeast growth and fermentation is completed. Furfural and HMF are furan derivatives with a furan ring composed of $C_5H_4O_2$ and $C_6H_6O_3$, respectively. Their conversion products FM and FDM contain the furan rings $C_5H_6O_2$ and $C_6H_8O_3$, respectively (see Fig. 8.2).

These furan elements remain intact during the inhibitor conversion process and persist in the medium until the end of the fermentation. The presence of FM and FDM did not affect yeast growth and ethanol yield. Apparently, the aldehyde functional group in furfural and HMF is toxic to yeast but not to the furan ring or the associated alcohol functional groups. Clearly, aldehyde reduction is a mechanism for *in situ* detoxification of furfural and HMF (Liu *et al.*, 2008b).

The toxicity of aldehyde to yeast has been recognized for many years (Leonard and Hajny, 1945). Any potential further reduction or degradation of the furan ring or alcohol groups may not play a significant role for the *in situ* detoxification of furfural and HMF by the yeast. Detoxification of the inhibitors by the yeast did not involve utilization or degradation of the furan compounds, and so terms such as 'furan conversion' or 'furan reduction' that have been used in the literature should not be used in the context of these inhibitor conversions (Liu *et al.*, 2008b). Furthermore, the use of 'furan derivatives' as a general term for 'inhibitors' such as furfural and HMF should be avoided, as FM and FDM are also furan derivatives, and these appear to be either less toxic or non-toxic to yeast.

NAD(P)H-dependent enzymatic activities were observed under furfural and HMF stress conditions (Larroy *et al.*, 2002; Nilsson, *et al.* 2005; Liu, 2006; Petersson *et al.*, 2006; Liu *et al.*, 2008b). Recent studies have shown that aldehyde reduction involves multiple genes with reductase activities and that reduction is not the result of a single gene (Liu *et al.*, 2008b). Numerous genes have been identified that were significantly induced and were responsible for the biotransformation of the inhibitors (Liu, 2006; Liu and Slininger, 2006; Liu *et al.*, 2008b). However, in a yeast culture growing under inhibitor stress, not all of these enzymes are able to function, particularly when the yeast is growing as a batch culture (Liu *et al.*, unpublished data). HMF reduction has been reported to have a cofactor preference for NADPH (Wahlbom and Hähn-Hägerdal, 2002); however, a later study found a different strain of *S. cerevisiae* had a preference for NADH rather than NADPH (Nilsson *et al.*, 2005). The tolerant ethanologenic yeast strain Y-50049, which has an enhanced biotransformation ability, showed enzymatic activities for furfural and HMF reduction with both cofactors (Liu *et al.*, 2008b).

Further studies of selected genes responsible for the de-toxification have shown that individual enzymes have different cofactor preferences for the reduction of the same inhibitor. For example, in HMF and furfural reduction, *GRE3*- and *ALD4*-encoding enzymes showed strong reductase activity with NADH, while *ADH6* and *ADH7* functioned better with NADPH. In addition, a single functional gene deletion mutant did not significantly change the tolerance of the yeast to the inhibitors. Therefore, the *in situ* detoxification is due to numerous genes rather than to a single one. As for the cofactor preference, certain single genes also have enzyme activities coupled with both cofactors. The whole cell response in detoxification of the inhibitors therefore reflected the collective activities of all functional enzymes (Liu *et al.*, 2008b), and the reduction of furfural and HMF by multiple enzymes with reductase activities is coupled by both NADH and NADPH. The reduction of furfural and HMF therefore compete for NADH and can inhibit glycolysis.

It was observed that in the presence of these inhibitors, glucose was not utilized until appropriate furfural and/or HMF reduction levels had been reached (Liu *et al.*, 2004), and synergistic inhibition by furfural and HMF has been recognized during the extended lag phase of cell growth (Larsson *et al.*, 1999; Taherzadeh *et al.*, 2000a; Wahlbom and Hähn-Hägerdal, 2002; Liu *et al.*, 2004). For normal cell growth, NAD^+ needs to be regenerated from NADH to enable continued glycolysis. In the presence of the inhibitors, furfural and/or HMF can dominate the competition for NADH when they are at higher concentrations and so delay glycolysis. Once the inhibitors have been converted, glucose utilization can occur (Liu *et al.*, 2004).

It is likely that, with the conversion of furfural into FM and HMF into FDM, NAD^+ regeneration becomes freely available and so allows glucose oxidation in glycolysis. Synergistic competition of NADPH also affects biosynthesis pathways and, as a result, the metabolic process can also be significantly altered and delayed in the presence of the inhibitors. In addition to the toxicity of the inhibitors causing cell damage, furfural and HMF will also affect the cellular redox balance.

Conclusion

Dose-dependent inhibition allows for the potential adaptation of ethanologenic yeast so that it is able to transform the inhibitors furfural and HMF into the less toxic compounds of FM and FDM, respectively. The isolation and identification of the HMF metabolic conversion end product as FDM has clarified existing knowledge and provides a basis for metabolic profiling studies of yeast during inhibitor stress tolerance. A genomic approach is needed for efficient improvement of ethanologenic yeast performance. For high-throughput genomic expression studies, the proper application of quality control measurement is critical in ensuring the reliability and reproducibility of expression data, and to confirm and compare data. Gene expression responses of the ethanologenic yeast to furfural and HMF stress during the fermentation were not transient, and the adaptation to furfural and HMF was a continued dynamic process involving multiple genes. A comprehensive and updated yeast database will allow better global transcriptome profiling of the ethanologenic yeast and provide additional insights into the complexity of adaptation to inhibitor stress.

However, a great deal of information remains unknown and there is only limited functional annotation for some of the significant genes involved in the adaptation. Challenges remain to assign complete functions, draw meaningful conclusions from the complex relationships and assess biological confirmations of gene regulatory networks. Global transcriptome profiling of the tolerant strains is under investigation, and key functional genes and the relevant regulatory components responsible for the biotransformation and de-toxification of the inhibitor will be characterized. A more accurate global account of the genomic mechanism on inhibitor detoxification and tolerance of ethanologenic yeast can be expected from computational modelling of gene regulatory networks. Multiple gene-mediated aldehyde reduction has been demonstrated as a mechanism for the *in situ* detoxification of furfural and HMF. Studies on genomic mechanism of stress tolerance to furfural, HMF and the inhibitory complex involved in bioethanol conversion will be further elucidated to aid more robust strain design and development in the future.

Directed evolutionary genomic adaptation focused on the improvement of specific molecular functions and metabolic dynamics is a powerful means for the improvement and development of desirable strains. Such technology, combined with traditional genetic studies, will bring us to a new horizon in the understanding of ethanologenic yeast. A comprehensive genomic engineering approach will allow us to meet the challenges for efficient lignocellulosic biomass conversion to ethanol into the next decade and beyond.

Acknowledgement

This study was supported by the National Research Initiative of the USDA Cooperative State Research, Education and Extension Service, grant number 2006-35504-17359.

References

Antal, M.J., Mok, W.S.L. and Richards, G.N. (1990) Mechanism of formation of 5-(hydroxymethyl)-2-furaldehyde from D-fructose and sucrose. *Carbohydrate Research* 199, 91–109.

Antal, M.J., Leesomboon, T., Mok, W.S. and Richards, G.N. (1991) Mechanism of formation of 2-furaldehyde from D-xylose. *Carbohydrate Research* 217, 71–85.

Badiee, A., Eiken, H.G., Steen, V.M. and Lovlie, R. (2003) Evaluation of five different cDNA labeling methods for microarrays using spike controls. *BMC Biotechnology* 3, 23.

Baker, S.C., Bauer, S.R., Beyer, R.P., Brenton, J.D., Bromley, B., Burrill, J. *et al.* (2005) The external RNA controls consortium: a progress report. *Nature Methods* 2, 731–734.

Bammler, T., Beyer, R.P., Bhattacharya, S., Boorman, G.A., Boyles, A., Bradford, B.U. *et al.* (2005) Standardizing global gene expression analysis between laboratories and across platforms. *Nature Methods* 2, 351–356.

Banerjee, N., Bhatnagar, R. and Viswanathan, L. (1981) Development of resistance in *Saccharomyces cerevisiae* against inhibitory effects of Browning reaction products. *Enzyme and Microbial Technology* 3, 24–28.

Bothast, R. and Saha, B. (1997) Ethanol production from agricultural biomass substrate. *Advances in Applied Microbiology* 44, 261–286.

Brazma, A., Hingamp, P., Quackenbush, J., Sherlock, G., Spellman, P., Stoeckert, C. *et al.* (2001) Minimum information about a microarray experiment (MIAME) – toward standards for microarray data. *Nature Genetics* 29, 365–371.

Brejning, J., Jespersen, L, and Arneborg, N. (2003) Genome-wide transcriptional changes during the lag phase of *Saccharomyces cerevisiae. Archives of Microbiology* 179, 278–294.

Causton, H.C., Ren, B., Koh, S.S., Harbison, C.T., Kanin, E., Jennings, E.G., Lee, T.I., True, H.L., Lander, E.S. and Young, R.A. (2001) Remodeling of yeast genome expression in response to environmental changes. *Molecular Biology of the Cell* 12, 323–337.

Chen, D., Toone, M.W., Mata, J., Lyne, R., Burns, G., Kivinen, K., Brazma, A., Jones, N. and Bahler, J. (2003) Global transcriptional responses of fission yeast to environmental stress. *Molecular Biology of the Cell* 14, 214–229.

Chung, I.S. and Lee, Y.Y. (1985) Ethanol fermentation of crude acid hydrolyzate of cellulose using high-level yeast inocula. *Biotechnology and Bioengineering* 27, 308–315.

Churchill, G.A. (2004) Using ANOVA to analyze microarray data. *BioTechniques* 37, 173–177.

Dallas, P.B., Gottardo, N.G., Firth, M.J., Beesley, A.H., Hoffmann, K., Terry, P.A., Freitas, J.R., Boag, J.M., Cummings, A.J. and Kees, U.R. (2005) Gene expression levels assessed by oligonucleotide microarray analysis and quantitative real-time RT-PCR – how well do they correlate? *BMC Genomics* 6, 59.

Devantier, R., Pedersen, S. and Olsson, L. (2005) Transcription analysis of *S. cerevisiae* in VHG fermentation. *Industrial Biotechnology* 1, 51–63.

Devaux, F., Carvajal, E., Moye-Rowley, S. and Jacq, C. (2002) Genome-wide studies on the nuclear PDR3-controlled response to mitochondrial dysfunction in yeast. *FEBS Letters* 515(1–3), 25–28.

Dunlop, A.P. (1948) Furfural formation and behavior. *Industrial Engineering and Chemistry* 40, 204–209.

Erasmus, D., van der Merwe, G. and van Vuure, H (2003) Genome-wide expression analysis: metabolic adaptation of *Saccharomyces cerevisiae* to high sugar stress. *Yeast Research* 3, 375–399.

Etienne, W., Meyer, M.H., Peppers, J. and Meyer Jr, R.A. (2005) Comparison of mRNA gene expression by RT-PCR and DNA microarray. *BioTechniques* 36, 618–626.

External RNA Controls Consortium (2005). Proposed methods for testing and selecting the ERCC external RNA controls. *BMC Genomics* 6, 150.

Fisk, D.G., Ball, C.A., Dolinski, K., Engel, S.R., Hong, E.L., Issel-Tarver, L., Schwartz, K., Sethuraman, A., Botstein, D. and Cherry, J.M. (2006) *Saccharomyces cerevisiae* S288C genome annotation: a working hypothesis. *Yeast 23*, 857–865.

Gasch, A. and Werner-Washburne, M, (2002) The genomics of yeast response to environmental stress and starvation. *Functional and Integrated Genomics* 2, 181–192.

Gasch, A.P., Spellman, P.T., Kao, C.M., Carmel-Harel, O., Eisen, M.B., Storz, G., Botstein, D. and Brown, P.O. (2000) Genomic expression program in the response of yeast cells to environmental changes. *Molecular Biology of the Cell* 11, 4241–4257.

Gorsich, S.W., Dien, B.S., Nichols, N.N., Slininger, P.J., Liu, Z.L. and Skory, C.D. (2005) Tolerance to furfural-induced stress is associated with pentose phosphate pathway genes

ZWF1, GND1, RPE1, and *TKL1* in *Saccharomyces cerevisiae. Applied Microbiology and Biotechnology* 71, 339–349.

Hähn-Hägerdal, B., Wahlbom, F., Gárdony, M., Van Zyl, W., Otero, R. and Jönsson, L. (2001) Metabolic engineering of *Saccharomyces cerevisiae* for xylose utilization. *Advances in Biochemical Engineering and Biotechnology* 73, 53–84.

Haugen, A.C., Kelley, R., Collins, J.B., Tucker, C.J., Deng, C., Afshari, C.A., Brown, J.M., Ideker, T. and van Houten, B. (2004) Integrating phenotypic and expression profiles to map arsenic-response networks. *Genome Biology* 5(12), R95.

Ho, N., Chen, Z. and Brainard, A. (1998) Genetically engineered *Saccharomyces* yeast capable of effective cofermentation of glucose and xylose. *Applied and Environmental Microbiology* 64, 1852–1859.

Irizarry, R.A., Warren, D., Spencer, F., Kim, I.F., Biswal, S., Frank, B.C. *et al.* (2005) Multiple-laboratory comparison of microarray platforms. *Nature Methods* 2, 345–350.

James, T.C., Campbell, S., Donnelly, D. and Bond, U. (2003) Transcription profile of brewery yeast under fermentation conditions. *Journal of Applied Microbiology* 94, 432–448.

Jeffries, T. and Shi, Q. (1999) Genetic engineering for improved fermentation by yeasts. *Advances in Biochemical Engineering and Biotechnology* 65, 117–161.

Kanehisa, M., Goto, S., Hattori, M., Aoki-Kinoshita, K.F., Itoh, M., Kawashima, S., Katayama, T., Araki, M. and Hirakawa, M. (2006) From genomics to chemical genomics: new developments in KEGG. *Nucleic Acids Research* 34, D354–D357.

Kang, S. and Okada, H. (1973) Alcohol dehydrogenase of *Cephalosporium* sp. induced by furfural alcohol. *Journal of Fermentation Technology* 51, 118–124.

Kerr, M.K. and Churchill, G.A. (2001) Statistical design and the analysis of gene expression microarray data. *Genetic Research* 77, 123–128.

Khan, Q. and Hadi, S. (1994) Inactivation and repair of bacteriophage lambda by furfural. *Biochemistry and Molecular Biology International* 32, 379–385.

Klinke, H.B., Thomsen, A.B. and Ahring, B.K. (2004) Inhibition of ethanol-producing yeast and bacteria by degradation products produced during pre-treatment of biomass. *Applied Microbiology and Biotechnology* 66, 10–26.

Kuyper, M., Toirkens, M.J., Diderich, J.A., Winkler, A.A., van Dijken, J.P. and Pronk, J.T. (2005) Evolutionary engineering of mixed-sugar utilization by a xylose-fermenting *Saccharomyces cerevisiae* strain. *FEMS Yeast Research* 5, 925–934.

Lage, J.M., Hamann, S., Gribanov, O., Leamon, J.H., Pejovic, T. and Lizardi, P.M. (2002) Microgel assessment of nucleic acid integrity and labeling quality in microarray experiments. *BioTechniques* 32, 312–314.

Larkin, J.E., Frank, B.C., Gavras, H., Sultana, R. and Quackenbush, J. (2005) Independence and reproducibility across microarray platforms. *Nature Methods* 2, 337–344.

Larroy, C., Fernadez, M.R., Gonzalez, E., Pares, X. and Biosca, J.A. (2002) Characterization of the *Saccharomyces cerevisiae* YMR318C (ADH6) gene product as a broad specificity NADPH-dependent alcohol dehydrogenase: relevance in aldehyde reduction. *Biochemical Journal* 361, 163–172.

Larsson, S., Palmqvist, E., Hähn-Hägerdal, B., Tengborg, C., Stenberg, K., Zacchi, G. and Nilvebrant, N. (1999) The generation of inhibitors during dilute acid hydrolysis of softwood. *Enzyme and Microbial Technology* 24, 151–159.

Lee, J., Godon, C., Lagniel, G., Spector, D., Garin, J., Labarre, J. and Toledano, M.B. (1999) *Yap1* and *Skn7* control two specialized oxidative stress response regulons in yeast. *Journal of Biological Chemistry* 274(23), 16040–16046.

Lee, T., Rinaldi, N., Robert, F., Odom, D., Bar-Joseph, Z., Gerber, G. *et al.* (2002) Transcriptional regulatory networks in *Saccharomyces cerevisiae. Science.* 298(5594), 763–764.

Leonard, R.H. and Hajny, G.J. (1945). Fermentation of wood sugars to ethyl alcohol. *Industrial Engineering and Chemistry* 37, 390–395.

Lewkowski, J. (2001) Synthesis, chemistry and applications of 5-hydroxymethylfurfural and its derivatives. *Arkivoc* 1, 17–54.
Liu, Z.L. (2006) Genomic adaptation of ethanologenic yeast to biomass conversion inhibitors. *Applied Microbiology and Biotechnology* 73, 27–36.
Liu, Z.L. and Blaschek, H.P. (2009) Biomass conversion inhibitors and *in situ* detoxification. In: Vertès, A.A., Blaschek, H.P., Yukawa, H. and Qureshi, N. (eds) *Biomass to Biofuels.* John Wiley & Sons, Chichester, UK <In press>.
Liu, Z.L. and Sinha, S. (2006) Transcriptional regulatory analysis reveals PDR3 and GCR1 as regulators of significantly induced genes by 5-hydroxymethylfurfrual stress involved in bioethanol conversion for ethanologenic yeast *Saccharomyces cerevisiae. Microarray Gene Expression Data Society Meeting* 9, 119.
Liu, Z.L. and Slininger, P.J. (2005) Development of genetically engineered stress-tolerant ethanologenic yeasts using integrated functional genomics for effective biomass conversion to ethanol. In: Outlaw, J., Collins, K. and Duffield, J. (eds) *Agriculture as a Producer and Consumer of Energy*. CAB International, Wallingford, UK, pp. 283–294.
Liu, Z.L. and Slininger, P.J. (2006) Transcriptome dynamics of ethanologenic yeast in response to 5-hydroxymethylfurfural stress related to biomass conversion to ethanol. In: Mendez-Vilas, A. (ed.) *Modern Multidisciplinary Applied Microbiology: Exploiting Microbes and Their Interactions*. Wiley-VCH, Weinheim, Germany, pp. 679–684.
Liu, Z.L. and Slininger, P.J. (2007) Universal external RNA controls for microbial gene expression analysis using microarray and qRT-PCR. *Journal of Microbiological Methods* 68, 486–496.
Liu, Z.L, Slininger, P.J., Dien, B.S,, Berhow, M.A., Kurtzman, C.P. and Gorsich, S.W. (2004) Adaptive response of yeasts to furfural and 5-hydroxymethylfurfural and new chemical evidence for HMF conversion to 2,5-bis-hydroxymethylfuran. *Journal of Industrial and Microbiological Biotechnology* 31, 345–352.
Liu, Z.L., Slininger, P.J. and Gorsich, S.W. (2005) Enhanced biotransformation of furfural and 5-hydroxy methylfurfural by newly developed ethanologenic yeast strains. *Applied Biochemistry and Biotechnology* 121–124, 451–460.
Liu, Z.L., Slininger, P.J. and Andersh, B.J. (2006) Induction of pleiotropic drug resistance gene expression indicates important roles of PDR to cope with furfural and 5-hydroxymethylfurufral stress in ethanologenic yeast. In: *27th Symposium on Biotechnology and. Fuels Chemistry*, Denver, Colorado, 1–4 May 2005, p. 169.
Liu, Z.L., Saha, B.C. and Slininger, P.J. (2008a) Lignocellulosic biomass conversion to ethanol by *Saccharomyces*. In: Wall, J., Harwood, C. and Demain, A. (eds) *Bioenergy*. ASM Press, Washington, DC.
Liu, Z.L., Moon, J., Andersh, B.J., Slininger, P.J. and Weber, S. (2008b) Multiple gene-mediated aldehyde reduction is a mechanism of *in situ* detoxification of furfural and HMF by ethanologenic yeast *Saccharomyces cerevisiae. Applied Microbiology and Biotechnology* 81(4), 743–753.
Lucau-Danila, A., Lelandais, G., Kozovska, Z., Tanty, V., Delaveau, T., Devaux, F. and Jacq, C. (2005) Early expression of yeast genes affected by chemical stress. *Molecular Cell Biology* 25, 1860–1868.
Luo, C., Brink, D. and Blanch, H. (2002) Identification of potential fermentation inhibitors in conversion of hybrid poplar hydrolyzate to ethanol. *Biomass and Bioenergy* 22, 125–138.
Martin, C. and Jonsson, L. (2003) Comparison of the resistance of industrial and laboratory strains of *Saccharomyces* and *Zygosaccharomyces* to lignocellulose-derived fermentation inhibitors. *Enzyme and Microbial Technology* 32, 386–395.
Martin, C., Marcelo, M., Almazan, O. and Jonsson, L.J. (2007) Adaptation of a recombinant xylose-utilizing *Saccharomyces cerevisiae* strain to a sugarcane bagasse hydrolysate with high content of fermentation inhibitors. *Bioresource Technology* 98, 1767–1773.

Martinez, A., Rodriguez, M.E., York, S.W., Preston, J.F. and Ingram, L.O. (2000) Effect of $Ca(OH)_2$ treatments ('Overliming') on the composition and toxicity of bagasse hemicellulose hydrolyzates. *Biotechnology and Bioengineering* 69, 526–536.

Modig, T., Liden, G. and Taherzadeh, M. (2002) Inhibition effects of furfural on alcohol dehydrogenase, aldehyde dehydrogenase and pyruvate dehydrogenase. *Biochemical Journal* 363, 769–776.

Mohsenzadeh, M., Saupe-Thies, W., Sterier, G., Schroeder, T., Francella, F., Ruoff, P. and Rensing, L. (1998) Temperature adaptation of house keeping and heat shock gene expression in *Neurospora crassa*. *Fungal Genetics and Biology* 25, 31–43.

Morimoto, S. and Murakami, M. (1967) Studies on fermentation products from aldehyde by microorganisms: the fermentative production of furfural alcohol from furfural by yeasts (part I). *Journal of Fermentation Technology* 45, 442–446.

Mussatto, S.I. and Roberto, I.C. (2004) Alternatives for detoxification of dilute-acid lignocellulosic hydrolyzates for use in fermentative processes: a review. *Bioresource Technology* 93, 1–10.

Nemirovskii, V. and Kostenko, V. (1991) Transformation of yeast growth inhibitors which occurs during biochemical processing of wood hydrolysates. *Gidroliz Lesokhimm Prom-st* 1, 16–17.

Nemirovskii, V., Gusarova, L., Rakhmilevich, Y., Sizov, A. and Kostenko, V. (1989) Pathways of furfurol and oxymethyl furfurol conversion in the process of fodder yeast cultivation. *Biotekhnologiya* 5, 285–289.

Nigam, J. (2001) Ethanol production from wheat straw hemicellulose hydrolysate by *Pichia stipitis*. *Journal of Biotechnology* 87, 17–27.

Nilsson, A., Gorwa-Grauslund, M.F., Hähn-Hägerdal, B. and Liden, G. (2005) Cofactor dependence in furan reduction by *Saccharomyces cerevisiae* in fermentation of acid-hydrolyzed lignocellulose. *Applied and Environmental Microbiology* 71, 7866-7871.

Olsson, L. and Hähn-Hägerdal, B. (1996) Fermentation of lignocellulosic hydrolysates for ethanol production. *Enzyme and Microbial Technology* 18, 312–331.

Ostergaard, S., Olsson, L. and Nielsen, J. (2000) Metabolic engineering of *Saccharomyces cerevisiae*. *Microbiology and Molecular Biology Reviews* 64, 34–50.

Outlaw, J., Collins, K.J. and Duffield, J.A. (2005) *Agriculture as a Producer and Consumer of Energy*. CABI Publishing, Wallingford, UK, 345 pp.

Palmqvist, E. and Hähn-Hägerdal, B. (2000) Fermentation of lignocellulosic hydrolysates II: inhibitors and mechanisms of inhibition. *Bioresource Technology* 74, 25–33.

Palmqvist, E., Almeida, J. and Hähn-Hägerdal, B. (1999) Influence of furfural on anaerobic glycolytic kinetics of *Saccharomyces cerevisiae* in batch culture. *Biotechnology and Bioengineering* 62, 447–454.

Peterson, J., Umayam, L.A., Dickinson, T., Hickey, E.K. and White, O. (2001) The comprehensive microbial resource. *Nucleic Acids Research* 29, 123–125.

Petersson, A., Almeida, J.R.M., Modig, T., Karhumaa, K., Hähn-Hägerdal, B., Gorwa-Grauslund, M.F. and Liden, G. (2006) A 5-hyfroxymethyl furfural reducing enzyme encoded by the *Saccharomyces cerevisiae ADH6* gene conveys HMF tolerance. *Yeast* 23, 455–464.

Saha, B. (2003) Hemicellulose bioconversion. *Journal of Industrial Microbiology and Biotechnology* 30, 279–291.

Sanchez, B. and Bautista, J. (1988) Effects of furfural and 5-hydroxymethylfurfrual on the fermentation of *Saccharomyces cerevisiae* and biomass production from *Candida guilliermondii*. *Enzyme and Microbial Technology* 10, 315–318.

Schena, M., Shalon, D., Davis, R.W. and Brown, P.O. (1995) Quantitative monitoring of gene expression patterns with a complementary DNA microarray. *Science* 270, 467–470.

Sonderegger, M. and Sauer, U. (2003) Evolutionary engineering *of Saccharomyces cerevisiae* for anaerobic growth on xylose. *Applied and Environmental Microbiology* 69, 1990–1998.

Song, M. and Liu, Z.L. (2007) A linear discrete dynamic system model for temporal gene interaction and regulatory network influence in response to bioethanol conversion inhibitor HMF for ethanologenic yeast. *Lecture Notes in Bioinformatics* 4532, 77–95.

Taherzadeh, M., Gustafsson, L. and Niklasson, C. (2000a) Physiological effects of 5-hydroxymethylfurfural on *Saccharomyces cerevisiae*. *Applied Microbiology and Biotechnology* 53, 701–708.

Taherzadeh, M., Niklasson, C. and Liden, G. (2000b) On-line control of fed-batch fermentation of dilute-acid hydrolysates. *Biotechnology and Bioengineering* 69, 330–338.

Teixeira, M., Monteirom C.P., Jain, P., Tenreiro, S., Fernandes, A.R., Mira, N.P., Alenquer, M., Freitas, A.T., Oliveira, A.L. and Sá-Correia, I. (2006) The YEASTRACT database: a tool for the analysis of transcription regulatory associations in *Saccharomyces cerevisiae*. *Nucleic Acids Research* 34, D446–D451.

The Gene Ontology Consortium (2000) *Nature Genetics* 25, 25–29.

Villa, G.P., Bartroli, R., Lopez, R., Guerra, M., Enrique, M., Penas, M., Rodriquez, E., Redondo, D., Iglesias, I. and Diaz, M. (1992) Microbial transformation of furfural to furfuryl alcohol by *Saccharomyces cerevisiae*. *Acta Biotechnologia* 12, 509–512.

Wahlbom, C.F. and Hähn-Hägerdal, B. (2002) Furfural, 5-hydroxymethylfurfrual, and acetone act as external electron acceptors during anaerobic fermentation of xylose in recombinant *Saccharomyces cerevisiae*. *Biotechnology and Bioengineering* 78, 172–178.

Wheals, A.E., Basso, L.C., Alves, D.M. and Amorim, H.V. (1999) Fuel ethanol after 25 years. *Trends in Biotechnology* 17, 482–487.

YEAst Search for Transcriptional Regulators and Consensus Tracking (YEASTRACT) (2006) http://www.yeastract.com (accessed 12 September 2006).

Zaldivar, J., Nielsen, J. and Olsson, L. (2001) Fuel ethanol production from lignocellulose: a challenge for metabolic engineering and process integration. *Applied Microbiology and Biotechnology* 56, 17–34.

Zhang, W., Needham, D.L., Coffin, M., Rooker, A., Hurban, P., Tanzer, M.M. and Shuster, J.R. (2003) Microarray analysis of the metabolic response of *Saccharomyces cerevisiae* to organic solvent dimethyl sulfoxide. *Journal of Industrial Microbiology and Biotechnology* 30, 57–69.

9 Spoilage Yeasts and Other Fungi: their Roles in Modern Enology

Manuel Malfeito-Ferreira and Virgílio Loureiro

Departamento de Botânica e Engenharia Biológica, Instituto Superior de Agronomia, Technical University of Lisbon, Portugal

Introduction

The basic principle of crushing grapes and converting the grape juice into wine has been known for millennia. Wine has been known as 'the safe beverage' since the time of early civilisations, and one factor in the historical success of wine (and beer) in the Mediterranean basin has been the poor sanitary quality of drinking water of early towns and cities, which was frequently contaminated by salmonella or other faecal pathogens. In the second half of the 19th century, Louis Pasteur delivered his famous statement: 'Wine is the most hygienic of the beverages', by confirming that pathogenic bacteria did not have the capacity to grow or survive in wine in their vegetative form.

However, recent attention has been directed to food safety issues, such as the occurrence of the toxic substances ethyl carbamate, biogenic amines and ochratoxin A in wines and beers. These provide new challenges due to their potential detrimental effects on consumers' health. Of these three compounds only ochratoxin A is produced by filamentous fungi. Yeasts have only a minor (if any) role in the metabolism of these molecules and, in wine, the only situation where yeasts may pose an indirect health threat is related to the explosion of bottles due to re-fermentation of sweet wines. This rare type of incident, which can lead to eye injuries, has been described mainly in carbonated beverages (Kuhn *et al.*, 2004). Therefore, the most relevant current safety concerns are related to the potential activities of filamentous fungi, while the spoilage of wine is mainly influenced by yeast activity. This chapter addresses the main scientific and technological threats related to these two groups of microorganisms in winemaking.

Fungal Threats in Modern Winemaking

Ochratoxin A

Some mycotoxins, including patulin, aflatoxins and trichothecin, have been reported in grape products, but ochratoxin A (OTA) is the main toxin of concern in the wine industry (Hocking *et al.*, 2007). OTA was first reported in wine in 1995 (Zimmerli and Dick, 1996), and considerable data have since been generated for wine, grape juice and raisins, especially in the new millennium (see Chapter 3, this volume).

Nature of ochratoxin A

Ochratoxin A was originally described as a metabolite of *Aspergillus ochraceus* in a laboratory screening for toxinogenic fungi (van der Merwe *et al.*, 1965). It is the most abundant and hence the most commonly detected member of the family of ochratoxins, a group of three mycotoxins (A, B and C) produced as secondary metabolites by fungi. All of these are weak organic acids consisting of an isocoumarin derivative (see Fig. 9.1), and differ slightly from each other in their chemical structure (van der Merwe *et al.*, 1965). These differences, however, have marked effects on their respective toxic potentials. OTA is a mycotoxin considered as being a possible carcinogen for humans (IARC, 1993), and it has been shown to be nephrotoxic, hepatotoxic, teratogenic, carcinogenic and immunotoxic to several species of animals, and to cause kidney and liver tumours in mice and rats (JECFA, 2001).

The European Food Safety Authority (EFSA) has advised a tolerable weekly intake for OTA of 120 ng/kg body weight (b.w.) (EFSA, 2006). This is higher than the average dietary exposure for adult Europeans, which reach to 60 ng/kg b.w. (Mateo *et al.*, 2007). The maximum allowable limit in the European Union (EU) has been set at 2.0 μg OTA/kg of wine. This limit does not apply to dessert or liquor wines of more than 15% (v/v) ethanol (Mateo *et al.*, 2007). An evaluation of OTA in wines worldwide has shown various levels of contamination, but relatively few had levels exceeding 2.0 μg OTA/kg (Mateo *et al.*, 2007). High concentrations were more frequent in wines

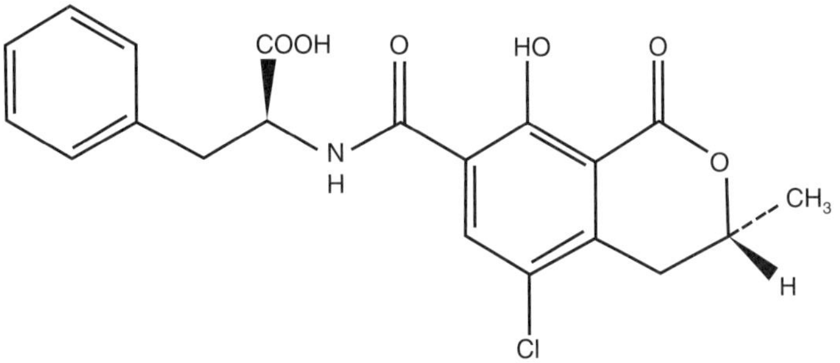

Fig. 9.1. Chemical structure of ochratoxin A.

produced from dried grapes and in raisins, and so wines may not represent a significant contribution to human exposure to OTA. Cereals are regarded as the major source of OTA intake in humans, due to their prevalence in diets.

Ochratoxin A-producing fungi on grapes

The main OTA-producing fungi that occur in grapes and, consequently, in grape juices, raisins, wine and wine derivatives, belong to the commonly termed 'black aspergilla', which are placed in *Aspergillus* section *Nigri*. The taxonomy of this group is somewhat uncertain, and OTA-producing strains have been identified to a number of taxa, including species, subspecies and varieties (for references see Samson *et al.*, 2004). In general, *A. carbonarius* is highly dominant, particularly in warmer regions, but in cooler regions there have been some reports of *A. ochraceus* and *Penicillium* species on grapes, although their significance is unclear.

Prevention and control of OTA production on grapes

The primary sources of *A. carbonarius* and *A. niger* on vines are from soil and soil remnants (Hocking *et al.*, 2007). Generally, colonization of grape bunches by black aspergilli and other fungi occurs when damage to the skin of the fruit allows entry to fruit tissues, where the low pH and high sugar content provide a favourable environment. As a rule, the competition among contaminant microorganisms is more favourable to *Botrytis cinerea*, the agent responsible for common grey rot. It has been reported that high temperatures (30°C) and high relative humidity (80–100%) give rise to higher amounts of OTA from *A. carbonarius* in grapes (Bellí *et al.*, 2007), suggesting that such conditions give competitive advantages to the black aspergilli population. The influence of high relative humidity seems to be less important than the influence of high temperature, and this may explain the occurrence of this species in hot and dry climates (Hocking *et al.*, 2007).

Zimmerli and Dick (1996) were the first authors to show that the OTA content of wines from southern wine-growing regions was higher than those of wines from northern areas, and these results were supported by the extensive work of Otteneder and Majerus (2000). More recently, OTA has been detected on grapes produced in many wine-growing countries, and the highest levels of OTA detected in each survey generally correlated to vines growing in the warmest zones of each geographical region. However, this observation was not seen in Australia and South Africa, where no correlation was found between OTA incidence and wine region or type (red or white) (Leong *et al.*, 2006a), although in Australia incidences were apparently lower in the cooler climate of Tasmania (Hocking *et al.*, 2007).

Given the ubiquity of black aspergilli in vineyards of warm regions, the wine industry must learn to live with them, and to minimize all biotic and abiotic factors that contribute to their growth on grapes (see Fig. 9.2). As a result, a code of good viticultural practices has been recommended by the

Office International de la Vigne et du Vin (OIV, 2005) for use mainly in areas with a high occurrence of *A. carbonarius*. The main measures considered within this publication are:

- Avoidance of all of the cropping practices that lead to an excessive vigour of vines and an exaggerated increase in yield, which makes bunches more compact and thereby susceptible to berry splitting (Leong *et al.*, 2006b).
- Avoidance of cultivars with thin-skinned berries and very compact bunches.
- Undertaking a plant sanitation programme through usage of efficient products directed against fungi – particularly powdery mildew and grey rot – and insects, mainly *Lobesia* spp., employing adequate dosage and timing while ensuring that the active ingredients reach all parts of the bunch, as well as penetrating its interior (Mínguez *et al.*, 2004).
- Controlling berry splitting due to rain just prior to harvest (JECFA, 2001).

OTA control strategies in wine

The presence of OTA in wines results exclusively from grape contamination. Thus, the principal measure in preventing grapes with high levels of OTA from entering wineries is to reject the rotten bunches, and particularly those that have brown or black rot. This is difficult to implement during harvest, as rot is not easily visible due to its development in the inner part of the bunch. Apart from South African and Australian wines (Leong *et al.*, 2006a), most studies indicate that red wines contain higher OTA concentrations than rosé and white wines from the same region (for references see the review of Varga and Kozakiewicz, 2006). Therefore, the winemaking processes could significantly influence OTA content.

Several studies have shown that the OTA content in wines increases with the maceration time and decreases with solid–liquid separation steps, such

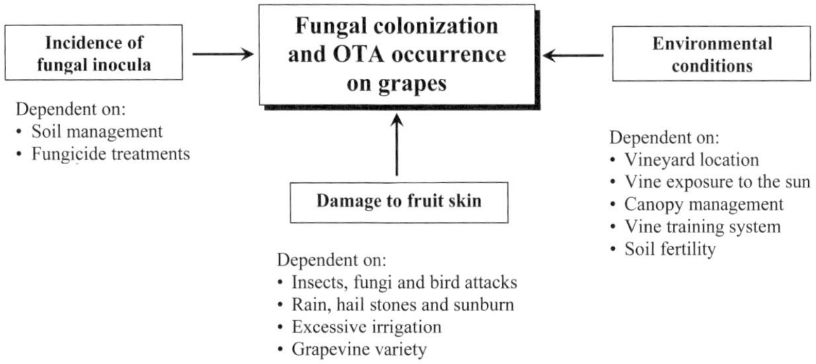

Fig. 9.2. Factors affecting grape colonization by OTA-producing fungi.

as pressing the juice or wine from the skins and decanting the wine from the precipitated solids (Fernandes *et al.*, 2003; Leong *et al.*, 2006a). Such decreases may be as much as 50–80% of the initial OTA concentration (Hocking *et al.*, 2007). Some studies have been undertaken to assess the effect of enological adjuvants (fining agents), such as bentonite, gelatin, charcoal and yeast cell wall preparations, on the removal of OTA from wines (Leong *et al.*, 2006a; Mateo *et al.*, 2007). Fernandes *et al.* (2007) showed that the mean carry-over of OTA from grapes to wine was 8.1% (w/w) after malolactic fermentation, even without the use of enological adjuvants. In addition, spontaneous reduction of up to 29% of OTA content has been observed during wine storage over 10–14 months (Hocking *et al.*, 2007).

The most important measures for a code of good enological practice to prevent or reduce the OTA content in wines are:

- Training of harvesters to reject rotten bunches, particularly those affected by dark brown or black moulds.
- In large wineries and cooperatives, monitoring of the sanitary quality of the grapes by FTIR (Fourier transform infrared spectroscopy) instruments and favouring the sanitary status of grapes and the crop by pricing strategies, particularly in cooperatives.
- Processing of mechanical harvested grapes separately when the sanitary quality of the crop is bad.
- Avoidance of long periods of maceration and inclusion of enological adsorbents, such as activated charcoal or yeast cell walls in red wine, and bentonite in white wine, when the crop has a relevant percentage of rotten grapes.
- Rapid drying of grapes used for producing dessert wines.
- Implemention of a complete hazard analysis and critical control point (HACCP) plan, from the vine to the bottled wine, in regions with high levels of OTA occurrence.

Yeast Threats to Wine Quality

The concept of spoilage yeasts

In practice only about a dozen yeast species are detrimental to food quality. Among the common yeast contaminants, those that can survive in foods but are unable to grow are called *adventitious*, *innocuous* or *innocent* yeasts. Those responsible for unwanted modifications of the final product, such as visual, textural or organoleptical (producing off-flavours or off-tastes) effects are called *spoilage* yeasts. However, for technologists the concept of spoilage yeast is narrower and includes only species that are able adversely to modify foods processed according to the standards of good manufacturing practices (GMPs) (Pitt and Hocking, 1985). These are the spoilage yeasts, *sensu stricto*, and comprise the species most resistant to the stresses produced by food or beverage processing (Loureiro and Querol, 1999).

In the wine industry the definition of spoilage is not always clear cut, particularly as microbial metabolites contribute to wine flavours and aroma, and the desirability of this is driven by many subjective factors (e.g. habits, fashions, opinion makers' choices) that influence consumer taste. This situation is clearly demonstrated by the presence of volatile phenols produced by the species *Dekkera bruxellensis* in red wines (see below).

Significance and occurrence of wine-related yeast species

The most commonly recognized symptoms of yeast spoilage are film formation in bulk wines, cloudiness, sediment formation, gas production in bottled wines and off-flavour production during all processing and storing stages (Loureiro and Malfeito-Ferreira, 2003a). The species that cause the most problems in the wine industry are listed in Table 9.1, together with the associated spoilage effects caused, risk of occurrence and remedial treatments.

Pre-fermentation and fermenting activities

GRAPES IN THE VINEYARD The natural microbial population of healthy grapes comprises fungi, yeasts and bacteria, and these generally do not cause any reduction in wine quality. Common yeast contaminants of grapes and grape juices before fermentation include the genera *Candida*, *Cryptococcus*, *Debaryomyces*, *Hansenula*, *Kloeckera/Hanseniaspora*, *Metchsnikowia*, *Pichia*, *Rhodotorula* and the yeast-like fungus *Aureobasidium pullulans* (Fleet *et al.*, 2002). These species are inhibited during wine fermentation carried out by *S. cerevisiae* or *S. bayanus*. If adequate fermentation conditions are provided (e.g. nutrient level, temperature) the final product is most likely to be achieved satisfactorily without any problems. The situation is, however, different when the grapes are of poor quality. It is common knowledge that grapes affected by plant-pathogenic fungi (rotten grapes) produce lower-quality wines, and the decrease in quality is dependent on both the type of rot and the percentage of affected grapes.

The main measures used in winemaking to overcome problems involve grape selection, increased sulfur dioxide usage and prompt inoculation with active starters. The selection of grapes can be based on chemical indicators that are related to the microbiological quality of the grapes. Such measures such as laccase activity (an indicator of grapes affected by grey rot) or volatile acidity and gluconic acid (indicators of grapes affected by sour rot) are already used in numerous wineries, particularly in cooperatives or large companies where the condition of the grapes is used to establish the price paid to growers. The use of FTIR instruments makes these determinations readily available, and this knowledge can be used to establish different processing procedures according to the quality of the raw material. In smaller wineries, grape selection can be undertaken by visual inspection and the removal or separate processing of poor-quality grapes. Microbiological quality control techniques are not necessary at this stage.

Table 9.1. Most frequent wine spoilage events and modes of prevention.

Yeast species	Spoilage event	Prevention	Risk
Pichia anomala, *Kloeckera apiculata*	Production of ethylacetate during juice settling or grape maceration	Addition of sulphur dioxide; decrease in temperature; prompt starter inoculation	Low
Pichia spp., *Candida* spp.	Formation of films with production of off-flavours	Remove oxygen by topping; addition of sulphur dioxide	Low
Saccharomyces cerevisiae	Production of sulphur-reduced compounds;	Correct juice nutritional status Treat wine when detected	Low/medium
	Re-fermentation of sweet wines; cloudiness and re-fermentation in bottled wine	Keep microbial load low; use sulphur dioxide, DMDC or thermal treatments	Medium/high
Saccharomyces ludwigii	Re-fermentation of sweet wines; cloudiness and re-fermentation in bottled wine	Keep microbial load low; use sulphur dioxide, DMDC or thermal treatments	Medium/high
Zygosaccharomyces bailii	Cloudiness and sediments in bottled wine	Keep microbial load low; use sulphur dioxide, DMDC or thermal treatments	Medium/high
Dekkera bruxellensis	Volatile phenol production in stored or bottled red wines	Keep microbial load low; use sulphur dioxide, DMDC or thermal treatments	Medium/high in stainless steel stored red wines; high in barrel-matured red wine

DMDC, dimethyl dicarbonate

GRAPE JUICES Spoilage events are rare in grape juices, particularly given the short time period before fermentation. However film-forming yeasts (e.g. *Pichia anomala*) or apiculate yeasts can grow rapidly in white grape juices, with long settling periods or long skin contact times, and in the lengthy red wine pre-fermentative maceration stage. These yeasts are easily controlled by normal winemaking measures (low temperature, sulfur dioxide, hygiene). In principle these species are inhibited during fermentation but, due to their fast growth, they can produce unwanted metabolites such as ethylacetate (which gives a vinegary smell) or acetaldehyde (which gives an oxidized taint) that can irreversibly spoil the wine in a short time period (Romano, 2005). Given the higher load of contaminating yeast species, the processing of juice from rotten grapes requires prompt induction of fermentation by active starters.

FERMENTATION Fermentation problems with sound grape juices are usually related to the activity of the fermenting yeasts (*S. cerevisiae* or *S. bayanus*). The production of off-flavours (reduced sulfur compounds) (Bell and Henschke, 2005) and acetic acid is due to nutritional imbalances or deficient fermenting conditions (e.g. high temperature) that may lead to stuck fermentations (Bisson and Butzke, 2000). These events are results of environmental conditions and not particular yeast spoilage characteristics. They can be overcome by the correct management of fermenting conditions. Stuck wines are highly susceptible to bacterial (lactic and acetic acid bacteria) spoilage, making it difficult to achieve wine stability during storage. The influence of contaminant yeast species during fermentation is limited or unknown, but spoilage species such as *Zygosaccharomyces bailii* survive in low numbers during fermentation and can cause problems in the later stages of wine production.

Post-fermentation activities

BULK STORAGE AND BOTTLING The most frequent and difficult problems with regard to yeast spoilage occur in wines after fermentation, during storage or bottling. During bulk wine storage, film-forming yeasts (e.g. *Candida* spp., *Pichia* spp.) may form pellicles on the surface and spoil the wine by the production of active odour compounds. The absence of oxygen and appropriate sulfur dioxide usage or hygiene measures will prevent the formation of films. These yeast species are frequently encountered in bottled wine, but they have low resistance to the stresses present in the bottled product and so tend to die or remain dormant when in low numbers.

The classical spoilage events due to spoilage yeasts, *sensu stricto*, occur in bottled wine, and the typical members of this group – *Z. bailii*, *S. cerevisiae*, *Schizosaccharomyces pombe* and *Saccharomycodes ludwigii* – can grow in this environment. Although these species are not frequent grape or winery contaminants, their stress resistance enables them to survive and proliferate under conditions that are not tolerated by other species. Re-fermentation problems in red wines have increased in recent years, due to the addition of concentrated grape juices to make softer wines and meet modern market demands. This problem is easily recognized by swollen packages in bag-in-box wines. In bottled wines, gas production and wine turbidity are also easily visible after the bottle has been opened and when the wine is poured. In red wines the effect of sediment formation or haziness is less obvious due to the darker colour.

4-Ethylphenol production

In the new millennium *Dekkera bruxellensis* has provided a new challenge in red wine spoilage, due to its ability to produce volatile phenols that are active odour compounds in bulk or bottled wines (Loureiro and Malfeito-Ferreira, 2006). Its effects are particularly problematic in premium red wines aged in costly oak barrels, and this considerably increases the economic losses from spoilage yeasts. At the current time this species is regarded as the main yeast threat to wine quality. In addition to the direct effect of volatile phenol production, there is also an indirect effect in the technological measures needed to control its activity. The onset of these problems has coincided with increased public exposure to wines and, while some consumers and opinion makers prefer wines tainted by volatile phenols, others consider that, even at low concentrations, these compounds diminish flavour complexity and reduce wine quality.

Wine spoilage by volatile phenols produced by *D. bruxellensis* causes the taint commonly described as 'horse sweat', and the processes involved have been understood only relatively recently. Production of volatile phenols is due to two enzymes, which decarboxylate hydroxycinnamic acids present in grape juices or wines (ferulic, *p*-coumaric and caffeic acids) into hydroxystyrenes (4-vinylguaiacol, 4-vinylphenol and 4-vinylcathecol, respectively). These hydroxystyrenes are then reduced to ethylphenols (4-ethylguaiacol, 4-ethylphenol and 4-ethylcathecol, respectively). This latter step is not frequent in wine-related yeasts but is particularly effective in *D. bruxellensis* and *Pichia guilliermondii*. *P. guilliermondii* is commonly present in grapes, grape juices and winery equipment (Dias *et al.*, 2003), but it does not grow and/or produce 4-ethylphenol in wines with average ethanol levels of 12% (v/v) (Barata *et al.*, 2006). Therefore, *D. bruxellensis* is the main agent of phenolic taint in wines.

A preference threshold can be used to measure the spoilage effect of compounds on the quality of wine. A preference threshold is defined as the minimum concentration at which 50% of the tasters, in a 70-person jury, reject the sample (Chatonnet *et al.*, 1992). The preference thresholds for 4-ethylphenol and for a mixture (10:1) of 4-ethylphenol and 4-ethylguaiacol in Bordeaux red wines are about 620 and 426 μg/l, respectively (Chatonnet *et al.*, 1992). Below these concentrations, volatile phenols may contribute favourably to the complexity of wine aroma by imparting flavours appreciated by many consumers, such as spices, leather, smoke or game. Concentrations above the preference thresholds result in wines that are clearly sub-standard for some consumers but remain acceptable to others. The definition of spoilage is further complicated, as thresholds are also dependent on both grapevine variety and the style of wine (Coulter *et al.*, 2003).

Analysis of volatile phenols from different countries showed that more than 25% of the red wines had levels of 4-ethylphenol higher than the preference threshold of 620 μg/l (see review of Loureiro and Malfeito-Ferreira, 2006). Although these results were not based on random sampling, they demonstrate that volatile phenols are a worldwide concern.

Prevention and Control of Spoilage Yeasts

Yeast monitoring

The conservative nature of the wine industry, together with the absence of microbiological safety hazards, has resulted in a general lack of strictness in implementing HACCP and control plans, processes that are mandatory in most food industries. In fact, the good microbial stability of most dry table wines that is attained when good winery practices are followed leads to the absence of microbiological control by most producers. Exceptionally, commercial contracts with modern distributors (supermarket chains and others) may require some routine microbiological analysis. As a result, microbiological control in the wine industry is, as a rule, synonymous with microbiological assessment (particularly of yeasts) in bottled sweet wine processing, where the risk of re-fermentation is high.

Following wine fermentation, most wineries use qualitative or quantitative chemical indicators to control the activity of lactic acid bacteria (malic acid assessment) and acetic acid bacteria (volatile acidity). It is not current practice to monitor the presence of spoilage yeasts. Microbiological analysis is not a requirement, but visual inspection of tank tops every two weeks is a simple and effective practice. In white wines, particularly those with residual sugar, the specific detection of *Z. bailii* or *S. cerevisiae* should be considered, as these may cause re-fermentation problems during storage.

Monitoring of D. bruxellensis

The relatively low demand for microbiological control during bulk wine storage is no longer advisable for red wines, particularly those retained for ageing. Presently, the detection of *D. bruxellensis* during all processing stages of premium red wines is a prerequisite for wineries. This species is frequently present at high levels just after the malolactic fermentation stage (Rodrigues *et al.*, 2001), leading to a premature 'horse sweat' taint. It is essential to monitor *D. bruxellensis* periodically during barrel ageing, irrespective of grape quality. This is particularly important when used barrels are in use, as these are a well-known ecological niche for these yeasts.

Microbiological criteria have been established that have been giving adequate results for many Portuguese wineries, and these are given here only as guidelines. In the first instance, bulk-stored wines should be screened for *D. bruxellensis* monthly, bimonthly or even every 3 months (according to the type of wine and container). Initial sample volumes of 1.0, 0.1, 0.01 and 0.001 ml should be taken from the air/liquid interface and from different depths in the storage container. When the result is positive for 1 ml, or less, and the level of 4-ethylphenol is higher than 150 μg/l, a fine filtration is recommended immediately, accompanied by sulfite addition. In subsequent analyses after filtration, it is generally sufficient to monitor the level of 4-ethylphenol. Prior to bottling the criteria are more stringent, and should involve 100, 10 and 1 ml aliquots of wine, sampled as above. When the result is positive in 1 or 10 ml, a very fine or sterilizing filtration is recommended. If only the 100 ml sample is positive then it is acceptable to control any viable yeast cells by the addition of preservatives (e.g. 1 mg/l of molecular sulfite). In this case, bottling must be technically correct and dissolved oxygen should be lowered to practically zero. Otherwise, a sterile filtration is recommended or, as an alternative, a light heat treatment of the wine to destroy viable cells.

Wine bottling

Wine bottling is the main part of the wine production process where conventional microbiological control may be implemented by wineries. Common procedures include the analysis of bottles, rinsing water, closures (corks, rip caps), bottling and corking machines and the atmosphere. When applied properly these controls can identify sources of contamination, and can help in determining corrective measures. Most frequently, the contamination sources are associated with the filling and corking machines (Loureiro and Malfeito-Ferreira, 2003b). The final analysis is the evaluation of any contamination in the bottled wine (Loureiro and Malfeito-Ferreira, 2003a). Common microbial contaminants do not survive for a long time after bottling, and if microbial counts are higher than the specifications the product can be retained until levels fall (see below).

Tools used in microbiological control in wineries

As a rule, yeast detection and enumeration methodologies are based on membrane filtration of wine samples or rinsing solutions, and on subsequent microbial growth on plates containing a general-purpose culture medium (Loureiro *et al.*, 2004). The use of the Most Probable Number (MPN) technique is not common, but it may prove useful when an estimation of yeast contamination in bulk wine is desired, or in wines with a high percentage of suspended solids.

Selective and/or differential culture media have only recently been used to any extent, mainly due to problems with *D. bruxellensis.* This situation is clearly distinct from the typical bacterial control of other food industries, where bacterial indicators based on a wide variety of selective/differential media play a central role. Accordingly, the use of zymological (zymo = yeast) indicators in wineries has been proposed in order to increase the utility of routine microbiological control (Loureiro and Querol, 1999).

The major purpose of zymological indicators is to measure the hygienic quality of surfaces that come into contact with wine and to assess the spoilage risks involved. Based on the wine yeast groups defined earlier, the hygienic quality of wine processing may be assessed by the detection and enumeration of film-forming yeasts by the MPN technique with a general-purpose culture medium. Selective/differential media have been developed to detect spoilage yeasts *sensu stricto*, and these are directed towards the most significant species, *Z. bailii* and *D. bruxellensis* (Schuller *et al.*, 2000; Rodrigues *et al.*, 2001). The presumptive results obtained from culture media can be further confirmed if necessary, by biomarkers (biomolecule indicators) such as long-chain fatty acids (Malfeito-Ferreira *et al.*, 1997) or through molecular biological identification. Chemical indicators can also provide a simple and rapid method for monitoring yeast activity. 4-ethylphenol is currently the most common indicator of *D. bruxellensis* activity, and should be used together with microbiological detection.

Acceptable levels of yeasts

The determination of acceptable levels of microorganisms in the final product is a common concern for many food industries. In foods harbouring pathogenic microorganisms, the law regulates the permitted levels and the technologist must comply. However there is very little existing legislation for yeasts, and the technologist must establish levels that are attainable under their industrial conditions and that ensure stability of the product during its shelf life (Loureiro and Malfeito-Ferreira, 2003a).

In the case of *Z. bailii*, one viable cell per bottle may cause spoilage (Davenport, 1986; Deak and Reichart, 1986; Thomas, 1993), but such a strict limit is difficult to attain in winery practice and is not at all appropriate when yeast counts may be due to other non-spoilage-causing contaminants. Occasionally, specifications are established as a function of the wine sugar content, assuming that sweet wines are more vulnerable than dry ones. However, Deak and Reichart (1986) demonstrated that product stability depends on the initial yeast population, and that there are no differences in the microbial stability of white, red, semi-dry and semi-sweet wines.

In the absence of a sound scientific background to establish appropriate specifications, the industry has established its own empirical limits that may be used for commercial purposes. Yeast counts as low as < 1 cell/500 ml or < 1 cell/ml are currently regarded as maximum acceptable levels (Andrews, 1992; Loureiro and Malfeito-Ferreira, 2003a), reflecting the level of caution necessary to prevent spoilage events. When levels are higher than acceptable, most wineries hold the product for enough time to meet specifications or to re-bottle it.

Microbial guidelines based on sampling plans defined by microbial attributes, similar to those applied in most food industries (Adams and Moss,

2000), could be an important improvement on wine microbiological control. Although these guidelines were devised for food pathogens, they may also be appropriate when dealing with spoilage situations. In an attribute sampling scheme, analytical results are assigned into two or three classes. In the two-class scheme, samples are either acceptable or unacceptable, and a sample fails if it contains more than a specified number of the targeted microorganism. In the three-class scheme samples may be acceptable, marginally acceptable or unacceptable. The two-class plans are more stringent and could be applied to the most significant and frequent species, such as *Z. bailli* and *D. bruxellensis*. Similar plans could be developed for other significant species, such as the frequent *S. cerevisiae* and the rare *Sch. pombe* and *S. ludwigii*, once appropriate culture media are developed.

When products are contaminated only by innocuous yeasts that will not grow in bottled wines, a higher number could be tolerated and a three-class plan is advised. The analysis should also be performed 24 h after bottling, to give time for any preservatives added to inactivate the microbial population. A number of guidelines based on experience and data provided by wineries, that could be applied industrially for the finished bottled product before shipment, are proposed in Table 9.2 (Loureiro and Malfeito-Ferreira, 2003a). A wine lot can be defined as the number of samples (bottles) per bottling day of a single wine. If a wine lot does not meet the specifications it should be held and analysed on a weekly basis until counts drop to acceptable levels. If counts in any of the indicators increase, the bottled wine should be re-processed.

Yeast control

The enologist may use a wide range of measures to prevent yeast spoilage in wine. These include inhibiting or killing microbes with chemical preservatives (sulfur dioxide, sorbic acid, DMDC-dimethyldicarbonate) and thermal treatments (flash pasteurization, hot bottling). Other physical operations involve the removal of the microbes from the wine, such as clarification, fining or filtration. All control operations must be accompanied by adequate hygiene procedures to prevent wine contamination by yeasts from other sources.

In the wine industry there are a number of hurdles associated with preservation methods, and one example is the need to decrease the use of sulfur dioxide, which has been associated with human allergies and subjected to stricter legal limits. However, decreases in sulfite usage are not always possible. In modern winemaking, oak barrels are used widely, particularly for high-quality red wines, and pose the main difficulty in preventing wine contamination by *D. bruxellensis*. Sterilizing agents with chlorine must be avoided to prevent the formation of the trichloroanisoles responsible for 'cork taint'. Most common treatments use hot water, sulfite solutions, steam and ozone as cleaning and disinfecting agents. However, their efficiency is very limited due to the porous nature of the wood. The contamination of the

Table 9.2. Proposed zymological guidelines (CFU/100 ml) for release of bottled wines to the market (from Loureiro and Malfeito-Ferreira, 2003a).

Wine style	Yeast indicator	Culture media[a]	Plan class	n[b]	m[c]	M[d]	c[e]
Sweet wines (sugar > 2 g/l)							
White, rosé and red	Total yeasts	WLN	3	5	10^0	10^1	2
	Z. bailii	ZDM	2	5	0	–	0
Dry wines (sugar < 2 g/l)							
White and light rosé	Total yeasts	WLN	3	5	10^1	10^2	2
	Z. bailii	ZDM	2	5	0	–	0
Red and dark rosé	Total yeasts	WLN	3	5	10^2	10^3	
Red (oak aged)	Total yeasts	WLN	3	5	10^1	10^3	2
	D. bruxellensis	DBDM	2	10	0	–	0

[a] WLN, Wallerstein laboratory nutrient; ZDM, *Zygosaccharomyces* differential medium; DBDM, *Dekkera/Brettanomyces* differential medium.
[b] Number of samples (bottles) to be taken from a lot.
[c] A measure that separates good quality from marginal quality.
[d] A measure which, if exceeded by any of the tested bottles, would lead to holding of the product.
[e] Maximum number of bottles that may fall into the marginally acceptable category before the lot is held.

outer layers of the wood may be significantly reduced but the inner layers, soaked by wine, can still harbour yeast populations that are able to re-contaminate wine after cleaning (Laureano *et al.*, 2004). When growing populations of *D. bruxellensis* are detected, it is necessary to use high doses of sulfur dioxide (e.g. 1 mg/l molecular form) (Barata *et al.*, 2007) together with DMDC, where legally authorized (Costa *et al.*, 2007), to inhibit yeast proliferation and prevent volatile phenol production.

Conclusions

In the new millennium there has been a rapid response from the scientific community to the problems of OTA and volatile phenols in wines. OTA studies have received increasing attention due to its effect on human health, and these have provided information about both its incidence and significance to human consumption. Volatile phenols are much more important to wine quality than OTA but, as they do not pose a human health problem, studies on wine spoilage by yeasts are relatively less developed.

Scientific studies on wine stability should be directed at establishing the vulnerability of different wines to microbial spoilage. Even without the use of preservatives, wine is a stressful environment for microbes, and many spoilage yeasts will not survive in the final product. As a result, some wines seem to be immune to spoilage without the need for any particular preventive measure, while others are readily spoiled if proper attention is not given to storage conditions. The concept of wine robustness could therefore be used as a measure of its resistance to microbial spoilage. The assessment of wine robustness could be used to establish an appropriate level of preventive measures, as well as to set the microbial guidelines to be used in microbiological control. Likewise, the evaluation of wine robustness could provide a useful tool in establishing predictive models of wine spoilage and to provide better support for risk assessment in HACCP plans in the wine industry. This type of plan requires prompt corrective measures when process deviations are detected.

Nevertheless, the wine industry does not have the routine analytical techniques required for rapid detection of either fungal metabolites or spoilage microorganisms. Therefore, the primary efforts to improve quality assurance in the future could be:

- Defining microbiological criteria in order to develop microbiological indicators.
- Standardizing microbiological criteria, such as attribute-based sampling, and developing standard analytical methods that enable the definition of appropriate microbiological specifications.
- Defining spoilage risks according to the concept of 'wine robustness' that determine the choice of appropriate microbiological specifications.
- Using rapid detection methods to enable prompt corrective measures.

References

Adams, M.R. and Moss M.D. (2000) *Food Microbiology*. The Royal Society of Chemistry Publishers, Cambridge, UK, pp. 395–409.

Andrews, S. (1992) Specifications for yeasts in Australian beer, wine and fruit juice products. In: Samson, R.A., Hocking, A.D., Pitt, J.I. and King, A.D. (eds) *Modern Methods in Food Mycology*. Elsevier, Amsterdam, pp. 111–118.

Barata, A., Correia, P., Nobre, A., Malfeito-Ferreira, M. and Loureiro, V. (2006) Growth and 4-ethylphenol production by the yeast *Pichia guilliermondii* in grape juices. *American Journal of Enology and Viticulture* 57, 133–138.

Barata, A., Caldeira, J., Botelheiro, R., Pagliara, D., Malfeito-Ferreira, M. and Loureiro, V. (2007) Survival patterns of *Dekkera bruxellensis* in wines and inhibitory effect of sulphur dioxide. *International Journal Food Microbiology* 121, 201–207.

Bell, S.-J. and Henschke, P.A. (2005) Implications of nitrogen nutrition for grapes, fermentation and wine. *Australian Journal of Grape and Wine Research* 11, 242–295.

Bellí, N., Marín, S., Coronas, I., Sanchis, V. and Ramos, A.J. (2007) Skin damage, high temperature and relative humidity as detrimental factors for *Aspergillus carbonarius* infection and ochratoxin A production in grapes. *Food Control* 18, 1343–1349.

Bisson, L.F. and Butzke, C.E. (2000) Diagnosis and rectification of stuck and sluggish fermentations. *American Journal of Enology and Viticulture* 51, 168–177.

Chatonnet, P., Dubourdieu, D., Boidron, J.N. and Pons, M. (1992) The origin of ethylphenols in wines. *Journal of Sciences and Food Agriculture* 60, 165–178.

Costa, A., Barata, A., Malfeito-Ferreira, M. and Loureiro, V. (2007) Evaluation of the inhibitory effect of dimethyl dicarbonate (DMDC) against microorganisms associated with wine. *Food Microbiology* 25, 422–427.

Coulter, A., Robinson, E., Cowey, G., Francis, I.L., Lattey, K., Capone, D., Gishen, M. and Godden, P. (2003) *Dekkera/Brettanomyces* yeast: an overview of recent AWRI investigations and some recommendations for its control. In: Bell, S.M., de Garis, K.A., Dundon, C.G., Hamilton, R.P., Partridge, S.J. and Wall, G.S. (eds) *Grapegrowing at the Edge; Managing the Wine Business; Impacts on Wine Flavour.* Proceedings of a Seminar, 10–11 July 2003, Barossa Convention Centre, Tanunda, Adelaide, Australia. Australian Society of Viticulture and Oenology, pp. 41–50.

Davenport, R.R. (1986) Unacceptable levels for yeasts. In: King, A., Pitt, J., Beuchat, L. and Corry, J. (eds) *Methods for the Mycological Examination of Food*. Plenum Press, New York, pp. 214–215.

Deak, T. and Reichart, O. (1986) Unacceptable levels of yeasts in bottled wine. In: King, A., Pitt, J., Beuchat, L. and Corry, J. (eds) *Methods for the Mycological Examination of Food.* Plenum Press, New York, pp. 215–218.

Dias, L., Dias, S., Sancho, T., Stender, H., Querol, A., Malfeito-Ferreira, M. and Loureiro, V. (2003) Identification of yeasts isolated from wine related environments and capable of producing 4-ethylphenol. *Food Microbiology* 20, 567–574.

EFSA (European Food Safety Authority) (2006) Opinion of the Scientific Panel on Contaminants in the Food Chain on a Request from the Commission related to ochratoxin A in food. Question No. EFSA-Q-2005-154, adopted on 4 April 2006. *The EFSA Journal* 365 (available at htpp://www. efsa.europa.eu/en/science/contam /contam. opinions/1521.html).

Fernandes, A., Venancio, A., Moura, F., Garrido, J. and Cerdeira, A. (2003) Fate of ochratoxin A during a vinification trial. *Aspects of Applied Biology* 68, 73–80.

Fernandes, A., Ratola, N., Cerdeira, A., Alves, A., and Venâncio, A. (2007) Changes in ochratoxin A concentration during winemaking. *American Journal of Enology and Viticulture* 58, 92–96.

Fleet, G., Prakitchaiwattana, C., Beh, A. and Heard, G. (2002) The yeast ecology of wine grapes. In: Ciani, M. (ed.) *Biodiversity and Biotechnology of Wine Yeasts.* Research Signpost, Kerala, India, pp. 1–17.

Hocking, A.D., Leong, S.L., Kazi, B.A., Emmett, R.W. and Scott, E.S. (2007) Fungi and mycotoxins in vineyards and grape products. *International Journal of Food Microbiology* 119, 84–88.

International Agency for Research on Cancer (IARC) (1993) *Some Naturally Occurring Substances: Food Items and Constituents, Heterocyclic Aromatic Amines and Mycotoxins* (Vol. 56). International Agency for Research on Cancer, Lyon, France, pp. 397–444; 445–466; 467–488.

JECFA (Joint FAO/WHO Expert Committee of Food Additives) (2001) Ochratoxin A. In: *Safety Evaluation of Certain Mycotoxins in Food.* Prepared by the 56th meeting of the JECFA. FAO Food and Nutrition Paper 74, Food and Agriculture Organization of the United Nations, Rome, pp. 281–415.

Kuhn, F., Mester, V., Morris, R. and Dalma, J. (2004) Serious eye injuries caused by bottles containing carbonated drinks. *British Journal of Ophthalmology* 88, 69–71.

Laureano, P., D'Antuono, I., Barata, A., Malfeito-Ferreira, M. and Loureiro, V. (2004). Effect of different sanitation treatments on the population of *Dekkera bruxellensis* recovered from the wood of barrels [in Portuguese]. *Enologia* 43/44, 3–8.

Leong, L.S., Hocking, A.D., Pitt, J.I., Kazi, B.A., Emmett, R.W. and Scott, E.S. (2006a) Black *Aspergillus* species in Australian vineyards: from soil to ochratoxin A in wine. *Advances in Experimental Medicine and Biology* 571, 153–171.

Leong, L.S., Hocking, A.D., Pitt, J.I., Kazi, B.A., Emmett, R.W. and Scott, E.S. (2006b) Australian research on ochratoxigenic fungi and ochratoxin A. *International Journal of Food Microbiology* 111, S10–S17.

Loureiro, V. and Malfeito-Ferreira, M. (2003a) Spoilage yeasts in the wine industry. *International Journal of Food Microbiology* 86, 23–50.

Loureiro, V. and Malfeito-Ferreira, M. (2003b) Yeasts in spoilage. In: Caballero, B., Trugo, L. and Finglas, P. (eds). *Encyclopedia of Food Sciences and Nutrition*, 2nd edn. Academic Press, London, pp. 5530–5536.

Loureiro, V. and Malfeito-Ferreira, M. (2006) Spoilage activities of *Dekkera/Brettanomyces* spp. In: Blackburn, C. (ed.) *Food Spoilage Microorganisms*. Woodhead Publishers, Cambridge, UK, pp. 354–398.

Loureiro, V. and Querol A. (1999) The prevalence and control of spoilage yeasts in foods and beverages. *Trends in Food Science and Technology* 10/11, 356–365.

Loureiro, V., Malfeito-Ferreira M. and Carreira A. (2004) Detecting spoilage yeasts. In: Steele, R. (ed.) *Understanding and Measuring the Shelf-life of Food*. Woodhead Publishers, Cambridge, UK, pp. 233–288.

Malfeito-Ferreira, M., Tareco, M. and Loureiro, V. (1997) Fatty acid profiling: a feasible typing system to trace yeast contaminations in wine bottling plants. *International Journal of Food Microbiology* 38, 143–155.

Mateo, R., Medina, A., Mateo, E.M., Mateo, F. and Jiménez, M. (2007) An overview of ochratoxin A in beer and wine. *International Journal of Food Microbiology* 119, 79–83.

Mínguez, S., Cantus, J.M., Pons, A., Margot, P., Cabañes, F.X., Masqué, C., Accensi, F., Elorduy, X., Giralt, L.L., Vilavella, M., Rico, S., Domingo, C., Blasco, M. and Capdevila, J. (2004) Influence of the fungus control strategy in the vineyard on the presence of Ochratoxin A in the wine. *Bulletin de L'OIV* 885–886, 821–831.

OIV (2005) *Code of Sound Vitivinicultural Practices in Order to Minimise Levels of Ochratoxin A in Vine-based Products*. Resolution Viti-Oeno 1/2005 (accessed at http://news.reseau-concept.net/images/oiv_uk/Client/VITI-OENO_1-2005_EN.pdf).

Otteneder, H. and Majerus, P. (2000) Occurrence of ochratoxin A (OTA) in wines: influence of the type of wine and its geographical origin. *Food Additives and Contaminants* 17, 793–798.

Pitt, J. and Hocking, A. (1985) *Fungi and Food Spoilage*. Academic Press, Sydney, Australia.

Rodrigues, N., Gonçalves, G., Pereira-da-Silva, S., Malfeito-Ferreira, M. and Loureiro, V. (2001) Development and use of a new medium to detect yeasts of the genera *Dekkera/Brettanomyces* spp. *Journal of Applied Microbiology* 90, 588–599.

Romano, P. (2005) Proprietà technologiche e di qualità delle specie di lieviti vinari. In: Vicenzini, M., Romano, P. and Farris, G. (eds) *Microbiologia del Vino*. Casa Editirice Ambrosiana, Milan, Italy, pp. 101–131.

Samson, R.A., Houbraken, J.A., Kuijpers, A.F., Frank, J.M. and Frisvad, J.C. (2004) New ochratoxin A- or sclerotium-producing species in *Aspergillus* section *Nigri*. *Studies in Mycology* 50, 45–61.

Schuller, D., Côrte-Real, M. and Leão, C. (2000) A differential medium for the enumeration of the spoilage yeast *Zygosaccharomyces bailii* in wine. *Journal of Food Protection* 63, 1570–1575.

Thomas, D.S. (1993) Yeasts as spoilage organisms in beverages. In: Rose, A.H. and. Harrison, J.S (eds) *The Yeasts*, Vol. 5, 2nd edn). Academic Press, London, pp. 517–561.

van der Merwe, K.J., Steyn, P.S., Fourie L., Scott, D.B. and Theron, J.J. (1965) Ochratoxin A, a toxic metabolite produced by *Aspergillus ochraceus* Wilh. *Nature* 205 (4976), 1112–1113.

Varga, J. and Kozakiewicz, Z. (2006) Ochratoxin A in grapes and grape-derived products. *Trends in Food Science Technology* 17, 72–81.

Zimmerli, B. and Dick, R. (1996) Ochratoxin A in table wine and grape-juice: occurrence and risk assessment. *Food Additives and Contaminants* 13, 665–668.

10 Medicinal Potential of *Ganoderma lucidum*

Daniel Sliva[1,2,3]

[1]*Cancer Research Laboratory, Methodist Research Institute, Indianapolis, USA;* [2]*Department of Medicine and* [3]*Indiana University Simon Cancer Center, School of Medicine, Indiana University, Indianapolis, USA*

Introduction

Ganoderma lucidum is one of the important Asian fungi that were collectively recognized in China and Korea as *ling zhi* (mushroom of immortality), and in Japan as *reishi* mushroom or *mannentake* (10,000 years mushroom), over 4000 years ago (Wasser, 2005). Ling zhi was recognized as a superior tonic in the most famous Chinese Materia Medica, the *Shen Nung Ben Cao Jing* (206 BC–AD 8) (Huang, 1993). Although ling zhi includes a variety of *Ganoderma* species with different colours and shapes, the red ling zhi (*G. lucidum*) was reported as treating binding in the chest, toning the heart, nourishing the centre, sharpening the wit and improving the memory (Wasser, 2005).

Ganoderma was held in high esteem due to its relative rarity and because it was reserved for use by only royalty. As a result of this esteem, *Ganoderma* was immortalized in the Chinese culture in literature, numerous paintings, statues, silk tapestries and on the robes of emperors (Wasser, 2005). Although *Ganoderma* was originally used to improve health and promote longevity, its potential therapeutic effects were recognized in traditional Chinese medicine for the treatment of a variety of diseases.

The two major groups of biologically active compounds isolated from *G. lucidum* are polysaccharides (mainly glucans and glycoproteins) and lanostane-type triterpenes (ganoderic acids, ganoderic alcohols and their derivatives) (Gao and Zhou, 2003). Some of these molecules have been shown to have specific effects in modulating particular signalling pathways, inhibiting specific enzymes or stimulating the immune response. *G. lucidum* has been shown to have the following effects:

- Anti-cancer activities (tumouricidal effects on cancer cells, induction of cytokines in immune cells).
- Antiviral effects (inhibition of viral replication).

- Hepatoprotective effects (antioxidative activity, modulation of activity of liver enzymes).
- Cardioprotective effects (anti-hypertensive effects, lowering of blood cholesterol).
- Hypoglycaemic effects (reduction of blood glucose).

The cumulative results of recent *in vitro* cell culture and animal studies have helped to elucidate the molecular mechanisms responsible for the medicinal effects of *G. lucidum*. Although extracts from *G. lucidum* are popularly used in the form of dietary supplements and in alternative medicine, the identification of specific compounds and knowledge of their therapeutic activities can be employed rationally for the future development of new drugs from this ancient medicinal mushroom.

Biologically Active Compounds

Polysaccharides

The earliest anti-tumour and immunomodulating activities described from medicinal mushrooms, including *G. lucidum*, were reported in 1957 (Ringler *et al.*, 1957). Polysaccharides were the first biologically active compounds isolated from *G. lucidum* in Japan, in the early 1980s (Miyazaki and Nishijima, 1981). In their original report Miyazaki and Nishijima identified arabinoxyglucan, consisting of a D-glucopyranosyl (1→4)-α- and -β- backbone with β-D-(1→3) and β-D-(1→6) linkages and branching sequences (Miyazaki and Nishijima, 1981). Although more than 200 types of polysaccharides have been isolated from the fruiting bodies, spores, mycelia and cultivation broth of *G. lucidum* (Zhou *et al.*, 2007), chemical analysis has shown that the most active polysaccharides are β-D-glucans.

Anti-tumour activity has been found mainly in the water-soluble branched (1→3)-β-D-glucans (see Fig. 10.1), which are usually isolated by extraction with hot water (Zhou *et al.*, 2007). In addition to the water-soluble β-glucan and glucurono-β-glucan, *G. lucidum* also contains a large amount of bioactive water-insoluble polysaccharides such as hetero β-glucan, xylo-β-glucan, xylomanno-β-glucan and manno-β-glucan. Although typical β-glucans have an average molecular weight of 1000 kDa, the molecular weights of other bioactive polysaccharides from *G. lucidum* are very varied. Examples include ganoderan C (5.8 kDa), ganoderan B (7.4 kDa), PL-1 (8.3 kDa), SP (10 kDa), ganoderan A (23 kDa), GL-1 (40 kDa), PL-3 (63 kDa), PL-4 (200 kDa) and PSGL-I-1A (718 kDa) (Zhou *et al.*, 2007).

Protein-bound polysaccharides have also been isolated from *G. lucidum*, including a neutral protein-bound polysaccharide (NPBP), an acidic protein-bound polysaccharide (APBP) and a fucose-containing glycoprotein (Wang *et al.*, 2002). A proteoglycan (GLIS) with a carbohydrate:protein ratio of 11.5:1.0 and a polysaccharide peptide (GLPP) have also been shown to be biologically active (Lin, 2005). Although the activity of polysaccharides from

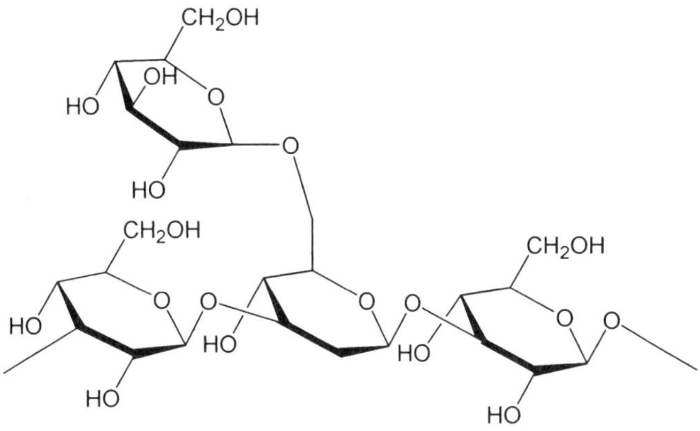

Fig. 10.1. Backbone structure of β-D-glucans from *Ganoderma lucidum.*

G. lucidum has been suggested as being mediated through the complement receptor-type 3 (CR3, $\alpha_M\beta_2$ integrin, CD11b/CD18), which binds β-glucan polysaccharides (Yan *et al.*, 1999), other mechanisms are also involved (see below).

Triterpenes

More than 20,000 triterpenes/triterpenoids have been identified in nature and some of them, such as oleanolic and ursolic acids, have been used to develop new multifunctional drugs for cancer therapy and prevention (Liby *et al.*, 2007). In 1982, ganoderic acids A and B were the first two triterpenes to be isolated from *G. lucidum* (Kubota *et al.*, 1982). Since then more than 130 triterpenes have been isolated from the fruiting bodies, spores, mycelia and culture media (Huie and Di, 2004), and new triterpenes continue to be isolated and identified. These bioactive molecules are predominantly oxygenated lanostane-type triterpenes that are grouped according to the number of carbons (C24, C27 and C30 compounds) and their functional groups (Gao, J. *et al.*, 2005). *Ganoderma lucidum* is reported to contain ganoderic and lucidenic acids and their alcohols, and aldehydes, ganodermic acids, ganoderenic acids, sterols and other oxygenated triterpenes (see Fig. 10.2; Lindequist *et al.*, 2005). Most of these triterpenes are biologically active and have been suggested as potential drugs for a variety of conditions.

Other compounds

In addition to containing small amounts of polyphenols, steroids, lignin, ganomycins, vitamins, lectin, nucleosides, nucleotides and organic germanium, *G. lucidum* also contains significant amounts of amino acids

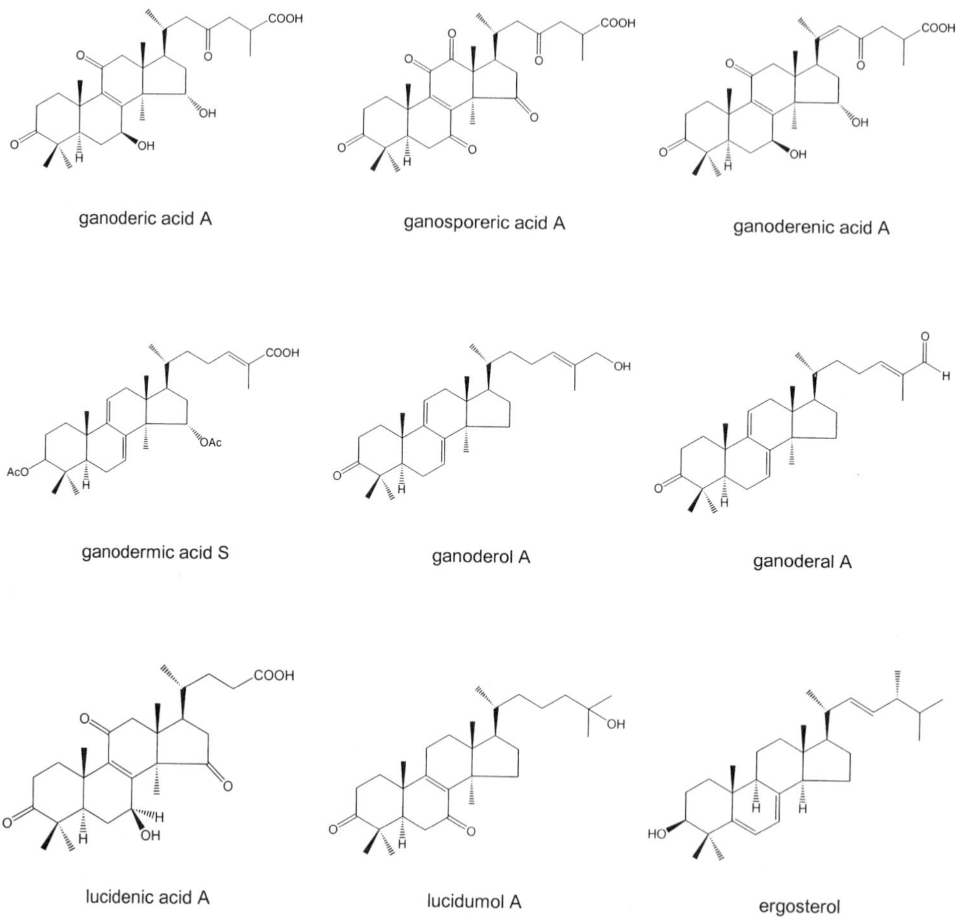

Fig. 10.2. Triterpenes identified from *Ganoderma lucidum.*

such as alanine, leucine, aspartic acid and glutamic acid (Gao and Zhou, 2003). A peptide isolated from fermented *G. lucidum* powder demonstrated a strong antioxidant activity in lipid peroxidation (Sun *et al.*, 2004).

Although some proteins in *G. lucidum* were isolated in complexes with polysaccharides (see above), others were isolated in their pure forms. An immunomodulatory protein LZ-8, (MW 12.4 kDa) has been obtained from a *G. lucidum* mycelial extract, and its complete amino acid sequence has been determined (Tanaka *et al.*, 1989). Ganodermin, an antifungal protein (MW 15 kDa), was isolated from fresh fruiting bodies of *G. lucidum* (Wang and Ng, 2006a), and a lectin (GLL-M, MW 18 kDa) was extracted from mycelia (Kawagishi *et al.*, 1997). In contrast, a novel proteinase A inhibitor purified from *G. lucidum* (MW 38 kDa) contained 70% carbohydrate (Tian and Zhang, 2005). A new laccase (MW 75 kDa) demonstrating anti-HIV activity has been isolated from fresh fruiting bodies (Wang and Ng, 2006b), and a novel 114 kDa hexameric lectin has also been isolated recently (Thakur *et al.*, 2007).

Mechanisms of Therapeutic Effects

Ganoderma lucidum and its extracts have been used for thousands of years to promote longevity and also to treat a variety of diseases, but *in vitro* cell culture and animal studies have only recently helped elucidate the molecular mechanisms responsible for these therapeutic effects.

Anti-tumour effects

Polysaccharides

Inhibition of the growth of cancer cells in animals was one of the first medicinal effects described for *G. lucidum*, and an early example was the inhibition of growth of subcutaneously transplanted sarcoma-180 ascites in mice after intraperitoneal injection of β-D-glucan (Miyazaki and Nishijima, 1981) (see Table 10.1). The *Ganoderma* polysaccharide fraction (PS-G) can stimulate human blood mononuclear cells to produce factors which are able to suppress proliferation and induce apoptosis of leukaemia cells (Wang *et al.*, 1997). PS-G can also induce the production of interleukin (IL)-1β, IL-6 and tumour necrosis factor (TNF)-α from macrophages and the secretion of interferon (IFN)-γ from T-lymphocytes (Wang *et al.*, 1997).

A *G. lucidum* isolated from a European forest (Slovenia) and cultivated in a liquid substrate was found to produce β-D-glucans, which induced TNF-α and IFN-γ synthesis in human peripheral blood mononuclear cells (Berovic *et al.*, 2003). The majority of the anti-cancer effects of *G. lucidum* polysaccharides have been attributed to the modulation of the immune system through the activation of macrophages, neutrophils, dendritic cells, natural killer cells and T- and B-lymphocytes (Lin, 2005; Zhu *et al.*, 2007). Mechanistically, *G. lucidum* polysaccharides (GLP) induce cytokine expression via Toll-like receptors, (TLR)-4, in macrophages and dendritic cells, whereas immunoglobulin (Ig) production is mediated through TLR-4/TLR-2 in B-lymphocytes (Hua *et al.*, 2007). Collectively, this GLP-dependent expression of cytokines and Ig employs signalling through mitogen-activated protein kinases (MAPKs) and transcription factor NF-κB (Lin *et al.*, 2006).

Oral administration of water-soluble polysaccharides extracted from *G. lucidum* (as the 'over-the counter' product Ganopoly) have been reported to enhance the immune response in patients with advanced-stage cancer, although the response in patients with advanced lung cancer was variable (Gao, J. *et al.*, 2005). In addition to stimulating the immune system, a polysaccharide extract from *G. lucidum* has also been suggested as a chemopreventative cancer agent due to its ability to inhibit cell transformation, as demonstrated by the suppression of foci formation induced by the *Ras* oncogene (Hsiao *et al.*, 2004).

Table 10.1. Anti-cancer effects of polysaccharides, peptides and proteins.

Fraction	Biological effects	Cell type(s) involved	Reference(s)
β-D-glucans	Inhibition of growth of sarcoma cells in mice; induction of TNF-α and IFN-γ synthesis	S-180	Miyazaki and Nashijima, 1981
Polysaccharides	Immunomodulating effects	Blood mononuclear cells Macrophages, neutrophils, dendritic cells, NK cells, T- and B-lymphocytes	Berovic *et al.*, 2003 Lin, 2005; Zhu *et al.*, 2007 Lin *et al.*, 2006; Hua *et al.*, 2007
PS (polysaccharides)	Inhibition of cell transformation by the *Ras* oncogene	R6 cells	Hsiao *et al.*, 2004
Ganopoly (polysaccharides)	Induction of immune response in cancer patients; increase in plasma IL-2, IL-6, IFN-γ; decrease in plasma IL-1, TNF-α	T-cells, NK cells	Gao, Y. *et al.*, 2005
G009 (amino-polysaccharides)	Antioxidative effects	HL-60	Lee *et al.*, 2001
GLPP (polysaccharide peptide)	Inhibition of growth of lung carcinoma cells in mice; induction of apoptosis and suppression of secretion of VEGF from vascular endothelial cells	PG HUVEC	Cao and Lin, 2006
GPS (polysaccharide–protein complex)	Induction of activity of glutathione S-transferase	NCTC-clone cells	Kim, H. *et al.*, 1999
GLIS (proteoglycans)	Stimulation of cell proliferation; induction of IL-2 secretion; stimulation of expression of PKCα and PKCγ	B-lymphocytes	Lin, 2005
LZ-8 (protein)	Induction of secretion of IL-1β, IL-2, TNF-α, IFN-γ, IL-2	Blood lymphocytes T-cells	Haak-Frendscho *et al.*, 1993 Hsu *et al.*, 2008

Amino-polysaccharides, glycoproteins and proteins

An amino-polysaccharide fraction isolated from *G. lucidum* has been considered as having a chemopreventative potential in that it inhibited lipid peroxidation (Lee *et al.*, 2001). Although the *G. lucidum* polysaccharide peptide (GLPP) did not inhibit proliferation of human lung carcinoma cells *in vitro*, GLPP markedly reduced the growth of these cells in mice (Cao and Lin, 2006). Additionally GLPP has been shown to suppress proliferation and induce apoptosis of human umbilical vascular endothelial cells by down-regulating the expression of anti-apoptotic protein Bcl-2 and up-regulating the expression of proapoptotic Bax-2 proteins, respectively (Cao and Lin, 2006). GLPP can also suppress the secretion of vascular endothelial growth factor (VEGF) from lung cancer cells, further confirming the anti-angiogenic properties of GLPP (Cao and Lin, 2006).

A polysaccharide–protein complex isolated from *G. lucidum* has been found significantly to increase the activity of glutathione S-transferase, a detoxification enzyme used to degrade carcinogens in the initiation stages of cancer (Kim, H. *et al.*, 1999). The anti-tumour activity of glycoprotein isolated from *G. lucidum* has been linked to its potency to stimulate production of IL-2, IL-4 and IFN-γ (Wang *et al.*, 2002). It has been suggested that the GLIS proteoglycan may improve the immune response of cancer patients by stimulating activation, proliferation and differentiation of B-lymphocytes and by the production of immunoglobulins (Lin, 2005).

A novel immunomodulatory protein has been purified from *G. lucidum* and named Ling Zhi-8 (LZ-8) (Tanaka *et al.*, 1989). The anti-tumour activity of LZ-8 was found to be associated with the stimulation of production of IL-1β, IL-2, TNF-α and IFN-γ by human peripheral blood lymphocytes (Haak-Frendscho *et al.*, 1993). Recently, recombinant LZ-8 protein was produced in different expression systems, and its capacity to modulate the production of Th1 and Th2 cytokines was evaluated (Yeh *et al.*, 2008). In addition, recombinant LZ-8 has been reported as modulating protein tyrosine kinase (PTK) and protein kinase C (PKC) signalling to produce IL-2 in human T-cells (Hsu *et al.*, 2008).

Triterpenes

As mentioned above, one of the major groups of compounds isolated from *G. lucidum* are the lanostane-type triterpenes. The anti-cancer activities of these biologically active terpenoids were originally demonstrated by their cytotoxic/killing effects on a variety of cancer cells. Ganoderic acids U, V, W, X and Y were found to be cytotoxic against hepatoma cells *in vitro* (Toth *et al.*, 1983), and 3-β-hydroxy-26-oxo-5-α-lanosta-8,24-dien-11-one was found to suppress nasopharyngeal carcinoma and hepatoma cells (Lin *et al.*, 1991).

Some of the triterpenes have shown a selective cytotoxicity against Lewis lung carcinoma (LLC) and mouse sarcoma (Meth-A) cells. For example, ganoderic acid C1, ganolucidic acid A and lucideric acid α were found to be cytotoxic against LLC cells; ganoderic acid γ, ganoderic acid ε and ganoderic acid G were cytotoxic against Meth-A cells; and ganoderic acid θ, lucidumol A, lucidumol B, ganodermanondiol, ganodermanontriol and ganoderiol F killed both LLC and Meth-A cells (Min *et al.*, 2000). In addition to demonstrating cytotoxicity against LLC and Meth-A cells, ganodermanondiol, lucialaldehyde B and lucialaldehyde C have been found to be cytotoxic to human breast cancer cells (Gao *et al.*, 2002). Ganoderic acid E, lucidenic acid A and lucidenic acid N are reported to be cytotoxic to human hepatoma and mouse leukaemia cells (Wu *et al.*, 2001).

Some studies have focused more on the specific mechanism(s) responsible for anti-cancer activity (see Table 10.2). Prenyltransferases such as farnesyl protein transferase (FPT) are essential for the cell-transforming activities of the Ras oncoprotein. Ganoderic acids A and C have been shown to inhibit the enzymatic activity of FPT (Lee *et al.*, 1998). Antioxidative activities have been identified in the triterpene fraction that contains ganoderic acids A, B, C and D, lucidenic acid B and ganodermanontriol as major compounds (Zhu *et al.*, 1999).

The triterpenes 26,27-dihydroxy-5-α-lanosta-7,9-(11),24-triene-3,22-dione and 26-hydroxy-5-α-lanosta-7,9-(11),24-triene-3,22-dione have shown some chemopreventative activity by inducing NAD(P)H:quinine oxidoreductase, a phase-2 drug-metabolizing enzyme (Ha *et al.*, 2000). Ganoderic acid DM and 5-α-lanosta-7,9(11),24-triene-15-α,26-dihydroxy-3-one can inhibit the activity of 5α-reductase, a pharmacological target in prostate cancer (Liu *et al.*, 2006), and 7-oxo-ganoderic acid Z and 15-hydroxy-ganoderic acid S can both suppress the activities of HMG-CoA reductase and acyl-CoA acyltransferase. These last two enzymes are in the mevalonate pathway, which is an important target for anti-cancer therapy (Li *et al.*, 2006).

The quantity and composition of triterpenes can vary between different specimens of *G. lucidum*, and this can be reflected in the amounts and types of purified triterpenes that can be obtained from them (Gao, J.-J. *et al.*, 2004). Therefore, some studies have evaluated therapeutic mechanism(s) against cancer cells with alcohol extracts of *G. lucidum* containing unidentified or partially identified triterpenes. These extracts can inhibit various carcinoma cells through cell cycle arrest at G0/G1 or G2/M by specific mechanisms (Zhu *et al.*, 2000; Hu *et al.*, 2002; Lin *et al.*, 2003; Lu *et al.*, 2004).

An alcohol extract from *G. lucidum* was found to inhibit proliferation of breast cancer cells MCF-7 by cell-cycle arrest at G1 by down-regulating expression of cyclin D1 and inducing expression of the cell-cycle inhibitor p21/Waf-1 (Hu *et al.*, 2002). This extract also induced expression of the proapoptotic Bax protein, resulting in the programmed cell death of these breast cancer cells (Hu *et al.*, 2002). Another alcohol extract from *G. lucidum* was found to induce proapoptotic caspase-3 activity and decrease expression of pro-inflammatory cyclooxygenase-2 (COX-2) in human colon cancer cells HT-29 (Hong *et al.*, 2004).

An ethanol extract from *G. lucidum* has been reported to inhibit the growth of human urothelial cells (HUC) by arresting the cell-cycle at G2/M

Table 10.2. Anti-cancer effects of triterpenes.

Triterpene	Biological effects	Reference
Ganoderic acid A, C	Inhibition of prenyltransferase (FPT) activity	Lee *et al.*, 1998
Ganoderic acid A, H	Suppression of cell proliferation and invasive behaviour of breast cancer cells through the inhibition of Cdk4 and uPA	Jiang *et al.*, 2008
Ganoderic acid D	Inhibition of proliferation and induction of apoptosis of cervical carcinoma cells	Yue *et al.*, 2008
Ganoderic acid Me	Suppression of tumour growth and lung metastasis through the induction of expression of IL-2 and IFN-γ	Wang *et al.*, 2007
Ganoderic acid T	Inhibition of proliferation and induction of apoptosis of lung cancer cells through the up-regulation of expression of p53	Tang *et al.*, 2006
Ganoderic acid X	Induction of apoptosis and inhibition of topoisomerases; activation of ERK, JNK and caspase-3 in hepatoma cells	Li *et al.*, 2005
7-oxo-ganoderic acid Z, 15-hydroxy-ganoderic acid Z	Inhibition of HMG CoA reductase; inhibition of acetyl CoA acyltransferase	Li *et al.*, 2006
Ganoderiol F	Activation of ERK and up-regulation of p16 in hepatoma cells	Chang *et al.*, 2006
Ganoderol B	Inhibition of prostate cancer growth through the inhibition of androgen receptor signalling	Liu *et al.*, 2007
Lucidenic acid A, B, C, N	Suppression of invasion of hepatoma cells through the inhibition of MMP-9 activity	Weng *et al.*, 2007
Lucidenic acid B	Inhibition of MAP/ERK signalling in hepatoma cells	Weng *et al.*, 2008
Ganoderic acid A, B, C, D, Lucidenic acid B, Ganodermanontriol	Antioxidative activity	Zhu *et al.*, 1999
Ganoderic acid DM, 5α-lanosta-7,9-(11),24-triene-15α,26dihydroxy-3-one	Inhibition of 5α-reductase	Liu *et al.*, 2006
Ganoderic acid A, F, DM, T-Q, lucidenic acid A, D2, E2, P, methyl lucidenate A, 2, E2, Q, 20-hydroxylucidenic acid N	Inhibition of carcinogen-induced oedema and inflammation	Akihisa *et al.*, 2007
26-hydroxy-5α-lanosta-7,9-(11),24-triene-3,22-dione 26,27-dihydroxy-5α-lanosta-7,9-(11),24-triene-3,22-dione	Induction of NAD(P)H:quinine oxidoreductase (QR)	Ha *et al.*, 2000

and to inhibit carcinogen-induced cell migration by increasing actin polymerization independently of matrix metalloproteinase-2 (MMP-2) and focal adhesion kinase (FAK) expression (Lu *et al.*, 2004). A triterpene-enriched extract has also been found to suppress proliferation of hepatoma cells by cell-cycle arrest at the G2 phase without any effect on the normal human liver cell line (Lin *et al.*, 2003). Inhibition of the hepatoma cells was associated with the suppression of protein kinase C (PKC) and the activation of p38 MAPK and c-Jun N-terminal kinase (JNK) (Lin *et al.*, 2003).

A triterpene-rich fraction combined with lovastatin, a liver-specific HMG-CoA reductase inhibitor, has been found to markedly inhibit tumour growth in nude mice inoculated with human hepatoma Hep 3B/T2 cells (Shiao, 2003). In addition, a further alcohol extract has been reported to inhibit the growth of human cervical carcinoma HeLa cells through cell-cycle arrest at G1 phase and to significantly decrease the level of intracellular calcium. It has been suggested that the triterpene-dependent changes in the calcium transport system might alter the signal transduction involved in the regulation of the cell cycle (Zhu *et al.*, 2000).

A dietary supplement, ReishiMax, which contains 6% triterpenes in addition to polysaccharides, was reported to suppress growth and metastatic potential of the highly invasive human breast cancer cells MDA-MB-231 by inhibiting the activity of AKT kinase and transcription factors AP-1 and NF-κB. This resulted in the down-regulation of cyclin D1 and urokinase plasminogen activator (uPA) (Sliva *et al.*, 2003; Jiang *et al.*, 2004a; Slivova *et al.*, 2004). ReishiMax had phytoestrogenic properties in that it modulated oestrogen receptor signalling in breast cancer cells MCF-7 and down-regulated c-myc oncogene (Jiang *et al.*, 2006). In addition, ReishiMax contained the triterpenes ganoderic acids A, F and H, and also inhibited the oxidative stress-induced invasiveness of breast cancer cells by suppressing the phosphorylation of extracellular signal-regulated protein kinases (Erk1/2), which resulted in the down-regulation of c-fos and suppressed secretion of interleukin-8 (IL-8) from MCF-7 cells (Thyagarajan *et al.*, 2006).

ReishiMax was found to inhibit proliferation of prostate cancer cells PC-3 by down-regulating cyclin B and Cdc2 and up-regulating p21 expression, which was shown also by cell-cycle arrest at G2/M phase, and the induction of apoptosis of PC-3 cells was associated with the up-regulation of expression of proapoptotic Bax (Jiang *et al.*, 2004b). Moreover, ReishiMax also inhibited angiogenesis of vascular endothelial cells by suppressing secretion of the proangiogenic factors VEGF and transforming growth factor-β1 (TGF-β1) from PC-3 cells (Stanley *et al.*, 2005).

A triterpenoid fraction from *G. lucidum* was found to inhibit primary tumour growth in the spleen, liver metastasis and secondary tumour growth in the liver in intrasplenically implanted LLC cells (a renal epithelial cell line) in mice. Ganoderic acid F was isolated from this fraction and inhibited Matrigel-induced angiogenesis (Kimura *et al.*, 2002). Ganoderic acids A and H have been reported to suppress cell proliferation, colony formation and invasive behaviour (cell adhesion, cell migration and cell invasion) of MDA-MB-231 cells (Jiang *et al.*, 2008). This suppression is mediated through the inhibition of transcription factors AP-1 and NF-κB, resulting in the down-regulation of Cdk4 and the suppression of uPA secretion (Jiang *et al.*, 2008).

Ganoderic acid T has been shown to inhibit proliferation of highly metastatic lung cancer cells 95-D through cell-cycle arrest at G1 phase, and induced apoptosis by up-regulating the expression of p53 and Bax proteins and stimulating caspase-3 activity (Tang *et al.*, 2006). Similarly, ganoderic acid D can inhibit proliferation of HeLa human cervical carcinoma cells through cell cycle arrest at G2/M phase, and induced apoptosis (Yue *et al.*, 2008). Proteomic analysis has shown that ganoderic acid D can modulate expression of 21 proteins, and down-regulate eukaryotic translation initiation factor 5A (eIF5A) and microtubule-associated protein RP/EB family member 1 (EB1). The up-regulation of 14-3-3-epsilon and thioredoxin-dependent peroxide reductase mitochondrial precursor (PRDX3) were confirmed by Western blot analysis (Yue *et al.*, 2008).

Lucidenic acids A, B, C and N are reported to suppress PMA-induced MMP-9 activity and invasion of HepG2 cells (Weng *et al.*, 2007). The inhibitory effect of lucidenic acid B was mediated through MAPK/ERK signalling and the inhibition of NF-κB and AP-1 (Weng *et al.*, 2008). Ganoderol B can show anti-androgenic activity by inhibiting androgen-induced growth of prostate cancer cells LNCaP, down-regulating androgen receptor (AR) signalling and suppressing the regrowth of the ventral prostate induced by testosterone in rats (Liu *et al.*, 2007).

In addition to the direct effects of triterpenes on cancer cells, triterpenes can also mediate their effects through the immune system, and an ethanol extract of *G. lucidum* containing terpenes and polyphenols was found to suppress carrageenan-induced acute and formalin-induced chronic inflammatory oedema in animals (Lakshmi *et al.*, 2003). Purified ganoderic acid Me can suppress both tumour growth and lung metastasis of LLC cells in mice and stimulate an immune response by increasing the expression of IL-2 and IFN-γ (Wang *et al.*, 2007). The ganoderic acids A, F, DM and T-Q, lucidenic acids A, D2, E2 and P, methyl lucidenates A, D2, E2 and Q and 20-hydroxylucidenic acid N can all inhibit TPA-induced aural oedema inflammation in mice (Akihisa *et al.*, 2007).

Although the majority of research on *Ganoderma* has been focused on *G. lucidum*, some 258 species have been described in the genus (http://zipcodezoo.com/Fungi/G/Ganoderma_amboinense.asp). Many of these species have been described on morphological data, which is often inadequate for accurate identification in these fungi, and many names may be synonyms or misapplied (Buchanan, 2001). Other biologically active triterpenes have been isolated from other members of the genus, and ganoderic acid X, isolated from *G. amboinenese*, can inhibit DNA synthesis in human hepatoma HuH-7 cells by inhibiting topoisomerases. Ganoderic acid X has also been reported to induce apoptosis in these cells by activating ERK and JNK kinases and stimulating caspase-3 activity (Li, C.-H. *et al.*, 2005). Ganoderiol F that has also been isolated from *G. amboinenese* is reported to induced cell-cycle arrest at G1 phase in HepG2 cells, by activating ERK kinase and up-regulating the cyclin-dependent kinase inhibitor p16 (Chang *et al.*, 2006). Long-term treatment of HepG2 cells with ganoderiol F resulted in cellular senescence (Chang *et al.*, 2006).

Antiviral effects

Both water and methanol extracts of *G. lucidum* have been found to markedly inhibit the cytopathic effects of the herpes simplex virus (HSV) and vesicular stomatitis virus (VSV) *in vitro* (Eo *et al.*, 1999a) (see Table 10.3). The acidic protein-bound polysaccharide (APBP) showed the most anti-herpetic activity in Vero and HepG2 cells (Eo *et al.*, 1999b), and the combination of APBP with an anti-herpetic drug such as acyclovir or interferon α, can have a synergistic effect on HSV activity (Kim, Y.S. *et al.*, 2000). The anti-herpetic activity of APBP has been associated with its binding to specific glycoproteins responsible for the attachment and penetration of HSV (Kim, Y.S. *et al.*, 2000). A proteoglycan isolated from *G. lucidum* has shown anti-HSV activity, and it has been suggested that this may be due to it inhibiting viral replication by suppressing viral adsorption and entry into target cells (Liu *et al.*, 2004; Li, Z. *et al.*, 2005).

The *G. lucidum* compounds lucidenic acid O and lucidenic lactone are reported to prevent the activity of reverse transcriptase in human immunodeficiency virus type 1 (Mizushina *et al.*, 1999). All of the following have been shown to significantly inhibit the induction of the Epstein–Barr virus early antigen (EBV-EA) by 12-O-tetradecanoylphorbol-13-acetate (TPA) in Raji cells: (i) lucidenic acids A, B, C, D, E, F, N and P; (ii) methyl lucidenates A, C, D, E, F, L, P and Q; (iii) 20(21)-dehydrolucidenic acid A; (iv) methyl 20(21)-dehydrolucidenate A; (v) 20-hydroxy-lucidenic acids D2, E2, F, N and P; (vi) ganoderic acids A, C1, C2, DM, E, F and T-Q; and (vii) ganodermanondiol and methyl ganoderate F (Iwatsuki *et al.*, 2003; Akihisa *et al.*, 2007).

In addition, an uncharacterised ganoderic acid isolated from *G. lucidum* has been reported to inhibit replication of the hepatitis B virus (HBV) (Li and Wang, 2006). Ganodermadiol, lucidadiol and applanoxidic acid G obtained from a European specimen of *G. pfeifferi* have all shown antiviral activity against influenza virus types A and HSV-1 (Mothana *et al.*, 2003). Ganoderone A, lucialdehyde B and ergosta-7,22-dien-3β-ol from *G. pfeifferi* have also shown potent inhibitory activity against the herpes simplex virus (Niedermeyer *et al.*, 2005).

Effects on the liver

Anti-hepatotoxic activity has been demonstrated with water extracts from *G. lucidum*, *G. formosanum* and *G. neo-japonicum* (Lin *et al.*, 1995; Table 10.4). All these extracts were found to suppress carbon tetrachloride (CCl_4)-induced toxicity in rat liver, leading to decreases in glutamic oxaloacetic transaminase (GOT) and lactate dehydrogenase (LDH) levels in serum (Lin *et al.*, 1995). Hepatoprotective and anti-fibrotic activities have been reported for a *G. tsugae* extract used in CCl_4-induced chronic liver injury in rats (Wu *et al.*,

Table 10.3. Antiviral effects of *Ganoderma* extracts and compounds.

Compound(s)	Biological effect(s)	Reference(s)
Water/methanol extract	Inhibition of herpes simplex virus (HSV) and vesicular stomatitis virus (VSV)	Eo *et al.*, 1999a
Acidic protein-bound polysaccharide (APBV)	Suppression of HSV activity through the inhibition of viral binding and its penetration	Kim, H. *et al.*, 2000
Proteoglycan	Suppression of HSV activity through the inhibition of viral replication	Liu *et al.*, 2004; Li, Z. *et al.*, 2005
Ganoderic acid A, C1, C2, DM,, E, F, T-Q, ganodermanondiol, methyl-ganoderate F, lucidenic acid A, B, C, D, E, F, N, P, methyl-lucidenate A, C, D, E, F, L, P, Q methyl-20(21)-dehydrolucidenic acid A, 20-hydroxy-lucidenic acid D2, E2, F, N, P	Inhibition of Epstein–Barr virus early antigen	Iwatsuki *et al.*, 2003; Akihisa *et al.*, 2007
Lucidenic acid O, lucidenic lactone	Inhibition of HIV reverse transcriptase	Mizushina *et al.*, 1999
Ganoderic acid (uncharacterized)	Inhibition of replication of hepatitis B virus	Li and Wang, 2006
Ganodermadiol, lucidadiol, applanoxidix acid G	Inhibition of influenza virus type A and HSV	Mothana *et al.*, 2003
Ganoderone A, lucialdehyde B, ergosta-7,22-dien-3β-ol	Inhibition of HSV	Niedermeyer *et al.*, 2005

Table 10.4. Hepatoprotective effects of *Ganoderma* extracts and compounds.

Compound	Biological effects	Reference
Water extract	Inhibition of of glutamic oxaloacetic transaminase (GOT) and lactate dehydrogenase (LDH)	Lin *et al.*, 1995
	Induction of superoxide scavenging activity	Lee *et al.*, 1998
	Inhibition of transaminases and methionine adenosyltransferase (MAT1); inhibition of transforming growth factor-β1 (TGF-β1)	Park *et al.*, 1997
Polysaccharides	Inhibition of aspartate transaminase (AST), alanine transaminase (ALT) and alkaline phosphatase (ALP)	Zhang *et al.*, 2002
Polysaccharide (GLPS)	Inhibition of ALT; down-regulation of iNOS expression	Wang *et al.*, 2007
	Inhibition of CYP2E1, CYP1A2, CYP3A	Yang *et al.*, 2006
Proteoglycan (GLPG)	Suppression of TNF-α in plasma	Lakshmi *et al.*, 2006
Methanol extract	Inhibition of glutamate oxaloacetate transaminase (SGOT), glutamate pyruvate transaminase (SGPT) and alkaline phosphatase (ALP)	Kim, D. *et al.*, 1999
Methanol extract	Induction of glutathione peroxidase (GPx), catalase (CAT), glutathione-S-transferase (GST) and superoxide dismutase (SOD)	Kim, D. *et al.*, 1999
Ganoderenic acid A	Inhibition of β-glucuronidase	Wachtel-Galor *et al.*, 2004

2004). A hepatoprotective effect on ethanol-induced liver damage has also been reported for a hot-water extract of *G. lucidum* (Lee *et al.*, 1998). It was suggested that this effect may be due to a superoxide scavenging activity, as animals treated with the extract had decreased levels of malonic dialdehyde (Lee *et al.*, 1998). A *G. lucidum* extract (GLE) has been reported to improve CCl_4-induced liver fibrosis, leading to increases in plasma albumin and the albumin/globulin ratio, a reduction in the hepatic hydroxyproline and decreased activities of transaminases (Lin and Lin, 2006). In addition, GLE treatment may decrease expression of transforming growth factor-β1 (TGF-β1) and methionine adenosyltransferase (MAT1) (Lin and Lin, 2006).

A polysaccharide isolated from *G. lucidum* has been reported to reduce the serum aspartate and alanine transaminases, alkaline phosphatase, total bilirubin and collagen content in the livers of rats with biliary obstruction-induced liver cirrhosis (Park *et al.*, 1997). The *G. lucidum* polysaccharide GLPS has been shown to have a hepatoprotective effect on an immune liver injury induced by *Mycobacterium bovis* BCG infection in mice (Zhang *et al.*,

2002). In this study GLPS decreased alanine transaminase release and nitrous oxide production, inhibited inducible nitric oxide synthase (iNOS) expression and improved the pathological changes of chronic and acute inflammation that had been induced by the BCG infection (Zhang *et al.*, 2002). Additionally, GLPS suppressed the cytochrome P450 activity of the enzymes CYP2E1, CYP1A2 and CYP3A in hepatic microsomes, a finding that could indicate that the GLPS hepatoprotective mechanism is mediated by P450 oxidative metabolism (Wang *et al.*, 2007). The anti-hepatotoxic mechanisms of *G. lucidum* proteoglycan (GLPG) in CCl_4-induced liver injury have been linked to the suppression of plasma TNF-α and free radical scavenging activity (Yang *et al.*, 2006).

A methanolic extract from *G. lucidum* has been shown to prevent hepatic damage by benzo[a]pyrene by inhibiting serum glutamate oxaloacetate transaminase, serum glutamate pyruvate transaminase and alkaline phosphatase (Lakshmi *et al.*, 2006). In addition, this extract enhanced the levels of reduced glutathione and activities of glutathione peroxidase, glutathione-S-transferase, superoxide dismutase and catalase (Lakshmi *et al.*, 2006). Ganoderenic acid A isolated from *G. lucidum* has been considered a potent hepatoprotective against CCl_4-induced liver injury, as it can inhibit β-glucuronidase activity (Kim, D. *et al.*, 1999). Uncharacterized ganoderic acid from *G. lucidum* has also been shown to protect the liver from CCl_4- or *Mycobacterium bovis* BCG infection-induced injuries (Li and Wang, 2006).

Although a controlled human supplementation study did not show any evidence of liver, renal or DNA toxicity from *G. lucidum* usage (Wachtel-Galor *et al.*, 2004), two recent reports have described some hepatotoxic effects (Yuen *et al.*, 2004; Wanmuang *et al.*, 2007). These effects may have been due to other components in the *G. lucidum* powder formulation, as the addition of other plants or ingredients in Chinese herbal medicine preparation is very common (Yuen *et al.*, 2004).

Effects on the cardiovascular system

A water extract from *G. lucidum* mycelia was found to decrease systolic and diastolic blood pressure in laboratory animals independent of heart rate, which suggests that hypotension induced by the extract may be mediated through the central inhibition of sympathetic nerve activity (Lee and Rhee, 1990; Table 10.5).

High cholesterol levels have been linked to atherosclerosis and heart disease, and a variety of different fungi have been reported to have cholesterol-lowering and other cardioprotective effects (Berger *et al.*, 2004). A *G. lucidum* extract has shown anti-atherosclerotic properties by inhibiting the accumulation of intracellular cholesterol in the cells of the human aortal *intima* (Li Khva *et al.*, 1989). *G. lucidum* polysaccharides have been found to decrease significantly levels of total cholesterol (TC), triglycerides (TG) and LDL cholesterol, and to increase levels of serum HDL cholesterol in hyperlipidemic rats (Chen *et al.*, 2005).

Table 10.5. Cardioprotective effects of *Ganoderma* extracts and compounds.

Compound	Biological effects	Reference(s)
Alcohol/water extract	Inhibition of cholesterol accumulation in aorta	Chen *et al.*, 2005
Water extract	Suppression of blood pressure	Li Khva *et al.*, 1989
Polysaccharides	Suppression of serum total cholesterol (TC), triglycerides (TG), LDL cholesterol and increased serum of HDL cholesterol	Frye and Leonard, 1999
Ganoderic acid C, ganoderol A, B, ganoderal A, ganoderic acid Y	Inhibition of lanosterol 14α-demethylase	Morigiwa *et al.*, 1986; Hajjaj *et al.*, 2005
7-oxo-ganoderic acid Z, 15-hydroxy-ganoderic acid S	Inhibition of HMG-CoA reductase	Li *et al.*, 2006
Ganoderic acid F	Inhibition of angiotensin-converting enzyme	Su *et al.*, 1999
Ganodermic acid S	Inhibition of platelet aggregation	Kino *et al.*, 1990

One of the possible mechanisms for reducing cholesterol is the inhibition of enzymes involved in cholesterol biosynthesis. Lanosterol analogues have been identified among *G. lucidum* triterpenes, and these compounds can act as dual inhibitors by suppressing lanosterol 14α-methyl demethylase (P-450DM), and as partial inhibitors of HMG-CoA reductase (Frye and Leonard, 1999). Ganoderic acid C, its derivatives and 26-oxygenosterols (ganoderol A and B, ganoderal A and ganoderic acid Y) can therefore potently inhibit lanosterol 14α-demethylase (Komoda *et al.*, 1989; Hajjaj *et al.*, 2005). As mentioned above, 7-oxo-ganoderic acid Z and 15-hydroxy-ganoderic acid S can suppress the activity of HMG-CoA reductase (Li *et al.*, 2006). Ganoderic acid F has been reported to give some atherosclerosis protection due to the inhibition of an angiotensin-converting enzyme (Morigiwa *et al.*, 1986), and ganodermic acid S has been shown to inhibit platelet aggregation (Su *et al.*, 1999).

Anti-diabetic effects

LZ-8, an immunomodulatory protein isolated from *G. lucidum* (Tanaka *et al.*, 1989), has been shown to be mitogenic *in vitro* towards spleen cells of non-obese diabetic mice (Kino *et al.*, 1990), and intraperitoneal administration of LZ-8 was reported to prevent insulitis and almost normalize the number of insulin-producing cells (Kino *et al.*, 1990; Table 10.6). In addition, LZ-8 has been found to prevent the incidence of autoimmune diabetes in non-obese diabetic mice and to delay the rejection process of transplanted allogeneic pancreatic rat islets, leading to prolonged survival without serious side

Table 10.6. Anti-diabetic effects of *Ganoderma* extracts and compounds.

Compound(s)	Biological effects	Reference
Polysaccharides (GLPS)	Increase in serum insulin; decrease in blood glucose;	He *et al.*, 2006
	decrease in serum triglycerides	Zhang and Lin, 2004
Polysaccharides (Ganopoly)	Decrease in blood glucose	Gao, Y. *et al.*, 2004
Protein (LZ-8)	Prevention of insulitis; normalization of insulin-producing cells;	van der Hem *et al.*, 1995
	delayed rejection of transplanted pancreatic islet cells	Zhang *et al.*, 2003

effects (van der Hem *et al.*, 1995).

Ganoderma lucidum polysaccharides (GLPS) have been reported to increase insulin and reduce glucose serum levels in a dose-dependent manner in alloxan-induced diabetic mice (Zhang *et al.*, 2003) and, interestingly, GLPS was also found to inhibit alloxan-induced activation of NF-κB in the pancreas (Zhang *et al.*, 2003). GLPS was able to decrease serum glucose and triglyceride levels in streptozotocin-induced diabetic mice, and improved the renal morphometric changes and oxidative stress state of diabetic mice. These observations suggest that GLPS may prevent or delay the progression of diabetic renal complications (He *et al.*, 2006). A hypoglycaemic effect has also been shown for GLPS in normal mice, through its insulin-releasing activity. This was due to a facilitation of Ca^{2+} inflow to the pancreatic beta cells (Zhang and Lin, 2004). The commercially available *G. lucidum* water-soluble polysaccharides (Ganopoly) have been shown clinically to decrease the postprandial blood glucose levels in type-2 diabetic patients (Gao, Y. *et al.*, 2004), although the major compounds responsible for the hypoglycaemic effects have been identified as peptidoglycans (Zhou *et al.*, 2007).

Conclusions

As described above, the Asian medicinal fungus *Ganoderma lucidum* appears to provide a wide range of compounds with a variety of different therapeutic activities. The increased interest in alternative therapies by Western medicine has helped to 'rediscover' some of the alternative/natural products (herbs/mushrooms), and this has led to the identification and evaluation of their biologically active compounds (Lindequist *et al.*, 2005). *Ganoderma lucidum* has been described as 'a therapeutic fungal biofactory' (Paterson, 2006), and this could provide new opportunities for chemopreventative or therapeutic approaches.

In conclusion, the identification of biologically active compounds from *G. lucidum* now allows the chemical synthesis of these compounds (or more active/stable analogues), and these may provide significant leads for the development of new drugs from this ancient medicinal fungus.

References

Akihisa, T., Nakamura, Y., Tagata M., Tokuda, H., Yasukawa, K., Uchiyama, E., Suzuki, T. and Kimura, Y. (2007) Anti-inflammatory and anti-tumour-promoting effects of triterpene acids and sterols from the fungus *Ganoderma lucidum. Chemistry and Biodiversity* 4, 224–231.

Berger, A., Rein, D., Kratky, E., Monnard, I., Hajjaj, H., Meirim, I., Piguet-Welsch, C., Hauser, J., Mace, K. and Niederberger, P. (2004) Cholesterol-lowering properties of *Ganoderma lucidum in vitro, ex vivo,* and in hamsters and minipigs. *Lipids Health Disease* 18, 2.

Berovic, M., Habijanic, J., Zore, I., Wraber, B., Hodzar, D., Boh, B. and Pohleven, F. (2003) Submerged cultivation of *Ganoderma lucidum* biomass and immunostimulatory effects of fungal polysaccharides. *Journal of Biotechnology* 103, 77–86.

Buchanan, P.K. (2001) A taxonomic overview of the genus *Ganoderma* with special reference to species of medicinal and nutraceutical importance. In: *Proceedings of the International Symposium on Ganoderma Science,* Auckland, New Zealand, pp. 1–8.

Cao, Q.-Z. and Lin, Z.-B. (2006) *Ganoderma lucidum* polysaccharides peptide inhibits the growth of vascular endothelial cell and the induction of VEGF in human lung cancer cell. *Life Sciences* 78, 1457–1463.

Chang, U.-M., Li, C.-H., Lin, L.-I., Huang, C.-P., Kan, L.-S. and Lin, S.-B. (2006) Ganoderiol F, a ganoderma triterpene, induces senescence in hepatoma HepG2 cells. *Life Sciences* 79, 1129–1139.

Chen, W.-Q., Luo, S.-H., LI, H.-Z. and Yang, H. (2005) Effects of *Ganoderma lucidum* polysaccharides on serum lipids and lipoperoxidation in experimental hyperlipidemic rats. *Zhongguo Zhong Yao Za Zhi/Zhongguo Zhongyao Zazhi* [China Journal of Chinese Materia Medica] 30, 1358–1360.

Eo, S.K., Kim, Y.S., Lee, C.K. and Han, S.S. (1999a) Antiviral activities of various water and methanol soluble substances isolated from *Ganoderma lucidum. Journal of Ethnopharmacology* 68, 129–136.

Eo, S.K., Kim, Y.S., Lee, C.K. and Han, S.S. (1999b) Antiherpetic activities of various protein-bound polysaccharides isolated from *Ganoderma lucidum. Journal of Ethnopharmacology* 68, 175–181.

Frye, L.L. and Leonard, D.A. (1999) Lanosterol analogs: dual-action inhibitors of cholesterol biosynthesis. *Critical Reviews in Biochemistry and Molecular Biology* 34, 123–140.

Gao, J.-J., Min, B.-S., Ahn, E.-M., Nakamura, N., Lee, H.-K. and Hattori, M. (2002) New triterpene aldehydes, lucialdehydes A–C, from *Ganoderma lucidum* and their cytotoxicity against murine and human tumour cells. *Chemical and Pharmaceutical Bulletin* 50, 837–840.

Gao, J.-J., Nakamura, N., Min, B.-S., Hirakawa, A., Zuo, F. and Hattori, M. (2004) Quantitative determination of bitter principles in specimens of *Ganoderma lucidum* using high-performance liquid chormatography and its application to the evaluation of *Ganoderma* products. *Chemical Pharmaceutical Bulletin* 52, 688–695.

Gao, J.L., Yu, Z.L., Li, S.P. and Wang, Y.T. (2005) Research advance on triterpenoids of *Ganoderma lucidum. Edible Fungi of China* 24, 6–11.

Gao, Y. and Zhou, S. (2003) Cancer prevention and treatment by *Ganoderma,* a mushroom with medical properties. *Food Reviews International* 19, 275–325.

Gao, Y., Lan, J., Dai, X., Ye, J. and Zhou, S. (2004) A phase I/II study of a *Ganoderma lucidum* (W. Curt.: Fr.) Lloyd (Aphyllophoromycetidae) extract in patients with type II diabetes mellitus. *International Journal of Medicinal Mushrooms* 6, 3–9.

Gao, Y., Tang, W., Dai, X., Gao, H., Chen, G., Ye, J., Chan, E., Koh, H.L., Li, X. and Zhou, S. (2005) Effects of water-soluble *Ganoderma lucidum* polysaccharides on the immune functions of patients with advanced lung cancer. *Journal of Medicinal Food* 8, 159–168.

Ha, T.B., Gerhauser, C., Zhang, W.D., Ho-Chong-Line, N. and Fouraste, I. (2000) New lanostanoids from *Ganoderma lucidum* that induce NAD(P)H:quinone oxidoreductase in cultured hepalcic7 murine hepatoma cells. *Planta Medica* 66, 681–684.

Haak-Frendscho, M., Kino, K., Sone, T. and Jardieu, P. (1993) Ling Zhi-8: a novel T cell mitogen induces cytokine production and upregulation of ICAM-1 expression. *Cellular Immunology* 150, 101–113.

Hajjaj, H., Mace, C., Roberts, M., Niederberger, P. and Fay, L.B. (2005) Effect of 26-oxygenosterols from *Ganoderma lucidum* and their activity as cholesterol synthesis inhibitors. *Applied and Environmental Microbiology* 71, 3653–3658.

He, C.Y., Li, W.D., Guo, S.X., Lin, S.Q. and Lin, Z.B. (2006) Effect of polysaccharides from *Ganoderma lucidum* on streptozotocin-induced diabetic nephropathy in mice. *Journal of Asian Natural Products Research* 8, 705–711.

Hong, K.-J., Dunn, D.M., Shen, C.-L. and Pence, B.C. (2004) Effects of *Ganoderma lucidum* on apoptotic and anti-inflammatory function in HT-29 human colonic carcinoma cells. *Phytotherapy Research* 18, 768–770.

Hsiao, W.L.W., Li, Y.Q., Lee, T.L., Li, N., You, M.M. and Chang, S.T. (2004) Medicinal mushroom extracts inhibit ras-induced cell transformation and the inhibitory effect requires the presence of normal cells. *Carcinogenesis* 25, 1177–1183.

Hsu, H.-Y., Hua, K.-F., Wu, W.-C., Hsu, J., Weng, S.-T., Lin, T.-L., Liu, C.-Y., Hseu, R.-S. and Huang, C.-T. (2008) Reishi immuno-modulation protein induces interleukin-2 expression via protein kinase-dependent signaling pathways within human T cells. *Journal of Cellular Physiology* 215, 15–26.

Hu, H., Ahn, N.-S., Yang, X., Lee, Y.-S. and Kang, K.-S. (2002) *Ganoderma lucidum* extract induces cell cycle arrest and apoptosis in MCF-7 human breast cancer cell. *International Journal of Cancer* 102, 250–253.

Hua, K.-F., Hsu, H.-Y., Chao, L.K., Chen,S.-T., Yang, W.-B., Hsu, J. and Wong, C.-H. (2007) *Ganoderma lucidum* polysaccharides enhance CD14 endocytosis of LPS and promote TLR4 signal transduction of cytokine expression. *Journal of Cellular Physiology* 212, 537–550.

Huang, K.C. (1993) *The Pharmacology of Chinese Herbs.* CRC Press, Boca Raton, Florida.

Huie, C.W. and Di, X. (2004) Chromatographic and electrophoretic methods for Lingzhi pharmacologically active components. *Journal of Chromatography B: Analytical Technologies in the Biomedical and Life Sciences* 812, 241–257 (http://zipcodezoo.com/Fungi/G/Ganoderma_amboinense.asp).

Iwatsuki, K., Akihisa, T., Tokuda, H., Ukiya, M., Oshikubo, M., Kimura, Y., Asano, T., Nomura, A. and Nishino, H. (2003) Lucidenic acids P and Q, methyl lucidenate P, and other triterpenoids from the fungus *Ganoderma lucidum* and their inhibitory effects on Epstein-Barr virus activation. *Journal of Natural Products* 66, 1582–1585.

Jiang, J., Slivova, V., Harvey, K., Valachovicova, T. and Sliva, D. (2004a) *Ganoderma lucidum* suppresses growth of breast cancer cells through the inhibition of Akt/NF-kappaB signalling. *Nutrition and Cancer* 49, 209–216.

Jiang J., Slivova, V., Valachovicova, T., Harvey, K. and Sliva, D. (2004b) *Ganoderma lucidum* inhibits proliferation and induces apoptosis in human prostate cancer cells PC-3. *International Journal of Oncology* 24, 1093–1099.

Jiang, J., Slivova, V. and Sliva, D. (2006) *Ganoderma lucidum* inhibits proliferation of human breast cancer cells by down-regulation of estrogen receptor and NF-kappaB signaling. *International Journal of Oncology* 29, 695–703.

Jiang, J., Grieb, B., Thyagarajan, A. and Sliva, D. (2008) Ganoderic acids suppress growth and invasive behavior of breast cancer cells by modulating AP-1 and NF-kappaB signaling. *International Journal of Molecular Medicine* 21, 577–584.

Kawagishi, H., Mitsunaga, S., Yamawaki, M., Ido, M., Shimada, A., Kinoshita, T., Murata, T., Usui, T., Kimura, A. and Chiba, S. (1997) A lectin from mycelia of the fungus *Ganoderma lucidum*. *Phytochemistry* 44, 7–10.

Kim, D.H., Shim, S.B., Kim, N.J. and Jang, I.S. (1999) Beta-glucuronidase-inhibitory activity and hepatoprotective effect of *Ganoderma lucidum*. *Biological and Pharmaceutical Bulletin* 22, 162–164.

Kim, H.S., Kacew, S. and Lee, B.M. (1999) *In vitro* chemopreventive effects of plant polysaccharides (*Aloe barbadensis, Lentinus edodes, Ganoderma lucidum* and *Coriolus versicolor*). *Carcinogenesis* 20, 1637–1640.

Kim, Y.S., Eo, S.K., Oh, K.W., Lee, C. and Han, S.S. (2000) Antiherpetic activities of acidic protein-bound polysacchride isolated from *Ganoderma lucidum* alone and in combinations with interferons. *Journal of Ethnopharmacology* 72, 451–458.

Kimura, Y., Taniguchi, M. and Baba, K. (2002) Antitumour and antimetastatic effects on liver of triterpenoid fractions of *Ganoderma lucidum*: mechanism of action and isolation of an active substance. *Anticancer Research* 22, 3309–3318.

Kino, K., Mizumoto, K., Sone, T., Yamaji, T., Watanabe, J., Yamashita, A., Yamaoka, K., Shimizu, K., Ko, K. and Tsunoo, H. (1990) An immunomodulating protein, Ling Zhi-8 (LZ-8), prevents insulitis in non-obese diabetic mice. *Diabetologia* 33, 713–718.

Komoda, Y., Shimizu, M., Sonoda, Y. and Sato, Y. (1989) Ganoderic acid and its derivatives as cholesterol synthesis inhibitors. *Chemical and Pharmaceutical Bulletin* 37, 531–533.

Kubota, T., Asaka, Y., Miura, I. and Mori, H. (1982) Structures of ganoderic Acid A and B, two new lanostane-type bitter triterpenes from *Ganoderma lucidum*. *Helvetica Chimica Acta* 65, 611–619.

Lakshmi, B., Ajith, T.A., Sheena, N., Gunapalan, N. and Janardhanan, K.K. (2003) Antiperoxidative, anti-inflammatory, and antimutagenic activities of ethanol extract of the mycelium of *Ganoderma lucidum* occurring in South India. *Teratogenesis, Carcinogenesis, and Mutagenesis*(Suppl. 1), 85–97.

Lakshmi, B., Ajith, T.A., Jose, N. and Janardhanan, K.K. (2006) Antimutagenic activity of methanolic extract of *Ganoderma lucidum* and its effect on hepatic damage caused by benzo[a]pyrene. *Journal of Ethnopharmacology* 107, 297–303.

Lee, J.M., Kwon, H., Jeong, H., Lee, J.W., Lee, S.Y., Baek, S.J. and Surh, Y.J. (2001) Inhibition of lipid peroxidation and oxidative DNA damage by *Ganoderma lucidum*. *Phytotherapy Research* 15, 245–249.

Lee, S., Park, S., Oh, J.W. and Yang, C. (1998) Natural inhibitors for protein prenyltransferase. *Planta Medica* 64, 303–308.

Lee, S.Y. and Rhee, H.M. (1990) Cardiovascular effects of mycelium extract of *Ganoderma lucidum*: inhibition of sympathetic outflow as a mechanism of its hypotensive action. *Chemical and Pharmaceutical Bulletin* 38, 1359–1364.

Li, C., Li, Y. and Sun, H.H. (2006) New ganoderic acids, bioactive triterpenoid metabolites from the mushroom *Ganoderma lucidum*. *Natural Product Research* 20, 985–991.

Li, C.-H., Chen, P.-Y., Chang, U.-M., Kan, L.-S., Fang, W.-H., Tsai, K.-S. and Lin, S.-B. (2005) Ganoderic acid X, a lanostanoid triterpene, inhibits topoisomerases and induces apoptosis of cancer cells. *Life Sciences* 77, 252–265.

Li, Y.-Q. and Wang, S.-F. (2006) Anti-hepatitis B activities of ganoderic acid from *Ganoderma lucidum*. *Biotechnology Letters* 28, 837–841.

Li, Z., Liu, J. and Zhao, Y. (2005) Possible mechanism underlying the antiherpetic activity of a proteoglycan isolated from the mycelia of *Ganoderma lucidum in vitro*. *Journal of Biochemistry and Molecular Biology* 38, 34–40.

Liby, K.T., Yore, M.M. and Sporn, M.B. (2007) Triterpenoids and rexinoids as multifunctional agents for the prevention and treatment of cancer. *Nature Reviews Cancer* 7, 357–369.

Li Khva, R., Vasil'ev, A.V., Orekhov, A.N., Tertov, V.V. and Tutel'ian, V.A. (1989) Anti-atherosclerotic properties of higher mushrooms (a clinico-experimental investigation). *Voprosy Pitaniia* 1, 16–19.

Lin, C.N., Tome, W.P. and Won, S.J. (1991) Novel cytotoxic principles of Formosan *Ganoderma lucidum*. *Journal of Natural Products* 54, 998–1002.

Lin, J.M., Lin, C.C., Chen, M.F., Ujiie, T. and Takada, A. (1995) Radical scavenger and antihepatotoxic activity of *Ganoderma formosanum*, *Ganoderma lucidum* and *Ganoderma neo-japonicum*. *Journal of Ethnopharmacology* 47, 33–41.

Lin, K.-I., Kao, Y.-Y., Kuo, H.-K., Yang, W.-B., Chou, A., Lin, H.-H., Yu, A.L. and Wong, C.-H. (2006) Reishi polysaccharides induce immunoglobulin production through the TLR4/TLR2-mediated induction of transcription factor Blimp-1. *Journal of Biological Chemistry* 281, 24111–24123.

Lin, S.-B., Li, C.-H., Lee, S.-S. and Kan, L.-S. (2003) Triterpene-enriched extracts from *Ganoderma lucidum* inhibit growth of hepatoma cells via suppressing protein kinase C, activating mitogen-activated protein kinases and G2-phase cell cycle arrest. *Life Sciences* 72, 2381–2390.

Lin, W.-C. and Lin, W.-L. (2006) Ameliorative effect of *Ganoderma lucidum* on carbon tetrachloride-induced liver fibrosis in rats. *World Journal of Gastroenterology* 12, 265–270.

Lin, Z.-B. (2005) Cellular and molecular mechanisms of immuno-modulation by *Ganoderma lucidum*. *Journal of Pharmacological Sciences* 99, 144–153.

Lindequist, U., Niedermeyer, T.H.J. and Julich, W.-D. (2005) The pharmacological potential of mushrooms. *Evidence-Based Complementary and Alternative Medicine: eCAM* 2, 285–299.

Liu, J., Yang, F., Ye, L.-B., Yang, X.-J., Timani, K.A., Zheng, Y. and Wang, Y.-H. (2004) Possible mode of action of antiherpetic activities of a proteoglycan isolated from the mycelia of *Ganoderma lucidum in vitro*. *Journal of Ethnopharmacology* 95, 265–272.

Liu, J., Kurashiki, K., Shimizu, K. and Kondo, R. (2006) 5α-Reductase inhibitory effect of triterpenoids isolated from *Ganoderma lucidum*. *Biological Pharmaceutical Bulletin* 29, 392–395.

Liu, J., Shimizu, K., Konishi, F., Kumamoto, S. and Kondo, R. (2007) The anti-androgen effect of ganoderol B isolated from the fruiting body of *Ganoderma lucidum*. *Bioorganic and Medicinal Chemistry* 15, 4966–4972.

Lu, Q.-Y., Jin, Y.-S., Zhang, Q., Zhang, Z., Heber, D., Go, V.L.W., Li, F.P. and Rao, J.Y. (2004) *Ganoderma lucidum* extracts inhibit growth and induce actin polymerization in bladder cancer cells *in vitro*. *Cancer Letters* 216, 9–20.

Min, B.S., Gao, J.J., Nakamura, N. and Hattori, M. (2000) Triterpenes from the spores of *Ganoderma lucidum* and their cytotoxicity against meth-A and LLC tumour cells. *Chemical and Pharmaceutical Bulletin* 48, 1026–1033.

Miyazaki, T. and Nishijima, M. (1981) Studies on fungal polysaccharides. XXVII. Structural examination of a water-soluble, antitumour polysaccharide of *Ganoderma lucidum*. *Chemical and Pharmaceutical Bulletin* 29, 3611–3616.

Mizushina, Y., Takahashi, N., Hanashima, L., Koshino, H., Esumi, Y., Uzawa, J., Sugawara, F. and Sakaguchi, K. (1999) Lucidenic acid O and lactone, new terpene inhibitors of eukaryotic DNA polymerases from a basidiomycete, *Ganoderma lucidum*. *Bioorganic and Medicinal Chemistry* 7, 2047–2052.

Morigiwa, A., Kitabatake, K., Fujimoto, Y. and Ikekawa, N. (1986) Angiotensin converting enzyme-inhibitory triterpenes from *Ganoderma lucidum*. *Chemical and Pharmaceutical Bulletin* 34, 3025–3028.

Mothana, R.A.A., Awadh Ali, N.A., Jansen, R., Wegner, U., Mentel, R. and Lindequist, U. (2003) Antiviral lanostanoid triterpenes from the fungus *Ganoderma pfeifferi*. *Fitoterapia* 74, 177–180.

Niedermeyer, T.H.J., Lindequist, U., Mentel, R., Gordes, D., Schmidt, E., Thurow, K. and Lalk, M. (2005) Antiviral terpenoid constituents of *Ganoderma pfeifferi*. *Journal of Natural Products* 68, 1728–1731.

Park, E.J., Ko, G., Kim, J. and Sohn, D.H. (1997) Antifibrotic effects of a polysaccharide extracted from *Ganoderma lucidum*, glycyrrhizin, and pentoxifylline in rats with cirrhosis induced by biliary obstruction [erratum appears in *Biol. Pharm. Bull.* (1998) 21(6), 649]. *Biological and Pharmaceutical Bulletin* 20, 417–420.

Paterson, R.R.M. (2006) Ganoderma – a therapeutic fungal biofactory. *Phytochemistry* 67, 1985–2001.

Ringler, R.L., Byerrum, R.U., Steven, T.A., Clark, P.P. and Stock, C.C. (1957) Studies on antitumour substances produced by basidiomycetes. *Antiobiotic Chemotherapy* 7, 1–5.

Shiao, M.-S. (2003) Natural products of the medicinal fungus *Ganoderma lucidum*: occurrence, biological activities, and pharmacological functions. *Chemical Record: an Official Publication of the Chemical Society of Japan* 3, 172–180.

Sliva, D., Sedlak, M., Slivova, V., Valachovicova, T., Lloyd, F.P., Jr. and Ho, N.W.Y. (2003) Biologic activity of spores and dried powder from *Ganoderma lucidum* for the inhibition of highly invasive human breast and prostate cancer cells. *Journal of Alternative and Complementary Medicine* 9, 491–497.

Slivova, V., Valachovicova, T., Jiang, J. and Sliva, D. (2004) *Ganoderma lucidum* inhibits invasiveness of breast cancer cell. *Journal of Cancer Integrative Medicine* 2, 25–30.

Stanley, G., Harvey, K., Slivova, V., Jiang, J. and Sliva, D. (2005) *Ganoderma lucidum* suppresses angiogenesis through the inhibition of secretion of VEGF and TGF-beta1 from prostate cancer cells. *Biochemical and Biophysical Research Communications* 330, 46–52.

Su, C.Y., Shiao, M.S. and Wang, C.T. (1999) Predominant inhibition of ganodermic acid S on the thromboxane A2-dependent pathway in human platelets response to collagen. *Biochimica et Biophysica Acta* 1437, 223–234.

Sun, J., He, H. and Xie, B.J. (2004) Novel antioxidant peptides from fermented mushroom *Ganoderma lucidum*. *Journal of Agricultural and Food Chemistry* 52, 6646–6652.

Tanaka, S., Ko, K., Kino, K., Tsuchiya, K., Yamashita, A., Murasugi, A., Sakuma, S. and Tsunoo, H. (1989) Complete amino acid sequence of an immunomodulatory protein, ling zhi-8 (LZ-8). An immunomodulator from a fungus, *Ganoderma lucidium*, having similarity to immunoglobulin-variable regions. *Journal of Biological Chemistry* 264, 16372–16377.

Tang, W., Liu, J.-W., Zhao, W.-M., Wei, D.-Z. and Zhong, J.-J. (2006) Ganoderic acid T from *Ganoderma lucidum* mycelia induces mitochondria-mediated apoptosis in lung cancer cells. *Life Sciences* 80, 205–211.

Tian, Y.-P. and Zhang, K.-C. (2005) Purification and characterization of a novel proteinase A inhibitor from *Ganoderma lucidum* by submerged fermentation. *Enzyme and Microbial Technology* 36, 357–361.

Thakur, A., Rana, M., Lakhanpal, T.N., Ahmad, A. and Khan, M.I. (2007) Purification and characterization of lectin from fruiting body of *Ganoderma lucidum*: lectin from *Ganoderma lucidum*. *Biochimica et Biophysica Acta* 1770, 1404–1412.

Thyagarajan, A., Jiang, J., Hopf, A., Adamec, J. and Sliva, D. (2006) Inhibition of oxidative stress-induced invasiveness of cancer cells by *Ganoderma lucidum* is mediated through the suppression of interleukin-8 secretion. *International Journal of Molecular Medicine* 18, 657–664.

Toth, J.O., Luu, B. and Ourisson, G. (1983) Les acides ganoderiques T and Z: triterpenes cytotoxiques de *Ganoderma lucidum* (Polyporacee). *Tetrahedron Letters* 24, 1081–1084.

van der Hem, L.G., van der Vliet, J.A., Bocken, C.F., Kino, K., Hoitsma, A.J. and Tax, W.J. (1995) Ling Zhi-8: studies of a new immunomodulating agent. *Transplantation* 60, 438–443.

Wachtel-Galor, S., Tomlinson, B. and Benzie, I.F.F. (2004) *Ganoderma lucidum* ('Lingzhi'), a Chinese medicinal mushroom: biomarker responses in a controlled human supplementation study [see comment]. *British Journal of Nutrition* 91, 263–269.

Wang, G., Zhao, J., Liu, J., Huang, Y., Zhong, J.-J. and Tang, W. (2007) Enhancement of IL-2 and IFN-gamma expression and NK cells activity involved in the anti-tumour effect of ganoderic acid Me *in vivo*. *International Immunopharmacology* 7, 864–870.

Wang, H.X. and Ng, T.B. (2006a) Ganodermin, an antifungal protein from fruiting bodies of the medicinal mushroom *Ganoderma lucidum*. *Peptides* 27, 27–30.

Wang, H.X. and Ng, T.B. (2006b) A laccase from the medicinal mushroom *Ganoderma lucidum*. *Applied Microbiology and Biotechnology* 72, 508–513.

Wang, S.Y., Hsu, M.L., Hsu, H.C., Tzeng, C.H., Lee, S.S., Shiao, M.S. and Ho, C.K. (1997) The anti-tumour effect of *Ganoderma lucidum* is mediated by cytokines released from activated macrophages and T lymphocytes. *International Journal of Cancer* 70, 699–705.

Wang, X., Zhao, X., Li, D., Lou, Y.-Q., Lin, Z.-B. and Zhang, G.-L. (2007) Effects of *Ganoderma lucidum* polysaccharide on CYP2E1, CYP1A2 and CYP3A activities in BCG-immune hepatic injury in rats. *Biological and Pharmaceutical Bulletin* 30, 1702–1706.

Wang, Y.-Y., Khoo, K.-H., Chen, S.-T., Lin, C.-C., Wong, C.-H. and Lin, C.-H. (2002) Studies on the immuno-modulating and antitumour activities of *Ganoderma lucidum* (Reishi) polysaccharides: functional and proteomic analyses of a fucose-containing glycoprotein fraction responsible for the activities. *Bioorganic and Medicinal Chemistry* 10, 1057–1062.

Wanmuang, H., Leopairut, J., Kositchaiwat, C., Wananukul, W. and Bunyaratvej, S. (2007) Fatal fulminant hepatitis associated with *Ganoderma lucidum* (Lingzhi) mushroom powder. *Journal of the Medical Association of Thailand* 90, 179–181.

Wasser, S.P. (2005) Reishi (*Ganoderma lucidum*). In: Coates P., Blackman, M.R., Cragg, G., Levine, M., Moss, J. and Whit, J. (eds) *Encyclopedia of Dietary Supplements*. Marcel Dekker, New York, pp. 603–622.

Weng, C.-J., Chau, C.-F., Chen, K.-D., Chen, D.-H. and Yen, G.-C. (2007) The anti-invasive effect of lucidenic acids isolated from a new *Ganoderma lucidum* strain. *Molecular Nutrition and Food Research* 51, 1472–1477.

Weng, C.-J., Chau, C.-F., Hsieh, Y.-S., Yang, S.-F. and Yen, G.-C. (2008) Lucidenic acid inhibits PMA-induced invasion of human hepatoma cells through inactivating MAPK/ERK signal transduction pathway and reducing binding activities of NF-kappaB and AP-1. *Carcinogenesis* 29, 147–156.

Wu, T.S., Shi, L.S. and Kuo, S.C. (2001) Cytotoxicity of *Ganoderma lucidum* triterpenes. *Journal of Natural Products* 64, 1121–1122.

Wu, Y.-W., Chen, K.-D. and Lin, W.-C. (2004) Effect of *Ganoderma tsugae* on chronically carbon tetrachloride-intoxicated rats. *American Journal of Chinese Medicine* 32, 841–850.

Yan, J., Vetvicka, V., Xia, Y., Coxon, A., Carroll, M.C., Mayadas, T.N. and Ross, G.D. (1999) Beta-glucan, a 'specific' biologic response modifier that uses antibodies to target tumours for cytotoxic recognition by leukocyte complement receptor type 3 (CD11b/CD18). *Journal of Immunology* 163, 3045–3052.

Yang, X.-J., Liu, J., Ye, L.-B., Yang, F., Ye, L., Gao, J.-R. and Wu, Z.-H. (2006) *In vitro* and *in vivo* protective effects of proteoglycan isolated from mycelia of *Ganoderma lucidum* on carbon tetrachloride-induced liver injury. *World Journal of Gastroenterology* 12, 1379–1385.

Yeh, C.M., Yeh, C.K., Hsu, X.Y., Luo, Q.M. and Lin, M.Y. (2008) Extracellular expression of a functional recombinant *Ganoderma lucidium* immunomodulatory protein by *Bacillus subtilis* and *Lactococcus lactis*. *Applied and Environmental Microbiology* 74, 1039–1049.

Yue, Q.X., Cao, Z.W., Guan, S.H., Liu, X.H., Tao, L., Wu, W.Y., Li, Y.X., Yang, P.Y., Liu, X. and Guo, D.A. (2008) Proteomic characterization of the cytotoxic mechanism of ganoderic acid D and computer-automated estimation of the possible drug–target network. *Molecular and Cellular Proteomics* 7, 949–961.

Yuen, M.-F., Ip, P., Ng, W.-K. and Lai, C.-L. (2004) Hepatotoxicity due to a formulation of *Ganoderma lucidum* (lingzhi). *Journal of Hepatology* 41, 686–687.

Zhang, G.-L., Wang, Y.-H., Ni,W., Ten, H.-L. and Lin, Z.-B. (2002) Hepatoprotective role of *Ganoderma lucidum* polysaccharide against BCG-induced immune liver injury in mice. *World Journal of Gastroenterology* 8, 728–733.
Zhang, H.-N. and Lin, Z.-B. (2004) Hypoglycemic effect of *Ganoderma lucidum* polysaccharides. *Acta Pharmacologica Sinica* 25, 191–195.
Zhang, H.-N., He, J.-H., Yuan, L. and Lin, Z.-B. (2003) *In vitro* and *in vivo* protective effect of *Ganoderma lucidum* polysaccharides on alloxan-induced pancreatic islets damage. *Life Sciences* 73, 2307–2319.
Zhou, X., Lin, J., Yin, Y., Zhao, J., Sun, X. and Tang, K. (2007) Ganodermataceae: natural products and their related pharmacological functions. *American Journal of Chinese Medicine* 35, 559–574.
Zhu, H.S., Yang, X.L., Wang, L.B., Zhao, D.X. and Chen, L. (2000) Effects of extracts from sporoderm-broken spores of *Ganoderma lucidum* on HeLa cells. *Cell Biology and Toxicology* 16, 201–206.
Zhu, M., Chang, Q., Wong, L.K., Chong, F.S. and Li, R.C. (1999) Triterpene antioxidants from *Ganoderma lucidum*. *Phytotherapy Research* 13, 529–531.
Zhu, X.-L., Chen, A.-F. and Lin, Z.-B. (2007) *Ganoderma lucidum* polysaccharides enhance the function of immunological effector cells in immunosuppressed mice. *Journal of Ethnopharmacology* 111, 219–226.

11 Current Advances in Dematiaceous Mycotic Infections

Sanjay G. Revankar

Department of Medicine, Division of Infectious Diseases, Wayne State University, Detroit, Michigan, USA

Introduction

Dematiaceous, or darkly pigmented fungi, are a heterogeneous group of organisms that have been associated with a variety of clinical syndromes (Revankar, 2004). They are uncommon causes of human disease, but can be responsible for life-threatening infections in both immunocompromised and immunocompetent individuals. In recent years these fungi have been increasingly recognized as important pathogens, and the spectrum of diseases with which they are associated has also expanded (Revankar, 2004).

The clinical syndromes are typically distinguished based on histologic findings into chromoblastomycosis, eumycetoma and phaeohyphomycosis. Chromoblastomycosis is caused by a small group of fungi that produce characteristic sclerotic bodies in tissue, and is usually seen in tropical areas (Pang *et al.*, 2004). Eumycetoma is a deep tissue infection, usually of the lower extremities, characterized by the presence of mycotic granules (Pang *et al.*, 2004). It is also associated with a relatively small group of fungi. Phaeohyphomycosis is a catch-all term generally reserved for the remainder of clinical syndromes caused by dematiaceous fungi (Rinaldi, 1996). For the purposes of this chapter, these will be divided into superficial infections, allergic disease, pneumonia, central nervous system (CNS) infection and disseminated disease. These are not necessarily mutually exclusive, and other groupings could be devised as well. For some infections in immunocompetent individuals, such as allergic fungal sinusitis and brain abscess, they are among the most common aetiologic fungi. Although some have questioned the use of the term 'dematiaceous' to describe this group of fungi (Pappagianis and Ajello, 1994), it remains widely used, and that convention will be continued in this chapter.

Mycology

Over 100 species and 60 genera of dematiaceous fungi have been implicated in human disease (Matsumoto *et al.*, 1994). They are typically soil organisms and are generally distributed worldwide. However, there are species that do appear to be geographically restricted, such as *Ramichloridium mackenziei*, which has only been seen in patients from the Middle East (Sutton *et al.*, 1998). Exposure is thought to be from inhalation or minor trauma, which may not even be noticed by the patient. Surveys of outdoor air for fungal spores routinely detect dematiaceous fungi (Shelton *et al.*, 2002). This suggests that most if not all individuals are exposed to them, although they remain uncommon causes of disease. As the number of patients immunocompromised from diseases and medical therapy increases, additional species are being reported as causes of human disease, expanding an already long list of potential pathogens. Common genera associated with specific clinical syndromes are listed in Table 11.1. The distinguishing characteristic common to all these various species is the presence of melanin in their cell walls, which imparts the dark colour to their conidia or spores and hyphae. The colonies are also typically brown to black.

Guidelines are available regarding the handling of potentially infectious fungi in the laboratory setting. It has been suggested that work with cultures of certain well-known pathogenic fungi, such as *Coccidioides immitis* and *Histoplasma capsulatum*, should be contained in a Biosafety Level 3 facility, which requires a separate negative-pressure room. Recently, certain agents of phaeohyphomycosis, in particular *Cladophialophora bantiana*, have been included in the list of fungi that should be kept under Biosafety Level 2 containment (Padhye *et al.*, 1998). This seems reasonable given their propensity, albeit rare, for causing life-threatening infection in normal individuals.

Pathogenesis

Few details are known regarding the pathogenic mechanisms by which these fungi cause disease, particularly in immunocompetent individuals. One of the likely virulence factors is the presence of melanin in the cell wall, which is common to all dematiaceous fungi. There are putative mechanisms by which melanin may act as a virulence factor (Butler and Day, 1998; Jacobson, 2000; Hamilton and Gomez, 2002). It may confer a protective advantage by scavenging free radicals and hypochlorite that are produced by phagocytic cells in their oxidative burst that would normally kill most organisms (Jacobson, 2000). In addition, melanin may bind to hydrolytic enzymes, thereby preventing their action on the plasma membrane (Jacobson, 2000). These multiple functions may help explain the pathogenic potential of some dematiaceous fungi, even in immunocompetent hosts.

Considerable work has been undertaken on several fungi that contain melanin (Casadevall *et al.*, 2000; Romero-Martinez *et al.*, 2000; Gomez *et al.*,

Table 11.1. Clinical syndromes, associated dematiaceous fungi and suggested therapy.

Clinical syndrome	Commonly associated fungal genera	Therapy
Onychomycosis	*Onychocola* *Alternaria*	Itra or terb +/– topical agents
Subcutaneous nodules	*Exophiala* *Alternaria* *Phialophora*	Surgery +/– itra
Chromoblastomycosis	*Fonsecaea* (*F. pedrosoi*) *Phialophora* *Rhinocladiella*	Itra
Eumycetoma	*Madurella* *Pyrenochaetae*	Itra
Keratitis	*Curvularia* *Bipolaris* *Exserohilum* *Lasiodiplodia*	Topical natamycin +/– itra
Allergic fungal sinusitis	*Bipolaris* *Curvularia*	Steroids +/– itra
Allergic bronchopulmonary mycosis	*Bipolaris* *Curvularia*	Steroids +/– itra
Pneumonia	*Ochroconis* *Exophiala* *Chaetomium*	Itra (AmB if severe)
Brain abscess	*Cladophialophora* (*C. bantiana*) *Ramichloridium* (*R. mackenziei*) *Ochroconis*	High dose azole + lipid AmB +5-FC
Disseminated disease	*Scedosporium (S. prolificans)* *Bipolaris* *Exophiala* (syn. *Wangiella*)	Lipid AmB + azole +/– echinocandin

Itra, itraconazole; Terb, terbinafine; AmB, amphotericin B; 5-FC, flucytosine; azole, itraconazole, voriconazole or posaconazole.

2001). Specifically, in the yeasts *Cryptococcus neoformans* and *Exophiala dermatitidis* (syn. *Wangiella dermatitidis*), disruption of melanin production leads to markedly reduced virulence in animal models (Kwon-Chung *et al.*, 1982; Dixon *et al.*, 1987). Melanin has also been shown to reduce the susceptibility of *C. neoformans* and *H. capsulatum* to amphotericin B and caspofungin, possibly by binding these drugs (van Duin *et al.*, 2002; Ikeda *et al.*, 2003). This effect is not apparent with azole drugs (van Duin *et al.*, 2002). Finally, melanin has even been found in the yeast *Candida albicans*, suggesting that it may be a more fundamental and common virulence factor than previously thought (Morris-Jones *et al.*, 2005).

It is interesting to note that almost all allergic disease and eosinophilia is caused by two genera, *Bipolaris* and *Curvularia* (Revankar, 2004). These organisms are very common in the environment, so exposure is practically universal. The virulence factors in these fungi that are responsible for eliciting allergic reactions are unclear at present.

Diagnosis

These ubiquitous fungi are often considered contaminants in cultures, making the determination of clinical significance problematic. A high degree of clinical suspicion – as well as correlation with appropriate clinical findings – is required when interpreting culture results. Unfortunately, there are no simple serologic or antigen tests available to detect these fungi in blood or tissue, unlike other common mycoses that cause human disease. Polymerase chain reaction (PCR) is being studied as an aid to the diagnosis of fungal infections, but as yet PCR methods are not widely available or reliable for this group of fungi. However, studies have begun to examine the potential of identifying species within this diverse group of fungi using PCR of highly conserved regions of ribosomal DNA (Abliz *et al.*, 2004).

The diagnosis of phaeohyphomycosis currently rests on pathological examination of clinical specimens and careful gross and microscopic examination of cultures, occasionally requiring the expertise of a mycology reference laboratory. In tissue, they will stain strongly with the Fontana–Masson stain, which is specific for melanin (Rinaldi, 1996). This can be helpful in distinguishing these fungi from other species, particularly *Aspergillus*. In addition, hyphae typically appear more fragmented in tissue than seen with *Aspergillus*, with irregular septate hyphae and beaded, yeast-like forms (Rinaldi, 1996).

In vitro Studies

In vitro antifungal testing has advanced considerably in recent years, especially when one considers that a standardized method for testing yeasts was not available until 1997 (National Committee for Clinical Laboratory Standards, 1997), and the first standardized method for filamentous fungi was not available until 2002 (National Committee for Clinical Laboratory Standards, 2002). Due to the relatively recent development of antifungal susceptibility testing, the available *in vitro* data for dematiaceous fungi are relatively sparse, and often rely on small numbers of isolates per species. An important issue is that much of the older literature is often inconsistent with regards to methodology, making reliable observations difficult. In addition, interpretive breakpoints are unavailable for most drugs and all the fungi. Clinical correlation data is practically non-existent, and therefore suggestions regarding susceptibility are guidelines only.

The newer azoles itraconazole and voriconazole demonstrate the most consistent *in vitro* activity against dematiaceous fungi, except for *Scedosporium prolificans* and *Scopulariopsis brumptii*, which are resistant to all azoles (McGinnis and Pasarell, 1998b; Carrillo and Guarro, 2001; Espinel-Ingroff *et al.*, 2001; Meletiadis *et al.*, 2002). Oral and intravenous (IV) formulations are available for both drugs. For itraconazole, the capsule form requires an acidic environment for absorption, while the suspension with cyclodextrin does not, and is absorbed more consistently. Itraconazole demonstrates good activity against the vast majority of dematiaceous fungi tested (Johnson *et al.*, 1998; McGinnis and Pasarell, 1998a,b; Espinel-Ingroff, 2001; Espinel-Ingroff *et al.*, 2001; van de Sande *et al.*, 2005).

Minimum inhibitory concentrations (MICs) are generally ≤ 0.125 μg/ml for this group of fungi. Voriconazole has become the treatment of choice for invasive aspergillosis, supplanting amphotericin B for this indication (Johnson and Kauffman, 2003). In addition, like itraconazole, voriconazole has a broad spectrum of activity that includes most dematiaceous fungi (McGinnis and Pasarell, 1998b; Espinel-Ingroff, 2001; Espinel-Ingroff *et al.*, 2001). However, MICs are usually slightly higher for voriconazole, although the clinical significance of this is unclear.

Other azoles have a much more limited role in the therapy of these infections. Ketoconazole was the first oral azole, and has a relatively broad spectrum. However, its current use has been significantly limited by a number of side effects and the availability of newer agents that are much better tolerated. Sparse *in vitro* data are available for dematiaceous fungi, but good activity is noted for the most common fungi causing chromoblastomycosis and mycetoma (Andrade *et al.*, 2004; van de Sande *et al.*, 2005). Fluconazole has negligible activity against dematiaceous fungi, and essentially no role in therapy given the variety of other options available.

The newest azole, posaconazole, has a broad spectrum similar to itraconazole, but with enhanced activity, particularly against *Aspergillus* and other fungi (Herbrecht, 2004). The published *in vitro* data are limited for dematiaceous fungi, but good activity has been demonstrated against most species tested, including *Bipolaris* spp., *C. bantiana* and *R. mackenziei* (Espinel-Ingroff, 1998; Al-Abdely *et al.*, 2000; Pfaller *et al.*, 2002). Ravuconazole is an investigational azole with activity against a wide variety of fungi (Fung-Tomc *et al.*, 1998).

Amphotericin B generally has good *in vitro* activity against most clinically important dematiaceous fungi. However, some species – including *S. prolificans* and *S. brumptii* – have been consistently resistant with a MIC ≥ 2 μg/ml *in vitro* (McGinnis and Pasarell, 1998b), while other species have occasionally been found to be resistant, including *Curvularia* spp., *Exophiala* spp. and *R. mackenziei* (McGinnis and Pasarell, 1998b; Sutton *et al.*, 1998). Significant nephrotoxicity often limits use of the standard formulation, but this has largely been reduced by the development of lipid-associated formulations. Use of lipid amphotericin B preparations allows for much higher doses than possible with standard amphotericin B, which may improve their efficacy against these fungi.

Flucytosine (5-FC) is unique in its mechanism of action, and inhibits DNA and RNA synthesis. Development of resistance during monotherapy has resulted in its use in combination therapy for systemic mycoses, most notably cryptococcal meningitis (Vermes *et al.*, 2000). *In vitro* studies are limited, although activity has been shown against a variety of dematiaceous fungi, including *C. bantiana*, *Exophiala* spp. and *Fonsecaea pedrosoi*, the major etiologic agent of chromoblastomycosis (Dixon and Polak, 1987; Caligiorne *et al.*, 1999).

Allylamines, like the azoles, also inhibit ergosterol synthesis, but act on squalene epoxidase, an enzyme two steps before the target of azoles. Their clinical role has been limited to treatment of dermatophyte infections, although there has been recent interest in potentially expanding their clinical spectrum (Hay, 1999). Terbinafine is the only oral allylamine available for systemic use; however, its extensive binding to serum proteins and distribution into skin and adipose tissue have diminished enthusiasm for its use in treating serious systemic fungal infections (Ryder and Frank, 1992; Hosseini-Yeganeh and McLachlan, 2002). *In vitro* studies against dematiaceous fungi are emerging and fairly broad-spectrum activity has been seen, including that against species of *Alternaria*, *Curvularia* and *Bipolaris* (McGinnis and Pasarell, 1998a; Jessup *et al.*, 2000).

The echinocandins are the latest group of antifungal agents to be developed, and they have a unique mode of action in inhibiting β-1,3 glucan synthesis and thereby disrupting the fungal cell wall (Deresinski and Stevens, 2003). They are available only in an intravenous formulation and are generally well tolerated. Caspofungin was the first of the class available for clinical use. *In vitro* studies with dematiaceous fungi are limited, although some activity has been demonstrated against species of *Curvularia*, *Bipolaris* and *F. pedrosoi* (Espinel-Ingroff, 1998). *C. bantiana* has higher MICs and *S. prolificans* appears resistant (Meletiadis *et al.*, 2002; Espinel-Ingroff, 2003). In general, MICs for dematiaceous fungi are higher than for *Aspergillus* sp. Only a few species have been evaluated *in vitro*, with activity shown against *C. bantiana*, *F. pedrosoi* and *Exophiala spinifera* (Espinel-Ingroff, 2003). MICs may be somewhat lower than for caspofungin. Micafungin and anidulafungin have also recently been approved for clinical use, but limited data are available for them.

The use of antifungal combinations is a potentially useful strategy for refractory infections, though it has not been studied extensively in dematiaceous fungi. The combination of itraconazole and terbinafine has been studied against *S. prolificans*, which is otherwise generally resistant to all agents. *In vitro*, synergistic activity was found against most isolates of this species, and no antagonism was noted (Meletiadis *et al.*, 2000). Voriconazole and terbinafine also display similar synergy *in vitro* (Meletiadis *et al.*, 2003). The mechanism is presumably through potent inhibition of ergosterol synthesis at two different steps of the pathway by these agents. However, this should be interpreted with caution, as terbinafine is not generally used for systemic infections. Another report has suggested synergy for *S. prolificans* with voriconazole and caspofungin (Steinbach *et al.*, 2003). Older literature also suggests synergy with ketoconazole and 5-FC for a variety of dematiaceous fungi (Corrado *et al.*, 1982), and this may be applicable to other azoles as well.

Animal Studies

There are relatively few animal studies with dematiaceous fungi. One of the earliest studies was a murine model of infection with *E. dermatitidis* and *F. pedrosoi* (Polak, 1984). Amphotericin B and 5-FC were active, although ketoconazole was not. In addition, the combination of amphotericin B with 5-FC appeared to be additive, but the combination of ketoconazole with 5-FC was not (Polak, 1984). In another study, 5-FC had the broadest activity against *C. bantiana*, *Ochroconis gallopava* and *E. dermatitidis* in mice, followed by amphotericin B and fluconazole (despite resistance *in vitro*) (Dixon and Polak, 1987). Terbinafine was inactive *in vivo*, despite good *in vitro* activity. More recent studies have found posaconazole to be effective in a variety of infections in mice (Al-Abdely *et al.*, 2000; Graybill *et al.*, 2004).

Clinical Syndromes and their Management

An overview of clinical syndromes, the associated fungi and suggested therapies is given in Table 11.1.

Superficial and deep local infections

Localized infections are the most common form of disease due to dematiaceous fungi. These cases are generally associated with minor trauma or other environmental exposure. Although many pathogens have been reported, relatively few are responsible for the majority of infections. Life-threatening disease is rare, although significant morbidity can occur depending on host immunity, site of infection and response to therapy.

Onychomycosis

Dematiaceous fungi are rare causes of onychomycosis. Clinical features may include a history of trauma, involvement of only one or two toenails or lack of response to standard systemic therapy (Gupta *et al.*, 2003). *Onychocola* and *Alternaria* species have been reported, with the former being highly resistant to therapy (Gupta *et al.*, 2003; Tosti *et al.*, 2003). Itraconazole and terbinafine are the most commonly used systemic agents, and may be combined with topical therapy for refractory cases (Tosti *et al.*, 2003). No published data are available for the newer azole agents.

Subcutaneous lesions

There are numerous case reports of subcutaneous infection due to a wide variety of fungi (Hiruma *et al.*, 1993; Clancy *et al.*, 2000; Kimura *et al.*, 2003; Sutton *et al.*, 2004), and species of *Exophiala, Alternaria, Phialophora* and *Bipolaris* are among the more common etiologic agents (Koga *et al.*, 2003). Minor trauma is the usual inciting factor, though it may be unrecognized by the patient. Lesions typically occur on exposed areas of the body and often appear as isolated cystic or papular lesions. Immunocompromised patients are at increased risk of subsequent dissemination, though rare cases have been described in apparently immunocompetent patients as well. Occasionally, infection may involve joints or bone, requiring more extensive surgery or prolonged antifungal therapy.

As for many of the infectious syndromes associated with dematiaceous fungi, therapy is not standardized. Surgical excision alone has been successful in a number of cases (Summerbell *et al.*, 2004). Oral systemic therapy with an azole antifungal agent in conjunction with surgery is frequently employed and has been used successfully (Clancy *et al.*, 2000; Sutton *et al.*, 2004).

Chromoblastomycosis

Chromoblastomycosis is a slowly progressive, chronic subcutaneous mycosis that is predominantly seen in tropical areas. Minor trauma typically precedes the lesions. Nodular lesions can progress over years to form large, verrucous plaques. Histopathology is characterized by the presence of sclerotic bodies in tissue. By far the most common species involved is *F. pedrosoi*, followed by *Phialophora verrucosa* and, less commonly, by *Cladophialophora carrionii* (syn. *Cladosporium carrionii*) and *Rhinocladiella aquaspersa* (Koga et *al.*, 2003).

There is no standard therapy, though in a large series cryotherapy, itraconazole or their combination were found to cure the largest number of cases (Bonifaz *et al.*, 2001). In developing countries, where systemic antifungals are not easily available or too expensive, the use of cryotherapy alone in a systematic manner over several months has also led to good cure rates (Castro *et al.*, 2003). The exact mechanism of this effect is unclear. As a single agent, itraconazole appears to be the most effective (Restrepo *et al.*, 1988; Kumarasinghe and Kumarasinghe, 2000; Bonifaz *et al.*, 2001; Castro *et al.*, 2003). A variety of other treatments have also been successful, including ketoconazole, flucytosine, local heat therapy and amphotericin B (Minotto *et al.*, 2001). However, the overall cure rate was only 57% in one large series of 100 cases from Brazil, despite use of multiple modalities (Minotto *et al.*, 2001). In refractory cases, the combination of itraconazole and terbinafine has been found to be useful (Gupta *et al.*, 2002).

Eumycetoma

Eumycetoma is another chronic subcutaneous fungal infection characterized by the presence of grains, or sclerotia, in tissue. These grains are usually white or black, depending on the fungal species involved, and are composed of fungal cells surrounded by a dense extracellular matrix containing a melanin compound, which gives it a dark colour and may have a role in protecting the organism from host defences (Ahmed *et al.*, 2004). Eumycetoma is common in many tropical and subtropical areas of the world, and the species involved are often associated with a particular geographic region. *Madurella mycetomatis* is one of the most common species, particularly in Africa (Ahmed *et al.*, 2004). Other species include *Pyrenochaeta romeroi* (South America) and *Leptosphaeria senegalensis* (Africa) (Ahmed *et al.*, 2004).

Unlike chromoblastomycosis and subcutaneous phaeohyphomycosis, which may be cured with surgical techniques alone, eumycetoma almost always requires prolonged systemic antifungal therapy in addition to surgery. Most experience has been with ketoconazole and itraconazole, although itraconazole appears to have more consistent clinical activity (Ahmed *et al.*, 2004; van de Sande *et al.*, 2005). Surgery can help to reduce the disease burden and occasionally cure small, localized lesions that do not involve bone. As with chromoblastomycosis, amphotericin B is largely ineffective (Rios-Fabra *et al.*, 1994).

Keratitis

Fungal keratitis is an important ophthalmological problem, particularly in tropical areas of the world. In one large series, 40% of all infectious keratitis was caused by fungi (Gopinathan *et al.*, 2002). The most common agents are species of *Fusarium* and *Aspergillus*, followed by dematiaceous fungi (up to 8–17% of cases) (Srinivasan, 2004). Approximately half the cases are associated with trauma; and previous eye surgery, diabetes and contact lens use have also been noted as important risk factors (Gopinathan *et al.*, 2002; Srinivasan, 2004). Diagnosis is based on the potassium hydroxide (KOH) smear test and culture.

Some of the largest case series with dematiaceous fungi have come from India (Garg *et al.*, 2000; Bharathi *et al.*, 2003). In a large review of keratitis due to dematiaceous fungi, 88 cases were examined (Garg *et al.*, 2000). The most common dematiaceous genus causing keratitis was *Curvularia*, followed by *Bipolaris*, *Exserohilum* and *Lasiodiplodia*. Almost half the cases were associated with trauma. Most patients received topical agents only (5% natamycin +/– azole), although more severe cases also received oral ketoconazole. Overall response was 72% in those available for follow-up. Surgery was needed in 13 patients, with an additional six requiring enucleation due to poor response.

In a US study of 43 cases of *Curvularia* keratitis, almost all were associated with trauma (Wilhelmus and Jones, 2001). Plants were the most common source, although several cases of metal injury were seen as well. Topical natamycin was used almost exclusively, with only a few severe cases requiring adjunctive therapy, usually with an azole. Of the oral agents, itraconazole had the best *in vitro* activity, though the majority of isolates were resistant to flucytosine. Surgery, including penetrating keratoplasty, was required in 19% of patients. At the end of therapy, only 78% had a visual acuity of 20/40 or better.

Topical polyenes, such as amphotericin B and natamycin, are commonly used, but oral and topical itraconazole has been found to be useful as well (Gopinathan *et al.*, 2002; Thomas, 2003). Voriconazole is a potentially useful agent, but published clinical experience is limited primarily to cases due to *S. apiospermum* (Hernandez *et al.*, 2004). However, many patients are left with residual visual deficits at the end of therapy, suggesting that further advances in therapy are needed for this debilitating disease.

Allergic disease

Fungal sinusitis

Patients with this condition usually present with chronic sinus symptoms that are not responsive to antibiotics. Previously, *Aspergillus* species were thought to be the most common fungi responsible for allergic sinusitis, but it is now appreciated that disease due to dematiaceous fungi actually comprises the majority of cases (Ferguson, 2000; Schubert, 2004). The most common fungi isolated are species of *Bipolaris* and *Curvularia*. The criteria that have been suggested for this disease include: (i) nasal polyps; (ii) presence of allergic mucin containing Charcot–Leyden crystals and eosinophils; (iii) hyphal elements in the mucosa without evidence of tissue invasion; (iv) a positive skin test to fungal allergens; and (v) characteristic areas of central hyperattenuation within the sinus cavity on computed tomography (CT) scans (Houser and Corey, 2000). Diagnosis generally depends on demonstration of allergic mucin, with or without actual culture of the organism. Therapy consists of surgery to remove the mucin, which is often tenacious, and systemic steroids. Antifungal therapy, usually in the form of itraconazole, may play a role in reducing the requirement for steroids, but this is not routinely recommended (Kuhn and Javer, 2000). Other azoles have only rarely been used for this disease.

Allergic bronchopulmonary mycosis (ABPM)

This is similar in presentation to allergic bronchopulmonary aspergillosis (ABPA), which is typically seen in patients with asthma or cystic fibrosis (Greenberger, 2002). There is a suggestion that allergic fungal sinusitis and allergic bronchopulmonary mycosis may actually be a continuum of disease and should be referred to as sinobronchial allergic mycosis (SAM) (Venarske and deShazo, 2002). Criteria for the diagnosis of ABPA in patients with asthma include: (i) asthma; (ii) a positive skin test for fungal allergens; (iii) elevated IgE levels; (iv) *Aspergillus*-specific IgE; and (v) proximal bronchiectasis (Greenberger, 2002). Similar criteria for ABPM are not established, but may include elevated IgE levels, positive skin tests and response to systemic steroids.

In reviewing cases of ABPM due to dematiaceous fungi, essentially all cases are due to species of *Bipolaris* or *Curvularia* (Lake *et al.*, 1991; Saenz *et al.*, 2001; Revankar, 2004). Asthma was common in these cases, but bronchiectasis was often not present, perhaps reflecting somewhat different pathogenicity mechanisms. All cases had either eosinophilia or elevated IgE levels. Therapy was primarily with systemic steroids, with a slow taper over two to three months or longer, if necessary. Itraconazole has been used as a steroid-sparing agent, but its efficacy is not clear and routine use of itraconazole is not generally recommended (Wark and Gibson, 2001).

Pneumonia

Non-allergic pulmonary disease is usually seen in immunocompromised patients, and may be due to a wide variety of species, in contrast to allergic disease (Brubaker *et al.*, 1988; Borges *et al.*, 1991; Manian and Brischetto, 1993; Yeghen *et al.*, 1996; Odell *et al.*, 2000; Greig *et al.*, 2001; Mazur and Judson, 2001; Revankar, 2004). Clinical manifestations include pneumonia, asymptomatic solitary pulmonary nodules and endobronchial lesions, which may cause haemoptysis. Therapy consists of systemic antifungal agents, usually amphotericin B or itraconazole initially, followed by itraconazole for a more prolonged period. Mortality rates are high in immunocompromised patients. Experience with voriconazole is anecdotal (Diemert *et al.*, 2001).

Brain abscess

This is a rare, but frequently fatal, manifestation of phaeohyphomycosis, often in immunocompetent individuals. In a review of 101 cases of central nervous system (CNS) infection due to dematiaceous fungi, 87 were found to be brain abscess (Revankar *et al.*, 2004). Over half the cases were in patients with no risk factor or immunodeficiency. The most common species involved was *C. bantiana,* which accounted for half of these cases. Symptoms included headache, neurological deficits and seizures. The pathogenesis may be haematogenous spread from an initial, presumably subclinical, pulmonary focus. It remains unclear why these fungi preferentially cause CNS disease in immunocompetent individuals.

Therapy varied depending on the case report. A retrospective analysis of reported cases suggested that the combination of amphotericin B, flucytosine and itraconazole may be associated with improved survival, though it was not frequently used. Complete excision of brain abscesses appeared to have better outcomes than aspiration or partial excision. Outcomes were poor, with an overall mortality of over 70%.

Only anecdotal reports are available for the newer azoles (voriconazole and posaconazole), which were not used in the above case series. Voriconazole was unsuccessful in treating three out of four cases of *C. bantiana* brain abscesses, although two of these patients were immunocompromised (Fica *et al.*, 2003; Trinh *et al.*, 2003; Levin *et al.*, 2004; Lyons *et al.*, 2005). Despite these reports, voriconazole may have a role in therapy of phaeohyphomycotic brain abscess, as it has been successfully used in cases of *Aspergillus* and *S. apiospermum* brain abscess (Nesky *et al.*, 2000; de Lastours *et al.*, 2003). Posaconazole has been reported to be effective in a case of a *R. mackenziei* brain abscess, which represents the first reported survival of infection due to this species (Al Abdely *et al.*, 2005).

Disseminated infection

This is the most uncommon manifestation of infection seen with dematiaceous fungi. In a recent review, most patients were immunocompromised, though occasional patients without known immunodeficiency or risk factors developed disseminated disease as well (Revankar *et al.*, 2002). In contrast to most invasive fungal infections, blood cultures were positive in over half of the cases. The most commonly isolated organism was *S. prolificans,* accounting for over one-third of cases. This species should be distinguished from *S. apiospermum* (teleomorph *Pseudallescheria boydii*), which some experts do not consider truly dematiaceous, and which has different antifungal susceptibilities. *Scedosporium prolificans* is generally resistant to all available antifungal agents. Peripheral eosinophilia was seen in 11% of cases, and these were generally due to species of *Bipolaris* and *Curvularia*. The mortality rate was over 70%, despite aggressive antifungal therapy.

There are no antifungal regimens associated with improved survival in disseminated infection and infection with *S. prolificans,* and this results in nearly 100% mortality in the absence of recovery from neutropenia. However, recent case reports have suggested that the combination of itraconazole or voriconazole with terbinafine may be synergistic against this species and improve outcomes (Meletiadis *et al.*, 2002; Howden *et al.*, 2003). Other combinations or therapies have not been shown to be effective, though clinical experience is limited, and will probably be confined to anecdotal reports, given the rarity of this infection. More recently, a case of disseminated *E. spinifera* infection was treated successfully with posaconazole after non-response to itraconazole and amphotericin B (Negroni *et al.*, 2004).

Future Perspectives

There are no new classes of drugs on the horizon that are expected to have significant activity against dematiaceous fungi. The current agents as a group have excellent activity, although some may have specific advantages for certain types of infections and fungal species. The further development of reliable animal models is needed better to study therapeutic methods and modes of action for uncommon and refractory infections. Progress in treating these infections is likely to come from accumulating clinical experience over time, as large, randomized trials are impractical for any of the clinical syndromes described. This means that the reporting of individual cases is important to advance our understanding, regardless of whether the therapy was successful or not. This relates both to the use of single agents and to combinations. It is likely that combination therapy will be utilized more often, and that additional clinical experience will drive future therapeutic options.

Conclusions

Infections due to dematiaceous fungi are uncommon, but have become increasingly recognized in a wide variety of clinical syndromes. Many species are associated with human infection, though relatively few are responsible for the majority of cases. Life-threatening infections are rare, but may even be seen in immunocompetent individuals, especially in cases of brain abscesses. As the fungi involved are typically soil organisms and common laboratory contaminants, they may be disregarded from clinical specimens as non-pathogenic. However, the clinical setting in which they are isolated should always be carefully considered before making decisions regarding therapy. *Bipolaris* and *Curvularia* species are often associated with allergic disease. Diagnosis depends on a high degree of clinical suspicion and appropriate pathological and mycological examination of clinical specimens. Much additional work is needed in order better to understand the pathogenic mechanisms underlying these infections.

Itraconazole, voriconazole and posaconazole demonstrate the most consistent *in vitro* activity against this group of fungi, although most clinical experience to date has been with itraconazole. Amphotericin B may be used for severe infections in unstable patients; high doses of lipid formulations may have a role in the treatment of refractory cases or in patients intolerant of standard amphotericin B. Once the infection is under control, longer-term therapy with a broad-spectrum oral azole is often reasonable until complete recovery is achieved. Terbinafine and flucytosine have limited roles in treating these fungi, and are generally used for superficial and/or subcutaneous infections, such as chromoblastomycosis. Terbinafine in particular appears to provide synergistic activity with azole antifungals, and this may be a useful strategy against refractory subcutaneous infections such as chromoblastomycosis and mycetoma that do not respond to conventional therapy. Use of this combination in serious systemic infections is unproven, limited to case reports and cannot be generally recommended. Echinocandins do not appear to be useful as single agents, but may be considered in combination therapy for difficult cases.

Combination therapy is a potentially useful therapeutic strategy for refractory infections, particularly brain abscesses and disseminated disease. However, it is not clear which antifungal drug combinations are most effective. Therapy is evolving for many of the clinical syndromes described, and randomized clinical trials to address this issue are impractical given the sporadic nature of cases. Ultimately, case reporting of both successful and unsuccessful clinical experiences will be important in attempting to define optimal therapy for infections caused by dematiaceous fungi.

References

Abliz, P., Fukushima, K., Takizawa, K. and Nishimura, K. (2004) Identification of pathogenic dematiaceous fungi and related taxa based on large subunit ribosomal DNA D1/D2 domain sequence analysis. *FEMS Immunology and Medical Microbiology* 40, 41–49.

Ahmed, A.O., van Leeuwen, W., Fahal, A., van de, S.W., Verbrugh, H. and van Belkum, A. (2004) Mycetoma caused by *Madurella mycetomatis*: a neglected infectious burden. *Lancet Infectious Disease* 4, 566–574.

Al-Abdely, H.M., Najvar, L., Bocanegra, R., Fothergill, A., Loebenberg, D., Rinaldi, M.G. and Graybill, J.R. (2000) SCH 56592, amphotericin B, or itraconazole therapy of experimental murine cerebral phaeohyphomycosis due to *Ramichloridium obovoideum* ('*Ramichloridium mackenziei*'). *Antimicrobial Agents and Chemotherapy* 44, 1159–1162.

Al Abdely, H.M., Alkhunaizi, A.M., Al Tawfiq, J.A., Hassounah, M., Rinaldi, M.G. and Sutton, D.A. (2005) Successful therapy of cerebral phaeohyphomycosis due to *Ramichloridium mackenziei* with the new triazole posaconazole. *Medical Mycology* 43, 91–95.

Andrade, T.S., Castro, L.G., Nunes, R.S., Gimenes, V.M. and Cury, A.E. (2004) Susceptibility of sequential *Fonsecaea pedrosoi* isolates from chromoblastomycosis patients to antifungal agents. *Mycoses* 47, 216–221.

Bharathi, M.J., Ramakrishnan, R., Vasu, S., Meenakshi, R. and Palaniappan, R. (2003) Epidemiological characteristics and laboratory diagnosis of fungal keratitis. A three-year study. *Indian Journal of Ophthalmology* 51, 315–321.

Bonifaz, A., Carrasco-Gerard, E. and Saul, A. (2001) Chromoblastomycosis: clinical and mycologic experience of 51 cases. *Mycoses* 44, 1–7.

Borges, M.C. Jr, Warren, S., White, W. and Pellettiere, E.V. (1991) Pulmonary phaeohyphomycosis due to *Xylohypha bantiana*. *Archives of Pathology and Laboratory Medicine* 15, 627–629.

Brubaker, L.H., Steele, J.C. Jr and Rissing, J.P. (1988) Cure of *Curvularia* pneumonia by amphotericin B in a patient with megakaryocytic leukemia. *Archives of Pathology and Laboratory Medicine* 112, 1178–1179.

Butler, M.J. and Day, A.W. (1998) Fungal melanins: a review. *Canadian Journal of Microbiology* 44, 1115–1136.

Caligiorne, R.B., Resende, M.A., Melillo, P.H., Peluso, C.P., Carmo, F.H. and Azevedo, V. (1999) *In vitro* susceptibility of chromoblastomycosis and phaeohyphomycosis agents to antifungal drugs. *Medical Mycology* 37, 405–409.

Carrillo, A.J. and Guarro, J. (2001) *In vitro* activities of four novel triazoles against *Scedosporium* spp. *Antimicrobial Agents and Chemotherapy* 45, 2151–2153.

Casadevall, A., Rosas, A.L. and Nosanchuk, J.D. (2000) Melanin and virulence in *Cryptococcus neoformans*. *Current Opinion in Microbiology* 3, 354–358.

Castro, L.G., Pimentel, E.R. and Lacaz, C.S. (2003) Treatment of chromomycosis by cryosurgery with liquid nitrogen: 15 years' experience. *International Journal of Dermatology* 42, 408–412.

Clancy, C.J., Wingard, J.R. and Hong, N.M. (2000) Subcutaneous phaeohyphomycosis in transplant recipients: review of the literature and demonstration of *in vitro* synergy between antifungal agents. *Medical Mycology* 38, 169–175.

Corrado, M.L., Kramer, M., Cummings, M. and Eng, R.H. (1982) Susceptibility of dematiaceous fungi to amphotericin B, miconazole, ketoconazole, flucytosine and rifampin alone and in combination. *Sabouraudia* 20, 109–113.

de Lastours, V., Lefort, A., Zappa, M., Dufour, V., Belmatoug, N. and Fantin, B. (2003) Two cases of cerebral aspergillosis successfully treated with voriconazole. *European Journal of Clinical Microbiology and Infectious Diseases* 22, 297–299.

Deresinski, S.C. and Stevens, D.A. (2003) Caspofungin. *Clinical Infectious Diseases* 36, 1445–1457.

Diemert, D., Kunimoto, D., Sand, C. and Rennie, R. (2001) Sputum isolation of *Wangiella dermatitidis* in patients with cystic fibrosis. *Scandinavian Journal of Infectious Diseases* 33, 777–779.

Dixon, D.M. and Polak, A. (1987) *In vitro* and *in vivo* drug studies with three agents of central nervous system phaeohyphomycosis. *Chemotherapy* 33, 129–140.

Dixon, D.M., Polak, A. and Szaniszlo, P.J. (1987) Pathogenicity and virulence of wild-type and melanin-deficient *Wangiella dermatitidis*. *Journal of Medical and Veterinary Mycology* 25, 97–106.

Espinel-Ingroff, A. (1998) Comparison of *in vitro* activities of the new triazole SCH56592 and the echinocandins MK-0991 (L-743,872) and LY303366 against opportunistic filamentous and dimorphic fungi and yeasts. *Journal of Clinical Microbiology* 36, 2950–2956.

Espinel-Ingroff, A. (2001) *In vitro* fungicidal activities of voriconazole, itraconazole, and amphotericin B against opportunistic moniliaceous and dematiaceous fungi. *Journal of Clinical Microbiology* 39, 954–958.

Espinel-Ingroff, A. (2003) *In vitro* antifungal activities of anidulafungin and micafungin, licensed agents and the investigational triazole posaconazole as determined by NCCLS methods for 12,052 fungal isolates: review of the literature. *Revista Iberoamericana de Micología* 20, 121–136.

Espinel-Ingroff, A., Boyle, K. and Sheehan, D.J. (2001) *In vitro* antifungal activities of voriconazole and reference agents as determined by NCCLS methods: review of the literature. *Mycopathologia* 150, 101–115.

Ferguson, B.J. (2000) Definitions of fungal rhinosinusitis. *Otolaryngologic Clinics of North America* 33, 227–235.

Fica, A., Diaz, M.C., Luppi, M., Olivares, R., Saez, L., Baboor, M. and Vasquez, P. (2003) Unsuccessful treatment with voriconazole of a brain abscess due to *Cladophialophora bantiana*. *Scandinavian Journal of Infectious Diseases* 35, 892–893.

Fung-Tomc, J.C., Huczko, E., Minassian, B. and Bonner, D.P. (1998) *In vitro* activity of a new oral triazole, BMS-207147 (ER-30346). *Antimicrobial Agents and Chemotherapy* 42, 313–318.

Garg, P., Gopinathan, U., Choudhary, K. and Rao, G.N. (2000) Keratomycosis: clinical and microbiologic experience with dematiaceous fungi. *Ophthalmology* 107, 574–580.

Gomez, B.L., Nosanchuk, J.D., Diez, S., Youngchim, S., Aisen, P., Cano, L.E., Restrepo, A., Casadevall, A. and Hamilton, A.J. (2001) Detection of melanin-like pigments in the dimorphic fungal pathogen *Paracoccidioides brasiliensis in vitro* and during infection. *Infection and Immunity* 69, 5760–5767.

Gopinathan, U., Garg, P., Fernandes, M., Sharma, S., Athmanathan, S. and Rao, G.N. (2002) The epidemiological features and laboratory results of fungal keratitis: a 10-year review at a referral eye care center in South India. *Cornea* 21, 555–559.

Graybill, J.R., Najvar, L.K., Johnson, E., Bocanegra, R. and Loebenberg, D. (2004) Posaconazole therapy of disseminated phaeohyphomycosis in a murine model. *Antimicrobial Agents and Chemotherapy* 48, 2288–2291.

Greenberger, P.A. (2002) Allergic bronchopulmonary aspergillosis. *Journal of Allergy and Clinical Immunology* 110, 685–692.

Greig, J.R., Khan, M.A., Hopkinson, N.S., Marshall, B.G., Wilson, P.O. and Rahman, S.U. (2001) Pulmonary infection with *Scedosporium prolificans* in an immunocompetent individual. *Journal of Infection* 43, 15–17.

Gupta, A.K., Taborda, P.R. and Sanzovo, A.D. (2002) Alternate week and combination itraconazole and terbinafine therapy for chromoblastomycosis caused by *Fonsecaea pedrosoi* in Brazil. *Medical Mycology* 40, 529–534.

Gupta, A.K., Ryder, J.E., Baran, R. and Summerbell, R.C. (2003) Non-dermatophyte onychomycosis. *Dermatologic Clinics* 21, 257–268.

Hamilton, A.J. and Gomez, B.L. (2002) Melanins in fungal pathogens. *Journal of Medical Microbiology* 51, 189–191.

Hay, R.J. (1999) Therapeutic potential of terbinafine in subcutaneous and systemic mycoses. *British Journal of Dermatology* 141, 36–40.
Herbrecht, R. (2004) Posaconazole: a potent, extended-spectrum triazole anti-fungal for the treatment of serious fungal infections. *International Journal of Clinical Practice* 58, 612–624.
Hernandez, P.C., Llinares, T.F., Burgos, S.J., Selva, O.J. and Ordovas Baines, J.P. (2004) Voriconazole in fungal keratitis caused by *Scedosporium apiospermum*. *Annals of Pharmacotherapy* 38, 414–417.
Hiruma, M., Kawada, A., Ohata, H., Ohnishi, Y., Takahashi, H., Yamazaki, M., Ishibashi, A., Hatsuse, K., Kakihara, M. and Yoshida, M. (1993) Systemic phaeohyphomycosis caused by *Exophiala dermatitidis*. *Mycoses* 36, 1–7.
Hosseini-Yeganeh, M. and McLachlan, A.J. (2002) Physiologically based pharmacokinetic model for terbinafine in rats and humans. *Antimicrobial Agents and Chemotherapy* 46, 2219–2228.
Houser, S.M. and Corey, J.P. (2000) Allergic fungal rhinosinusitis: pathophysiology, epidemiology, and diagnosis. *Otolaryngologic Clinics of North America* 33, 399–409.
Howden, B.P., Slavin, M.A., Schwarer, A.P. and Mijch, A.M. (2003) Successful control of disseminated *Scedosporium prolificans* infection with a combination of voriconazole and terbinafine. *European Journal of Clinical Microbiology and Infectious Diseases* 22, 111–113.
Ikeda, R., Sugita, T., Jacobson, E.S. and Shinoda, T. (2003) Effects of melanin upon susceptibility of *Cryptococcus* to antifungals. *Microbiology and Immunology* 47, 271–277.
Jacobson, E.S. (2000) Pathogenic roles for fungal melanins. *Clinical Microbiology Reviews* 13, 708–717.
Jessup, C.J., Ryder, N.S. and Ghannoum, M.A. (2000) An evaluation of the *in vitro* activity of terbinafine. *Medical Mycology* 38, 155–159.
Johnson, E.M., Szekely, A. and Warnock, D.W. (1998) *In vitro* activity of voriconazole, itraconazole and amphotericin B against filamentous fungi. *Journal of Antimicrobial Chemotherapy* 42, 741–745.
Johnson, L.B. and Kauffman, C.A. (2003) Voriconazole: a new triazole antifungal agent. *Clinical Infectious Diseases* 36, 630–637.
Kimura, M., Goto, A., Furuta, T., Satou, T., Hashimoto, S. and Nishimura, K. (2003) Multifocal subcutaneous phaeohyphomycosis caused by *Phialophora verrucosa*. *Archives of Pathology and Laboratory Medicine* 127, 91–93.
Koga, T., Matsuda, T., Matsumoto, T. and Furue, M. (2003) Therapeutic approaches to subcutaneous mycoses. *American Journal of Clinical Dermatology* 4, 537–543.
Kuhn, F.A. and Javer, A.R. (2000) Allergic fungal rhinosinusitis: perioperative management, prevention of recurrence, and role of steroids and antifungal agents. *Otolaryngologic Clinics of North America* 33, 419–433.
Kumarasinghe, S.P. and Kumarasinghe, M.P. (2000) Itraconazole pulse therapy in chromoblastomycosis. *European Journal of Dermatology* 10, 220–222.
Kwon-Chung, K.J., Polacheck, I. and Popkin, T.J. (1982) Melanin-lacking mutants of *Cryptococcus neoformans* and their virulence for mice. *Journal of Bacteriology* 150, 1414–1421.
Lake, F.R., Froudist, J.H., McAleer, R., Gillon, R.L., Tribe, A.E. and Thompson, P.J. (1991) Allergic bronchopulmonary fungal disease caused by *Bipolaris* and *Curvularia*. *Australian and New Zealand Journal of Medicine* 21, 871–874.
Levin, T.P., Baty, D.E., Fekete, T., Truant, A.L. and Suh, B. (2004) *Cladophialophora bantiana* brain abscess in a solid-organ transplant recipient: case report and review of the literature. *Journal of Clinical Microbiology* 42, 4374–4378.
Lyons, M.K., Blair, J.E. and Leslie, K.O. (2005) Successful treatment with voriconazole of fungal cerebral abscess due to *Cladophialophora bantiana*. *Clinical Neurology and Neurosurgery* 107, 532–534.

Manian, F.A. and Brischetto, M.J. (1993) Pulmonary infection due to *Exophiala jeanselmei*: successful treatment with ketoconazole. *Clinical Infectious Diseases* 16, 445–446.

Matsumoto, T., Ajello, L., Matsuda, T., Szaniszlo, P.J. and Walsh, T.J. (1994) Developments in hyalohyphomycosis and phaeohyphomycosis. *Journal of Medical and Veterinary Mycology* 32(Suppl. 1), 329–349.

Mazur, J.E. and Judson, M.A. (2001) A case report of a *Dactylaria* fungal infection in a lung transplant patient. *Chest* 119, 651–653.

McGinnis, M.R. and Pasarell, L. (1998a) *In vitro* evaluation of terbinafine and itraconazole against dematiaceous fungi. *Medical Mycology* 36, 243–246.

McGinnis, M.R. and Pasarell, L. (1998b) *In vitro* testing of susceptibilities of filamentous ascomycetes to voriconazole, itraconazole, and amphotericin B, with consideration of phylogenetic implications. *Journal of Clinical Microbiology* 36, 2353–2355.

Meletiadis, J., Mouton, J.W., Rodriguez-Tudela, J.L., Meis, J.F. and Verweij, P.E. (2000) *In vitro* interaction of terbinafine with itraconazole against clinical isolates of *Scedosporium prolificans*. *Antimicrobial Agents and Chemotherapy* 44, 470–472.

Meletiadis, J., Meis, J.F., Mouton, J.W., Rodriquez-Tudela, J.L., Donnelly, J.P. and Verweij, P.E. (2002). *In vitro* activities of new and conventional antifungal agents against clinical *Scedosporium* isolates. *Antimicrobial Agents and Chemotherapy* 46, 62–68.

Meletiadis, J., Mouton, J.W., Meis, J.F. and Verweij, P.E. (2003) *In vitro* drug interaction modeling of combinations of azoles with terbinafine against clinical *Scedosporium prolificans* isolates. *Antimicrobial Agents and Chemotherapy* 47, 106–117.

Minotto, R., Bernardi, C.D.V., Mallmann, L.F., Edelweiss, M.I.A. and Scrofernekr, M.L. (2001) Chromoblastomycosis: a review of 100 cases in the state of Rio Grande do Sul, Brazil. *Journal of the American Academy of Dermatology* 44, 585–592.

Morris-Jones, R., Gomez, B.L., Diez, S., Uran, M., Morris-Jones, S.D., Casadevall, A., Nosanchuk, J.D. and Hamilton, A.J. (2005) Synthesis of melanin pigment by *Candida albicans in vitro* and during infection. *Infection and Immunity* 73, 6147–6150.

National Committee for Clinical Laboratory Standards (1997) *Reference Method for Broth Dilution Antifungal Susceptibility Testing of Yeasts*, Approved standard M27-A. National Committee for Clinical Laboratory Standards, Wayne, Pennsylvania.

National Committee for Clinical Laboratory Standards (2002) *Reference Method for Broth Dilution Antifungal Susceptibility Testing of Conidium-forming Filamentous Fungi*, Approved M38-A. National Committe for Clinical Laboratory Standards, Wayne, Pennsylvania.

Negroni, R., Helou, S.H., Petri, N., Robles, A.M., Arechavala, A. and Bianchi, M.H. (2004) Case study: posaconazole treatment of disseminated phaeohyphomycosis due to *Exophiala spinifera*. *Clinical Infectious Diseases* 38, e15–e20.

Nesky, M.A., McDougal, E.C. and Peacock Jr, J.E. (2000) *Pseudallescheria boydii* brain abscess successfully treated with voriconazole and surgical drainage: case report and literature review of central nervous system pseudallescheriasis. *Clinical Infectious Diseases* 31, 673–677.

Odell, J.A., Alvarez, S., Cvitkovich, D.G., Cortese, D.A. and McComb, B.L. (2000) Multiple lung abscesses due to *Ochroconis gallopavum*, a dematiaceous fungus, in a non-immunocompromised wood pulp worker. *Chest* 118, 1503–1505.

Padhye, A.A., Bennett, J.E., McGinnis, M.R., Sigler, L., Fliss, A. and Salkin, I.F. (1998) Biosafety considerations in handling medically important fungi. *Medical Mycology* 36(Suppl. 1), 258–265.

Pang, K.R., Wu, J.J., Huang, D.B. and Tyring, S.K. (2004) Subcutaneous fungal infections. *Dermatologic Therapy* 17, 523–531.

Pappagianis, D. and Ajello, L. (1994) Dematiaceous – a mycologic misnomer? *Journal of Medical and Veterinary Mycology* 32, 319–321.

Pfaller, M.A., Messer, S.A., Hollis, R.J. and Jones, R.N. (2002) Antifungal activities of posaconazole, ravuconazole, and voriconazole compared to those of itraconazole and amphotericin B against 239 clinical isolates of *Aspergillus* spp. and other filamentous fungi: report from SENTRY Antimicrobial Surveillance Program, 2000. *Antimicrobial Agents and Chemotherapy* 46, 1032–1037.

Polak, A. (1984) Antimycotic therapy of experimental infections caused by dematiaceous fungi. *Sabouraudia* 22, 279–289.

Restrepo, A., Gonzalez, A., Gomez, I., Arango, M. and de Bedout, C. (1988) Treatment of chromoblastomycosis with itraconazole. *Annals of the New York Academy of Sciences* 544, 504–516.

Revankar, S.G. (2004) Dematiaceous fungi. *Seminars in Respiratory and Critical Care Medicine* 25, 183–190.

Revankar, S.G., Patterson, J.E., Sutton, D.A., Pullen, R. and Rinaldi, M.G. (2002) Disseminated phaeohyphomycosis: review of an emerging mycosis. *Clinical Infectious Diseases* 34, 467–476.

Revankar, S.G., Sutton, D.A. and Rinaldi, M.G. (2004) Primary central nervous system phaeohyphomycosis: a review of 101 cases. *Clinical Infectious Diseases* 38, 206–216.

Rinaldi, M.G. (1996) Phaeohyphomycosis. *Dermatologic Clinics* 14, 147–153.

Rios-Fabra, A., Moreno, A.R. and Isturiz, R.E. (1994) Fungal infection in Latin American countries. *Infectious Disease Clinics of North America* 8, 129–154.

Romero-Martinez, R., Wheeler, M., Guerrero-Plata, A., Rico, G. and Torres-Guerrero, H. (2000) Biosynthesis and functions of melanin in *Sporothrix schenckii*. *Infection and Immunity* 68, 3696–3703.

Ryder, N.S. and Frank, I. (1992) Interaction of terbinafine with human serum and serum proteins. *Journal of Medical and Veterinary Mycology* 30, 451–460.

Saenz, R.E., Brown, W.D. and Sanders, C.V. (2001) Allergic bronchopulmonary disease caused by *Bipolaris hawaiiensis* presenting as a necrotizing pneumonia: case report and review of literature. *American Journal of the Medical Sciences* 321, 209–212.

Schubert, M.S. (2004) Allergic fungal sinusitis: pathogenesis and management strategies. *Drugs* 64, 363–374.

Shelton, B.G., Kirkland, K.H., Flanders, W.D. and Morris, G.K. (2002) Profiles of airborne fungi in buildings and outdoor environments in the United States. *Applied and Environmental Microbiology* 68, 1743–1753.

Srinivasan, M. (2004) Fungal keratitis. *Current Opinion in Ophthalmology* 15, 321–327.

Steinbach, W.J., Schell, W.A., Miller, J.L. and Perfect, J.R. (2003) *Scedosporium prolificans* osteomyelitis in an immunocompetent child treated with voriconazole and caspofungin, as well as locally applied polyhexamethylene biguanide. *Journal of Clinical Microbiology* 41, 3981–3985.

Summerbell, R.C., Krajden, S., Levine, R. and Fuksa, M. (2004) Subcutaneous phaeohyphomycosis caused by *Lasiodiplodia theobromae* and successfully treated surgically. *Medical Mycology* 42, 543–547.

Sutton, D.A., Slifkin, M., Yakulis, R. and Rinaldi, M.G. (1998) U.S. case report of cerebral phaeohyphomycosis caused by *Ramichloridium obovoideum* (*R. mackenziei*): criteria for identification, therapy, and review of other known dematiaceous neurotropic taxa. *Journal of Clinical Microbiology* 36, 708–715.

Sutton, D.A., Rinaldi, M.G. and Kielhofner, M. (2004) First U.S. report of subcutaneous phaeohyphomycosis caused by *Veronaea botryosa* in a heart transplant recipient and review of the literature. *Journal of Clinical Microbiology* 42, 2843–2846.

Thomas, P.A. (2003) Fungal infections of the cornea. *Eye* 17, 852–862.

Tosti, A., Piraccini, B.M., Lorenzi, S. and Iorizzo, M. (2003) Treatment of nondermatophyte mold and *Candida* onychomycosis. *Dermatologic Clinics* 21, 491–497.

Trinh, J.V., Steinbach, W.J., Schell, W.A., Kurtzberg, J., Giles, S.S. and Perfect, J.R. (2003) Cerebral phaeohyphomycosis in an immunodeficient child treated medically with combination antifungal therapy. *Medical Mycology* 41, 339–345.

van de Sande, W.W., Luijendijk, A., Ahmed, A.O., Bakker-Woudenberg, I.A. and van Belkum, A. (2005) Testing of the *in vitro* susceptibilities of *Madurella mycetomatis* to six antifungal agents by using the Sensititre system in comparison with a viability-based 2,3-bis(2-methoxy-4-nitro-5-sulfophenyl)-5-[(phenylamino)carbonyl]-2H-tetrazolium hydroxide (XTT) assay and a modified NCCLS method. *Antimicrobial Agents and Chemotherapy* 49, 1364–1368.

van Duin, D., Casadevall, A. and Nosanchuk, J.D. (2002) Melanization of *Cryptococcus neoformans* and *Histoplasma capsulatum* reduces their susceptibilities to amphotericin B and caspofungin. *Antimicrobial Agents and Chemotherapy* 46, 3394–3400.

Venarske, D.L. and deShazo, R.D. (2002) Sinobronchial allergic mycosis: the SAM syndrome. *Chest* 121, 1670–1676.

Vermes, A., Guchelaar, H.J. and Dankert, J. (2000) Flucytosine: a review of its pharmacology, clinical indications, pharmacokinetics, toxicity and drug interactions. *Journal of Antimicrobial Chemotherapy* 46, 171–179.

Wark, P.A.B. and Gibson, P.G. (2001) Allergic bronchopulmonary aspergillosis: new concepts of pathogenesis and treatment. *Respirology* 6, 1–7.

Wilhelmus, K.R. and Jones, D.B. (2001) *Curvularia* keratitis. *Transactions of the American Ophthalmological Society* 99, 111–130.

Yeghen, T., Fenelon, L., Campbell, C.K., Warnock, D.W., Hoffbrand, A.V., Prentice, H.G. and Kibbler, C.C. (1996) *Chaetomium* pneumonia in a patient with acute myeloid leukaemia. *Journal of Clinical Pathology* 49, 184–186.

12 Biotechnological Aspects of *Trichoderma* spp.

A.M. Rincón,[1] T. Benítez,[1] A.C. Codón[1] and M.A. Moreno-Mateos [2]

[1]*Departamento de Genética, Facultad de Biología, Universidad de Sevilla, Spain;* [2]*Dpto. Señalización Celular, Centro Andaluz de Medicina Regenerativa y Biología Molecular, Edif. CABIMER, Seville, Spain*

Introduction

The genus *Trichoderma* consists of a heterogeneous group of fungi largely classified as anamorphic Hypocreales. They have a wide range of economically useful features that have applications for multiple biotechnological uses, mainly in agriculture, industry and environmental biotechnology.

All *Trichoderma* species are soil-born aerobic fungi that are often associated with root ecosystems. They have a wide metabolic capacity which makes them easily adaptable, and the genus is practically ubiquitous (Monte, 2001). *Trichoderma* is present in soils and other organic matter collected at all latitudes, some species being widely distributed while others are geographically limited (Samuels, 2006). Due to their metabolic versatility, many *Trichoderma* species are naturally resistant to many toxic compounds, including herbicides, fungicides and pesticides – and antibiotics produced by other microorganisms. This latter feature allows them to grow easily under extremely competitive conditions (Hjeljord and Tronsmo, 1998). Some species of *Trichoderma* are able to degrade some of these inhibitory compounds including hydrocarbons, chloro-phenolic compounds, polysaccharides and xenobiotics and pesticides, and these abilities makes them a potential tool in bioremediation of contaminated soils and water (Harman *et al.*, 2004b).

Most *Trichoderma* species have been described as saprophytic fungi, although some of them are, or have the capacity to be, mycoparasites (Barnett and Binder, 1973); and these may attack other fungi directly in order to utilize their cell walls and cytoplasmic components as nutritional resources. This mechanism, together with competition for nutrients and antibiotic production, allows *Trichoderma* species to be antagonists of a wide range of fungal plant pathogens (Benítez *et al.*, 1998a). *Trichoderma* species are not generally considered to be pathogens or otherwise detrimental to plant growth, and contrariwise their presence in the rhizosphere (biofertilization) can promote plant growth and development. In addition, *Trichoderma* species have recently been reported to be able to induce systemic and localized resistance in plants, conferring protection against a wide variety of pathogens (reviewed in Harman *et al.*, 2004a). These relationships with plants have resulted in *Trichoderma* species being used as biological control agents (BCAs) of plant diseases.

Trichoderma species produce high levels of extracellular proteins, and they are best known for their ability to produce enzymes that degrade cellulose and chitin, and other enzymes that may have commercial applications. Extensive reviews of these have been provided by Harman and Kubicek (1998) and Pentillä *et al.* (2004). More recently, cellulases from *Trichoderma* have been shown to contribute to the bioconversion of municipal solid waste to glucose for bio-ethanol production (Li *et al.*, 2007). This chapter will focus on some of the biotechnological applications of the genus in agriculture and the environment.

Several species of *Trichoderma* have been widely used as BCAs against phytopathogenic fungi, among them *Gliocladium virens* (syn. *Trichoderma virens*), *T. viride* and, most commonly, *T. harzianum*. There has been considerable research into the complex mechanisms involved in this control, and this has contributed to the subsequent improvement of *Trichoderma* strains as BCAs (Benítez *et al.*, 2004). Despite this, our knowledge of the biocontrol mechanisms remains somewhat incomplete, and it has become increasingly clear that induction of systemic and localized plant resistance is an important factor (Harman *et al.*, 2004a; Howell, 2006).

The *Trichoderma*–Pathogen Interaction

Competition

One of the advantages of most species of *Trichoderma* is their wide metabolic versatility that allows them to use different and complex carbon and nitrogen sources. This allows *Trichoderma* to limit the growth and proliferation of other competing microorganisms (Hjeljord and Tronsmo, 1998). Moreover, *Trichoderma* species have some high-affinity membrane transporters which allow them to take up nutrients at very limited concentrations (Delgado-Jarana *et al.*, 2003), and this has proved to be an efficient mechanism of biological control. Strain T-35 of *T. harzianum* has been reported to produce high-affinity siderophores and to inhibit growth of *Fusarium oxysporum*, mainly due to the lack of iron (Chet and Inbar, 1994). *Trichoderma* species are very resistant to a wide variety of toxic metabolites produced by other organisms (Hjeljord and Tronsmo, 1998), and it has been recently suggested that this might be due to the numerous ABC transporters detected in the different *Trichoderma* spp. genomes (Marra *et al.*, 2006).

Antibiosis

Many species of *Trichoderma* produce low-molecular weight secondary metabolites with antibiotic activity that can inhibit the growth of various microorganisms (Howell, 1998). These include, among others, gliovirin, several types of peptaibols, harzianic acid and 6-pentyl-pyrone (Howell, 2003). In some cases, there has been a good correlation between antibiotic production and inhibition ability, although this effect seems to be host specific (Howell, 2006). One example of this is gliotoxin, and production of this is neccesary for the biocontrol of *Pythium* but not for the control of *Rhizoctonia solani* (Howell and Stipanovic, 1995).

Combinations of antibiotic compounds and hydrolytic enzymes can have a synergistic effect, with the different components involved at different times during the interaction (Lorito *et al.*, 1996b). A mutant strain (PF1) that overproduces cell wall-degrading enzymes and a yellow pigment related to alpha-pyrones has been isolated from *T. harzianum* CECT 2413. The PF1 strain is capable of overgrowing *Rhizoctonia solani* better and faster than the wild type (Rey *et al.*, 2001). This mutant has also been shown to protect grapes against *Botrytis cinerea* both *in vitro* and in field trials (Rey *et al.*, 2001).

Mycoparasitism

Mycoparasitism by *Trichoderma* is a sequential process that involves the detection of the fungal host, chemotrophic growth towards the target, lysis and assimilation of the intracellular content (Benhamou and Chet, 1997). Several types of hydrolases are secreted in the course of the mycoparasitic response. Basal levels of exochitinases seem to be responsible for the release of low-molecular weight oligomers from the cell walls of target fungi (Harman, 2006). These oligomers may be elicitors and act as a signal that induces the expression of genes that encode for the cell wall-degrading enzymes (CWDEs), together with other genes related to mycoparasitism, and to morphological changes related to antifungal activity (Viterbo *et al.*, 2002).

It has recently been suggested that these elicitors might be short oligosaccharides bound to an amino acid residue (Woo *et al.*, 2006). The *Trichoderma* species attaches to the target fungus by binding carbohydrates of their cell walls to the lectins of the host. Once the contact has taken place, most *Trichoderma* species coil around the fungal target and form appresoria (Inbar and Chet, 1996). The coordinated action of these components (CWDEs, coiling and appresoria) produces cell wall degradation and lysis of the target fungus (Harman *et al.*, 2004a).

Improved Biocontrol Strains of *Trichoderma*

Several strategies have been used to obtain new and better antagonistic strains of *Trichoderma* with enhanced properties. Since the success of *Trichoderma* strains as biological control agents is partly due to their capacity to produce CWDEs, early attempts at strain improvement were focused on the over-expression of different CWDEs. In recent years efforts have been directed towards obtaining strains with altered intracellular signalling pathways. These pathways seem to control the broader aspects of mycoparasitism, and so mutants show more interesting and informative phenotypes than those that over-express CWDEs. In addition, some strains that have been isolated by classical genetics have proved to have good antagonistic properties.

Role of cell wall-degrading enzymes: homologous over-expression

Fungal cell walls consist mainly of β-1,4-N-acetyl-glucosamine (chitin), β-1,3 linked homopolymers of D-glucose with β-1,6 linked branches (glucans) and, in some fungi, there may be protein components associated with the polysaccharides (Peberdy, 1990). Therefore, early studies considered the extracellular hydrolytic systems used by *Trichoderma* against fungal cell walls.

Chitinases

The chitinolytic system of *Trichoderma* is composed of 1,4-β-N-acetylglucosaminidases, endochitinases and exochitinases (Sahai and Manocha, 1993). Most of these enzymes have been characterized and the associated genes have been isolated from various *Trichoderma* species (Limón and Benítez, 2002). The *chit42* and *chit33* genes that encode for endochitinases of 42 kDa and 33 kDa, respectively, have been over-expressed in strain CECT 2413 of *T. harzianum*, resulting in an improvement of its antifungal activity (Limón *et al.*, 1999). Other homologous chitinases have been over-expressed in different strains with similar results (Baek *et al.*, 1999; Viterbo *et al.*, 2001).

Molecular techniques and protein engineering have allowed new proteins with enhanced activity to be designed. This approach has been used to improve the specific activity of some *Trichoderma* chitinases that do not have a chitin-binding domain, by adding a cellulose-binding domain from cellobiohydrolase II of *T. reesei* (Limón *et al.*, 2004). Mutant strains over-expressing these modified enzymes show higher antifungal activities than those over-expressing native chitinases (Limón *et al.*, 2004). The over-expression of other types of hydrolases (glucanases and proteases) has resulted in similar phenotypes, but with some specific advantages (see below).

Strains that show a loss of function in some chitinase-encoding genes are interesting. A variable antifungal phenotype (with reduced, similar or even enhanced activity) has been observed in the same mutant depending on the fungal host (Woo *et al.*, 1999). The higher antagonistic ability observed in this mutant might be due to an over-expression of other chitinases, or to other hydrolytic enzymes in order to balance the loss of activity of one of them. This is a situation that has been suggested for other strains with a similar genetic background (Limón *et al.*, 2004).

Glucanases and proteases

The β-glucanolytic antifungal system of *Trichoderma* consists mainly of two enzymatic activities: endo- and exoglucanases. During mycoparasitism glucanases may break the β-1,3 or β-1,6 bonds in the cell wall β-glucan. Several endoglucanases have been characterized in *T. harzianum* CECT 2413, and their associated genes have been identified and cloned (Benítez *et al.*, 1998b; Montero *et al.*, 2005). Over-expression of some of these enzymes can result in strains with enhanced antifugal activity (Delgado-Jarana, 2001; Rincón, 2004). The specific effect of such strains against organisms which have glucan and cellulose, but not chitin in their cell walls, such as *Pythium ultimum* (Migheli *et al.*, 1998), is probably the most interesting aspect of over-expression of glucanases.

Proteases from *Trichoderma* have been used to inhibit the growth of fungal pathogens and also to reduce the negative effects of nematodes on plants (Sharon *et al.*, 2001; Suárez *et al.*, 2004). Therefore, different strains that over-express specific appropriate hydrolases can be used to control different pathogens. However, most of these single gene-based approaches have involved the isolation of particular strains specific to individual hosts or diseases. Over recent years various attempts have been made to identify more global regulators of mycoparasitism that could be used to obtain new strains with activity against a wide range of phytopathogens.

Global regulators of mycoparasitism

Carbon and nitrogen source regulation

Most CWDEs and other mechanisms related to mycoparasitism are regulated by carbon and nitrogen sources. Genes encoding CWDEs are usually repressed when easily assimilable carbon and/or nitrogen sources such as glucose and ammonium are present. The expression of these genes is de-repressed under carbon or nitrogen starvation, and it is strongly induced when a poorly assimilable carbon (such as chitin or glucan) or nitrogen (such as peptides or chitin) source is available (Benítez *et al.*, 1998b; Lorito, 1998; Donzelli and Harman, 2001).

Several transcription factors for both regulatory systems, e.g. CreA/Cre1 (carbon) or AreA (nitrogen), have been characterized in various fungi, including some *Trichoderma* species (Kudla *et al.*, 1990; Ilmen *et al.*, 1996). It has been shown that several genes related to mycoparasitism are subjected to control by *cre1* (Lorito *et al.*, 1996a), and a carbon catabolite-de-repressed cre1 mutant of *Hypocrea jecorina* has recently been shown to have increased antifungal activity against *P. ultimum*. However, the high cellulase activity in this mutant did not appear to be involved in its antagonistic ability (Seidl *et al.*, 2006).

Environmental stress and pH regulation

Several studies have suggested that mycoparasitism might be mediated by different stress signals, and that antagonism itself is a stress situation (Olmedo-Monfil *et al.*, 2002). The Seb1 protein that is homologous to the MSN2/MSN4 transcription factor related to stress response has been cloned and characterized in *T. atroviride* (Peterbauer *et al.*, 2002; Seidl *et al.*, 2004). However, Seb1 only partially controls the osmotic response in *T. atroviride* and does not seem to be related to the mycoparasitic response. pH regulation in *T. harzianum* is mediated by the Pac1 transcription factor (Moreno-Mateos *et al.*, 2007). Pac1 and pH can regulate antifugal activity, as some genes related to mycoparasitism that encode for CWDEs, a cell wall protein and a high-affinity glucose transporter are controlled by pH and Pac1 (Moreno-Mateos *et al.*, 2007).

Two different mutants with opposing phenotypes have been obtained from *T. harzianum*. A loss of function mutant of *pac1* has been shown to have an enhanced ability to inhibit growth and germination of several fungi (see Fig. 12.1), whereas a mutant with an allele of *pac1* that was functional under any pH condition showed better mycoparasitic behaviour against *R. solani* and *Phytophthora citrophthora* than the wild type (see Fig. 12.1). The latter mutant also had an increased resistance to saline and osmotic stress (Moreno-Mateos, 2006), which could be an advantage if these were used in high-salt content soils.

Cell-signalling pathways

Mycoparasitism, in common with other host–pathogen interactions, is regulated by highly conserved signal transduction pathways (Bölker, 2002). Some components of different signalling pathways that control mycoparasitic responses and antifungal activity in *Trichoderma* spp have been isolated and characterized. G-protein subunits and mitogen-activated protein kinases (MAPKs) have been studied, and different mutants for these genes have been generated. However, only some of these have shown increased biological control activity.

G-alpha proteins seem to have many different roles in mycoparasitism in *Trichoderma* species, and they appear to be involved in the specific interaction and recognition of different fungal phytopathogens (Mukherjee *et al.*, 2004; Zeilinger *et al.*, 2005). Several loss of function mutants have been obtained for the three main groups of G-alpha proteins. All of these resulted in a complete lack of mycoparasitic response but, in some cases, this appeared to be host specific. Research on the G proteins in *Trichoderma* has been useful to study the different signalling pathways that positively control mycoparasitic response. Unfortunately, from an applied point of view, only one of the mutants obtained so far seems to have an enhanced capacity to inhibit growth of host fungi, due in that case to the overproduction of low-molecular weight antifungal metabolites (Reithner *et al.*, 2005).

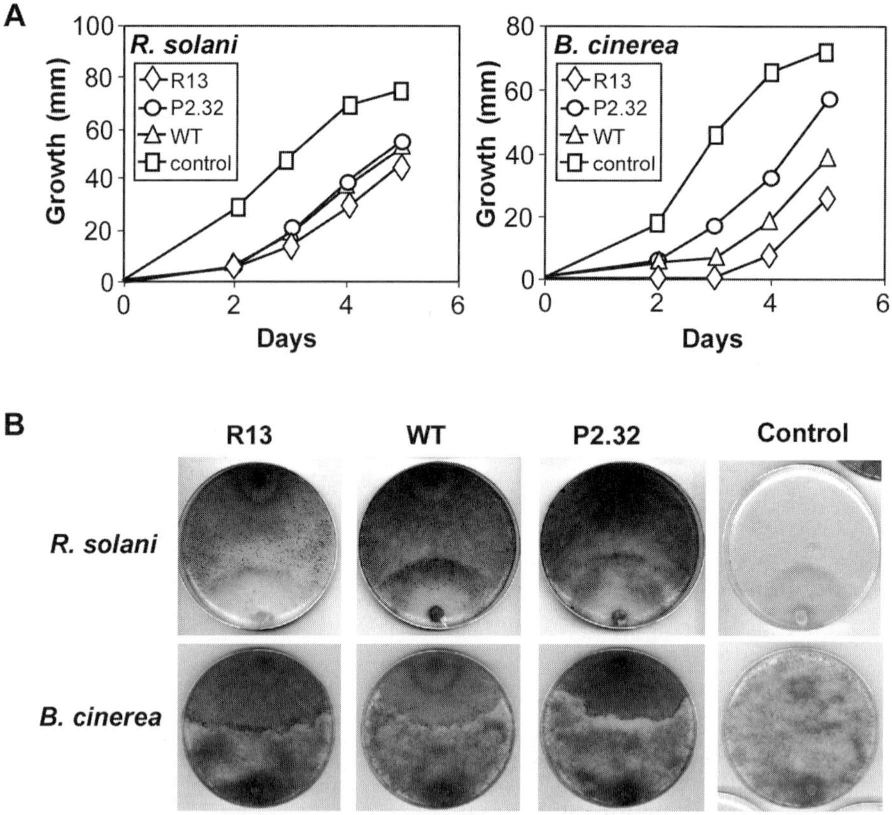

Fig. 12.1. Antagonistic ability of *T. harzianum* strains over-producing (P2.32) or lacking (R13) pH regulator PAC1. A, effect on enzymes and metabolites secreted by different strains. Wild type (WT), P2.32 and R13 cultures were grown on PPG plates, buffered at pH 5.5 and previously covered with cellophane, which was removed after 3 days; 5 mm mycelial discs of *Rhizoctonia solani* strain or spores of *Botrytis cinerea* were placed on plates. Radial growth of phytopathogen was measured at 48 h and then every 24 h for 5 days. Control: PPG plates where *T. harzianum* had not been previously grown. B, effect on mycoparasitism. Spores from WT, P2.32 and R13 strains were inoculated on to MM with 0.2% glucose as the only carbon source and buffered at pH 5.5. Mycelial discs (5 mm) of *R. solani* or *B. cinerea* were placed on the opposite side of Petri dishes. Control: plates without *T. harzianum* (from Moreno-Mateos *et al.*, 2007, with permission).

Some genes encoding for MAPKs have been cloned and characterized in *T. harzianum* (Mendoza-Mendoza *et al.*, 2003; Delgado-Jarana *et al.*, 2006). A mutant with a non-functional allele of the MAPK-encoding gene *tvk1* has been generated in *T. virens* strain Gv29-8 (Mendoza-Mendoza *et al.*, 2003). Use of this mutant as a BCA was found to reduce the effects of various fungal phytopathogens on cotton. At a molecular level, this MAPK seems to increase expression of several genes, among them those encoding for CWDEs. However, a similar MAPK mutant obtained from another *T. virens* strain failed to control fungal pathogens in a host-specific manner, similar to the G-protein loss of function mutants (Mukherjee *et al.*, 2003). This last phenotype has also been obtained in a *T. atroviride* loss of function mutant with a homologous gene. In contrast to the attenuated mycoparasitism obtained in direct confrontation assays, this mutant was able to protect bean plants against *R. solani* infection better than the wild type (Reithner *et al.*, 2007).

An adenylate cyclase-encoding gene (*tac1*) involved in cAMP signalling in *T. virens* has recently been cloned and studied (Mukherjee *et al.*, 2007). Deletion of *tac1* caused a total loss of mycoparasitic capacity against various phytopathogens and reduced production of secondary metabolites. One possible route for generating improved antagonistic strains could be the isolation of a mutant based on this gene with a hyperactive allele.

Most studies on signal transduction pathways carried out in *Trichoderma* have focused on the genes that control mycoparasitic response, and on the interactions with phytopathogens and plants. Only some of the mutants isolated seem to be better biological control agents than the wild type. However, all of these studies demonstrate that the different signal transduction pathways can have a major influence on the antagonistic properties of *Trichoderma*, and that these pathways may have a role in the future development of improved biocontrol strains.

The *Trichoderma*–Plant Interaction

Stimulation of plant defence mechanisms

The effect of *Trichoderma* strains against phytopathogenic fungi in plant protection has long been attributed mainly to their antagonist properties against pathogens. However, there is now enough evidence to demonstrate that *Trichoderma* can also induce systemic and localized resistance against a variety of phytopathogens in plants. This ability has been demonstrated by the observation of protection at various parts of the plant after specific application of *Trichoderma* on roots, and one example of this is the resistance induced by *T. virens* to green mottle mosaic virus in cucumber leaves (Lo *et al.*, 1998). In many cases, this resistance has been correlated with an increase in phytoalexins or pathogenesis-related (PR) proteins in the plants, including sites and tissues distant from the roots (see Fig. 12.2; Yedidia *et al.*, 2003).

In addition, synthesis of signal molecules such as salicylic acid has been detected in systemic resistance in *Trichoderma*-treated plants (de Meyer *et al.*, 1998), although localized resistance is not always accompanied by systemic resistance (Howell, 2003). From a biocontrol point of view, it is particulary interesting that *Trichoderma*-mediated induced systemic resistance (ISR) may provide long-lasting disease control, and in some cases this has been observed up to 100 days after *Trichoderma* application (Harman *et al.*, 2004a).

Several biochemical elicitors of disease resistance responses have been identified in plants after *Trichoderma* treatments and, for example, ethylene

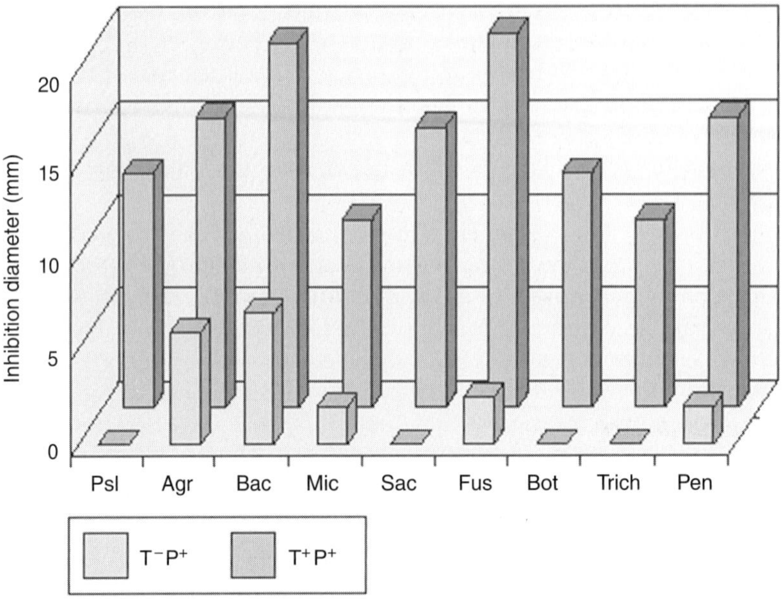

Fig. 12.2. Antimicrobial activity of aglycones in phenolic fraction from leaves of cucumber seedlings in which roots were either not colonized (light grey) or colonized by T-203 (dark grey) and then challenged with *Pseudomonas syringae* pv. *lachrymans* (Psl). Seedlings were of the same age and were extracted 2 days after challenge with Psl. These aglycones have antimicrobial activity against Psl, *Agrobacterium tumefaciens* (Agr), *Bacillus megaterium* (Bac), *Micrococcus luteus* (Mic), *Saccharomyces cerevisiae* (Sac), *Fusarium oxysporum* (Fus), *Botrytis cinerea* (Bot), *Trichoderma asperellum* (Trich) and *Penicillium italicum* (Pen) (from Harman *et al.*, 2004a, with permission).

synthesis has been related to the production of xylanases and cellulases in many cases (Dean *et al.*, 1989). A 22 kDa xylanase has also been shown to induce the opening of ionic chanels, biosynthesis of PR proteins, glycosylation and lipid acylation of phytoesterols (Hananaia and Avni, 1997). The activity of CWDEs of *Trichoderma* on plant or on fungal cell wall polymers could release oligosaccharides and low-molecular weight compounds that might act as inducers of plant defence mechanisms (Woo *et al.*, 2006). However, it has been shown that the activity of some of these enzymes is not a prerequisite for the induction of the aforementioned mechanisms, and both active and heat-denatured cellulases elicit salicylic acid or ethylene pathways, respectively, in melon cotyledons (Martínez *et al.*, 2001). In addition, Enkerli *et al.*, 1999 demonstrated that a mutation in the catalytic site of xylanase II in *T. reesei* did not impede induction of ISR by the fungus.

The elicitor ability of the 22 kDa xylanase appears to be mediated by membrane receptors that bind to the protein (Hananaia and Avni, 1997). It is interesting to note that this small protein is translocated through the vascular system of tobacco when introduced through cut petioles (Bailey *et al.*, 1991). In a similar way, *T. virens* produces several peptides that act as elicitors of phytoalexin biosynthesis and peroxidase activity in cotton. One of these is similar to the 22 kDa xylanase mentioned above, and another is similar to a *Fusarium sporotrichioides* serine proteinase (cited in Harman *et al.*, 2004a).

In pathogen–host interactions, the products of avirulence genes (*avr*) function as elicitors that induce a defence response in those plants that have the corresponding resistance gene. Avirulence genes have not yet been identified in *Trichoderma*, but proteomic analyses of strain T22 have shown the presence of proteins homologous to avr4 and avr9 from *Cladosporium fulvum* (Woo *et al.*, 2006), and a protein (Epl1) belonging to the cerato-platanin virulence factors family has been identified in *Hypocrea atroviridis* (Djonovic *et al.*, 2006).

Plant root colonization

Some *Trichoderma* strains are able to colonize root surfaces permanently and penetrate up to a few cell layers below the epidermis (see Fig. 12.3; Chacón *et al.*, 2007). Other strains can colonize only local sites in plant roots (Metcalfe and Wilsom, 2004), whereas a strain of *T. stromaticum* has been reported to colonize the vascular system in cocoa (Harman *et al.*, 2004a).

Colonization implies an ability to adhere to and recognize plant roots, to penetrate plant tissues and to withstand toxic metabolites produced by the

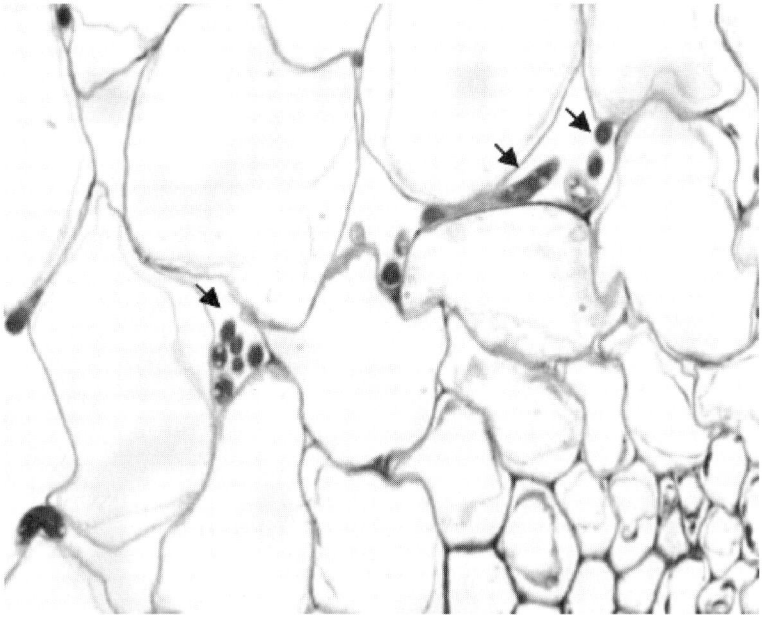

Fig. 12.3. Light micrography of ultra-thin cross-section of tomato root colonized by *Trichoderma harzianum* magnified × 2000. Arrows show hyphae stained by toluidine blue (from Chacón *et al.*, 2007, with permission).

plant. It has been reported that *Trichoderma* develops penetration-specific morphological structures similar to those formed at mycoparasitism, such as appresoria and papillae (Chacón *et al.*, 2007), and some of the molecules involved in *Trichoderma*–plant association are identical to those that induce a mycoparasitic response (Woo *et al.*, 2006). This process resembles that of mycorrhizal fungi, which penetrate the plant cell wall and then branch out extensively to form a highly branched haustorium without extensive disruption of plant tissues (Franken *et al.*, 2002).

Trichoderma spp. used as biocontrol agents are much more resistant to phytoalexins than are most other fungi. This resistance could explain their ability to colonize roots and establish a long-lasting relationship with plants, and it has also been associated with the presence of ABC transport systems (Harman *et al.*, 2004a). ABC protein members are ATP-dependent permeases that mediate the transport of different substrates through biological membranes. They have been linked with many important processes in *Trichoderma* such as spatial colonization, the secretion of antibiotics and CWDEs, and antagonistic ability (Woo *et al.*, 2006). They may also be involved in resistance mechanisms, as ABC knock-out mutants of *T. atroviride* are inhibited by antifungal compounds from *Botrytis cinerea*, *Rhizoctonia solani* and *Pythium ultimum*, unlike the wild type (Woo *et al.*, 2006).

It is not clear whether or not ISR against pathogens is always dependent on root colonization. However, in the few cases studied, a correlation has been observed between colonization and the induction of enzymatic activities related to plant defence, such as peroxidases, chitinases and β-1,3-glucanases (Yedidia *et al.*, 2003). In cucumber plants where the roots had been colonized by *Trichoderma asperellum* T-203, leaf extracts were found to possess more antimicrobial activity than non-treated controls (Harman *et al.*, 2004a). Colonization in those plants led to a transitory increase in the production of phenylalanine ammonia lyase (PAL) in both shoots and roots, which subsequently decreased within 2 days.

There is little information in the literature concerning fungal genes specifically expressed during *Trichoderma*–root interaction. A recent report has demonstrated that *T. virens* TmkA MAPK is involved in inducing systemic resistance in the plant, although it is not necessary for root colonization (Viterbo *et al.*, 2005). A broader analysis of fungal gene expression after plant root colonization by *T. harzianum* CECT 2413 has suggested the importance of genes involved in other metabolic pathways such as respiratory function, morphological changes, detoxification (ABC transporters), membrane and vesicle synthesis, sugar transport and lipid and protein metabolism (Chacón *et al.*, 2007). These findings are very similar to gene-profiling results described during plant–fungus interactions by mycorrhizal and pathogenic fungi (Duplessis *et al.*, 2002).

It has been observed that some *Trichoderma* strains can colonize leaf surfaces after spray application of conidia (Lo *et al.*, 1998), and *Trichoderma* spp. have also been successfully applied to control diseases in fruit, flowers and foliage (Benítez *et al.*, 1998b). There are, however, no reports of ISR induction in these cases.

Biofertilization

Frequent application of *Trichoderma* to plant soil can have a positive effect on plant development, and this has been demonstrated for seed germination, root development, the production of abundant early blossoms and increases in plant height and weight (Chang and Baker, 1986). Application of *T. hamatum* or *T. koningii* has been reported to result in a nearly 300% plant yield increase in some greenhouse experiments (Hornby, 1990). *T. harzianum* T22 has been shown to increase root development in maize and other plants, and this may last for the entire lifespan of annual plants and can be induced by adding small amounts of fungal conidia to seeds. Such addition can result in an increased yield and biomass, as well as an increased drought tolerance (Harman, 2000). However in maize, susceptibility to increasing yield by *T. harzianum* T22 seemed to have a strong genetic component (Harman *et al.*, 2004a). Interestingly, some synergy has been demonstrated between *T. harzianum* T22 and mycorrhizal fungi in root proliferation (Datnoff *et al.*, 1995).

The mechanisms responsible for the stimulation of plant growth and development have not yet been identified, but they are likely to be due to a combination of several factors. Some *Trichoderma* strains acidify the surrounding environment by secreting organic acids that can solubilize phosphates, micronutrients and mineral cations, including Fe^{3+}, Mn^{4+} and Mg^{2+} (Grondona *et al.*, 1997). This may help in the assimilation of these factors and assist promotion of plant growth (Altomare *et al.*, 1999). This might also explain why such stimulation is normally more evident under stress conditions in field experiments. Some authors have suggested that *Trichoderma* is also able to diminish, and even revert, oxidative damage in roots (Björkman *et al.*, 1998).

Another indirect effect of *Trichoderma* on growth promotion could be the elimination of minor pathogens or deleterious microorganisms in the rhizosphere, thus allowing the plants to reach their maximal developmental potential (Harman *et al.*, 1989). However, increases in plant yield have also been reported in axenic systems, where diffusible factors that might act as plant growth stimulators have been detected (Yedidia *et al.*, 2001). Similar yield increases were observed when seeds were separated from *Trichoderma* by a cellophane membrane, which may indicate that the fungus produces diffusible growth factors (Windham *et al.*, 1986). Molecules similar to auxins, cytokinins, zeatyn, gibberellin, GA3 and ethylene have been detected and identified in the laboratory (Osiewacz, 2002), although their effect on plants has not yet been demonstrated.

Improvement of Plant Defence: the Heterologous Expression of *Trichoderma* Genes

CWDE genes have been used to develop transgenic plants with enhanced antifungal abilities. *Nicotiana tabacum* and *Solanum tuberosum* plants that expressed the *chit42* gene from *T. harzianum* CECT 2413 showed high resistance to *R. solani* and *Alternaria alternata* (Lorito *et al.*, 1998). In broccoli, expression of *chit42* resulted in levels of resistance to *Alternaria brassicicola* similar to those obtained with the commercial fungicide Bayleton® (Mora and Earle, 2001). Expression of a gene homologous to *chit42* has given similar results in other crops, including apple and barley (Bolar *et al.*, 2000; Klemsdal *et al.*, 2004). Other CWDE genes, such as those encoding the β-1,3-glucanase I from *T. harzianum* CECT 2413, have been expressed in tobacco plants and this resulted in a high rate of survival from *R. solani* infections (see Fig. 12.4; Rincón, 2004).

There is very little information as to how over-expression of CWDEs affects the ability of *Trichoderma* to induce systemic resistance in plants. It has been demonstrated that in tobacco plants the expression of chitinases results in an induced systemic response, characterized by a higher resistance to bacterial foliar pathogens such as *P. syringae* and to fungal root pathogens such as *R. solani*. These plants had higher levels of activities associated with defence mechanisms than control plants. In addition, transgenic plants can show higher resistance to diverse abiotic stress conditions such as high concentrations of NaCl or heavy metals (de las Mercedes Dana *et al.*, 2006). Similar results have been observed in tobacco plants containing a β-1,3-glucanase gene from *T. harzianum* CECT 2413 (Rincón, 2004).

Recent findings on ISR stimulation in plants have widened the range of characteristics that could be considered for improvement in BCAs. For example, the selection of BCAs for plant root colonization would be enhanced by the isolation of strains that over-express ABC transporters, hydrophobins and repellents. In *T. harzianum* CECT 2413, over-expression of the hydrophobin gene *qid3* results in spores with improved adherence properties (Rosado *et al.*, 2007). Finally, heterologous expression of genes such as those that encode for *Trichoderma* avr proteins may be useful for engineering plants to increase disease resistance. *Trichoderma* gene expression may possibly become more effective at inducing plant resistance than the expression of

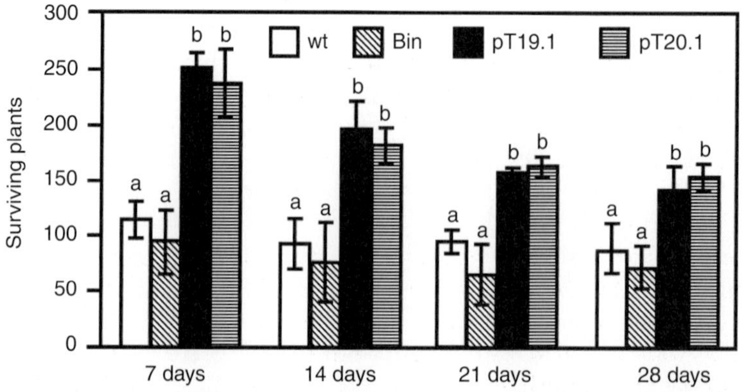

Fig. 12.4. Survival of tobacco plantlets after incubation for 1–4 weeks with *Rhizoctonia solani*. Wt, untransformed control; Bin, plants transformed with the empty vector; pT19.1 and pT20.1, transformants carrying the β-1,3-glucanase gene from *T. harzianum*. Figures bearing the same superscript letters have no significant differences (from Rincón, 2004, with permission).

homologue *avr* genes from pathogenic fungi, as genes from *Trichoderma* may be active in different plant species and cultivars.

Bioremediation

It has been suggested that *Trichoderma* species may have some potential in the alleviation of xenobiotics (Harman *et al.*, 2004b). The enzymatic activities of *Trichoderma* as a traditional biocontrol agent have been widely studied and, as many *Trichoderma* strains are naturally resistant to various toxic and xenobiotic compounds, they could be expected also to have enzyme systems that would degrade them.

Degradation of pollutants

Some strains of *Trichoderma* have been shown to be resistant to as much as 2000 ppm cyanide, and these isolates were able to detoxify the cyanide by using formamide hydrolase or rhodanase (Ezzi and Lynch, 2002, 2005). Wheat and pea seeds that had been coated with *T. harzianum* T22 have been shown to germinate and grow in soil treated with 50 or 100 ppm cyanide, where untreated seeds failed to germinate. It has also been shown that a *Trichoderma* strain can rapidly absorb a solution of the metallocyanide Prussian blue using cyanide-metabolizing enzymes (Harman *et al.*, 2004b).

Polyphenolics are toxic and very resistant to microbial degradation, making processing of such waste waters difficult. Adequate aeration and dilution of waste water can allow growth of *Trichoderma* spp., and these may metabolize the polyphenolic compounds. The enzymes responsible for the polyphenol catabolism have not been isolated, but a comparison of the activity among different fungi used to treat waste water showed the highest efficiency for the *T. harzianum* strain TC3 (Kissi *et al.*, 2001).

Many potential environmental contaminants (petroleum, coal tar, oil, shale) contain complex polycyclic aromatic hydrocarbons (PAHs). Some strains of *T. harzianum* and *T. viride* have been shown effectively to degrade pristane (Ravelet *et al.*, 2000) and hexadecane (Harman *et al.*, 2004b). Saraswathy and Hallberg (2002) reported that up to 75% of pyrene (4-ring PAH) was removed in axenic cultures of *Trichoderma* spp. when pyrene was the sole carbon source.

Organochlorides are contaminants that are commonly produced in the paper industry. *Trichoderma harzianum* and *T. viride* strains have been shown to degrade organochloride *in vitro* (Espósito and da Silva, 1998), and *T. harzianum* can also reduce concentrations of free chloroguaiacoles and adsorbable organic halogens (AOX) in a mineral salt medium (Van Leeuwen *et al.*, 1996).

Trichoderma may also be useful in the remediation of pesticide-contaminated environments. *T. harzianum* can effectively degrade a wide range of such compounds including ciliate, glyphosate, DDT, dieldrin, endosulfan, pentachloronitrobenzene and pentachlorophenol (Krzysko-Lupicka *et al.*, 1997). *Trichoderma harzianum* strain CCT-4790 has been reported to degrade 60% of the herbicide Diuron in soil after 24 h, and may have good potential for bioremediation of soil contaminated with residual phenyl urea herbicide (Espósito and da Silva, 1998).

A strain of *T. viride* has been used to degrade 14C photodieldrin, and this resulted in an appreciable conversion of the pesticide into water-soluble and non-insecticidal compounds after 4 to 5 weeks (Tabet and Lichtenstein, 1976). Similar results have been reported for synthetic pyrethroids, including β-cyfluthrin (Saikia and Gopal, 2004). A combination of *T. viride* with *Aspergillus carneus* and *F. oxysporum* has been found to degrade the herbicide trifluralin, resulting in less than 10% of the pesticide remaining after treatment (Zayed *et al.*, 1983).

Phytobial remediation

Phytoremediation is a form of bioremediation that uses living plants to remove, degrade, immobilize or retain contaminants *in situ*. Phytoremediation can be encouraged by using selected microorganisms to manipulate the rhizosphere, in a process defined as phytobial remediation (Lynch and Moffat, 2005). This process uses plants whose roots are colonized by symbiotic microbes, such as rhizobacteria and mycorrhizae that efficiently create stable microbial communities that either degrade toxic materials or assist in their uptake by the plant. *Trichoderma harzianum* has been used in this process, and work in this area has largely been carried out with strain T22, an isolate that has already been exploited as a biocontrol agent against root disease, and that is considered to have no deleterious effects on plants, animals or the environment when used at recommended rates. Strain T22 is rhizosphere competent, stable and robust and has been effective on all plants tested so far, ranging from ferns to conifers, monocotyledons to dicotyledons, and across a wide range of pH and soil conditions (Harman, 2000). An added advantage of T22 is that it can increase seedling vigour and deep rooting long after application (Adams *et al.*, 2007).

There are few low-cost and effective solutions for the clean-up of arsenic-polluted soils where arsenic is present at relatively low but toxic levels. However, ferns have been identified that hyperaccumulate arsenic in their fronds at levels between 2 and 200 times the level in soils (Ma *et al.*, 2001). Application of *T. harzianum* T22 to the hyperaccumulating fern *Pteris vittata* can increase arsenic uptake by up to 140% compared with non-treated plants (Lynch and Moffat, 2005).

As strain T22 is able to detoxify cyanide (see above), combinations of this strain with the cyanide-hyperaccumulating willow *Salix eriocephala* (Ebbs *et al.*, 2003) are expected to provide very effective methods of bioremediation. The phytobial association of T22 with both *P. vittata* and *Salix fragilis* has been reported to increase the absorption of essential elements and heavy metals, including cadmium, lead, manganese, nickel and zinc (see Table 12.1; Lynch and Moffat, 2005; Adams *et al.*, 2007). Moreover, *S. fragilis* grown with *T. harzianum* T22 in metal-contaminated soil produced a 30% increase in biomass dry weight; and the plants were 16% taller than non-inoculated controls. These results suggest that T22 may increase the rate of revegetation and the phytostabilization of metal-contaminated sites (Adams *et al.*, 2007).

Trichoderma may also be suitable for use in the remediation of organically contaminated materials. Agricultural systems and many industrial processes place large quantities of nitrates and phosphorus into waterways, and these contribute to hypoxia and eutrophization of water ecosystems. Maize plants treated with *T. harzianum* T22 have shown improved nitrogen utilization in soils treated with fertilizers with different nitrogen levels, although it is not clear whether this effect was due to a more efficient use of fertilizer or to mining of organic nitrogen pools in the soil (Harman, 2000).

Conclusion

Strains of *Trichoderma* species have long been used as biological control agents, and many formulations have already been commercialized. Research on the direct mechanisms of biocontrol used by *Trichoderma* species against phytopathogenic fungi has led to the improvement of a first generation of strains based on overproduction of antibiotics and hydrolytic enzymes. A

Table 12.1. Effect of *Trichoderma harzianum* T22 on the average uptake of cadmium (Cd), manganese (Mn), nickel (Ni), lead (Pb) and zinc (Zn) into leaves, stems, roots and whole saplings of crack willow (*Salix fragilis*) (μg metal absorbed by all biomass produced). Saplings were grown for 12 weeks at 25°C (n = 5). Values in brackets show the standard error of the mean.

	Plant section							
	Leaves		Stems		Roots		Whole sapling	
Metal	Control	T22	Control	T22	Control	T22	Control	T22
Cd	1.7	1.9	1.6	2.2[a]	1.3[b]	1.0	4.6	5.1
	(0.2)	(0.3)	(0.1)	(0.3)	(0.1)	(0.1)	(0.3)	(0.6)
Mn	24.3	26.6	12.3	15.3	8.7	7.9	45.3	49.7
	(4.3)	(3.5)	(2.9)	(3.6)	(0.9)	(1.1)	(6.4)	(7.7)
Ni	4.1	4.9	3.4	4.2[a]	3.7	3.6	11.2	12.7
	(0.5)	(0.4)	(0.3)	(0.3)	(0.2)	(0.3)	(0.5)	(0.8)
Pb	9.8	14.9	8.0	10.4[a]	42.8	37.8	60.6	63.0
	(0.9)	(3.5)	(0.6)	(1.1)	(2.6)	(5.1)	(2.2)	(8.1)
Zn	389.2	466.4	213.4	298.4[a]	118.1	150.2	720.8	915.0
	(52.8)	(70.6)	(23.6)	(34.7)	(7.8)	(21.9)	(72.3)	(112.7)
Total	429.1	514.7	238.7	330.5[a]	174.6	200.4	842.4	1,045.5
	(57.3)	(74.2)	(26.9)	(39.2)	(11.2)	(28.1)	(78.3)	(125.7)

[a] T22 saplings took up significantly more metal.
[b] Control saplings took up significantly more metal (P = 0.05).

second generation of broad-spectrum biocontrol agents is emerging from the recent research on the mechanisms that control mycoparasitism.

The recent work on the association of *Trichoderma* with plant roots and the associated effects on plant defence and development have given new insights into the agricultural application of this fungus. These findings may result in a transition from current specific, ecosystem-dependent application strategies to more generalistic strategies based more on plant response. In addition, plant–fungus association could overcome some of the problems associated with the instability of BCAs in the environment and could be an additional advantage in soil bioremediation mediated by plants.

Acknowledgements

This work was supported by the Ministry of Science and Technology (projects AGL2000-0524, BIO2003-03679, AGL2006-03947 and PETRI-95-1010.90.01) and by the Junta de Andalucía (PAI CVI-107, PO6-CVI-01546 and JA 65U0130118).

References

Adams, P., De-Leij, F.A. and Lynch, J.M. (2007) *Trichoderma harzianum* Rifai 1295-22 mediates growth promotion of crack willow (*Salix fragilis*) saplings in both clean and metal-contaminated soil. *Microbial Ecology* 54, 306–313.

Altomare, C., Norvell, W.A., Bjorkman, T. and Harman, G.E. (1999) Solubilization of phosphates and micronutrients by the plant growth-promoting and biocontrol fungus *Trichoderma harzianum* rifai 1295-22. *Applied and Environmental Microbiology* 65, 2926–2933.

Baek, J.M., Howell, C.R. and Kenerley, C.M. (1999) The role of an extracellular chitinase from *Trichoderma virens* Gv29-8 in the biocontrol of *Rhizoctonia solani*. *Current Genetics* 35, 41–50.

Bailey, B.A., Taylor, R., Dean, J.F. and Anderson, J.D. (1991) Ethylene biosynthesis-inducing endoxylanase is translocated through the xylem of *Nicotiana tabacum* cv Xanthi plants. *Plant Physiology* 97, 1181–1186.

Barnett, H.L. and Binder, F.L. (1973) The fungal host–parasite relationship. *Annual Review of Phytopathology* 11, 273–292.

Benhamou, N. and Chet, I. (1997) Cellular and molecular mechanisms involved in the interaction between *Trichoderma harzianum* and *Pythium ultimum*. *Applied and Environmental Microbiology* 63, 2095–2099.

Benítez, T., Delgado-Jarana, J., Rincón, A.M., Rey, M. and Limón, C.M. (1998a) Biofungicides: *Trichoderma* as a biocontrol agent against phytopathogenic fungi. *Recent Research Developments in Microbiology* 2, 129–150.

Benítez, T., Limón, M.C., Delgado-Jarana, J. and Rey, M. (1998b) Glucanolyitc and other enzymes and their genes. In: Harman, G.E. and Kubicek C.P. (eds) *Trichoderma & Gliocladium*, Vol. 2. Taylor & Francis Ltd, London, pp. 101–127.

Benítez, T., Rincón, A.M., Limón, M.C. and Codón, A.C. (2004) Biocontrol mechanisms of *Trichoderma* strains. *International Microbiology* 7, 249–260.

Björkman, T., Blanchard, L.M. and Harman, G.E. (1998) Growth enhancement of shrunken-2 sweet corn by *Trichoderma harzianum* 1295-22, effect of environmental stress. *Journal of the American Society for Horticultural Science* 123, 35–40.

Bolar, J.P., Norelli, J.L., Wong, K.W., Hayes, C., Harman, G.E. and Aldwinckle, H.S. (2000) Expression of endochitinase from *Trichoderma harzianum* in transgenic apple increases resistance to apple scab and reduces vigor. *Phytopathology* 90, 72–77.

Bölker, M. (2002) Signal transduction pathways in phytopathogenic fungi. In: Esser K. and Bennett J.W. (eds) *The Mycota (Agricultural Applications)*, Vol. XI. Springer-Verlag, Berlin, pp. 273–288.

Chacón, M.R., Rodríguez-Galán, O., Benítez, T., Sousa, S., Rey, M., Llobell, A. and Delgado-Jarana, J. (2007) Microscopic and transcriptome analyses of early colonization of tomato roots by *Trichoderma harzianum*. *International Microbiology* 10, 19–27.

Chang, Y.-C. and Baker, R. (1986) Increased growth in the presence of the biological control agent *Trichoderma harzianum*. *Plant Disease* 76, 60–65.

Chet, I. and Inbar, J. (1994) Biological control of fungal pathogens. *Applied Biochemistry and Biotechnology* 48, 37–43.

Datnoff, L.E., Nemec, S. and Pernezy, K. (1995) Biological control of *Fusarium* crown and root rot of tomato in Florida using *Trichoderma harzianum* and *Glomus* intraradices. *Biological Control* 5, 427–431.

Dean, J.F.D., Gamble, H.R. and Anderson, J.D. (1989) The ethylene biosynthesis-inducing xylanase, its induction in *Trichoderma viride* and certain plant pathogens. *Phytopathology* 79, 1071–1078.

de las Mercedes Dana, M., Pintor-Toro, J.A. and Cubero, B. (2006) Transgenic tobacco plants overexpressing chitinases of fungal origin show enhanced resistance to biotic and abiotic stress agents. *Plant Physiology* 142, 722–730.

de Meyer, G., Bigirimana, J., Elad, Y. and Höfte, M. (1998) Induced systemic resistence in *Trichoderma harzianum* T39 biocontrol of *Botrytis cinerea*. *European Journal of Plant Pathology* 104, 279–286.

Delgado-Jarana, J. (2001) Producción de beta-1,6-glucanasa II y genes regulados por pH en *Trichoderma harzianum*. MSc thesis, Universidad de Sevilla, Spain.

Delgado-Jarana, J., Moreno-Mateos, M.A. and Benítez, T. (2003) Glucose uptake in *Trichoderma harzianum*, role of *gtt1*. *Eukaryotic Cell* 2, 708–717.

Delgado-Jarana, J., Sousa, S., González, F., Rey, M. and Llobell, A. (2006) ThHog1 controls the hyperosmotic stress response in *Trichoderma harzianum*. *Microbiology* 152, 1687–1700.

Djonovic, S., Pozo, M.J., Dangott, L.J., Howell, C.R. and Kenerley, C.M. (2006) Sm1, a proteinaceous elicitor secreted by the biocontrol fungus *Trichoderma virens* induces plant defense responses and systemic resistance. *Molecular Plant–Microbe Interactions* 19, 838–853.

Donzelli, B.G. and Harman, G.E. (2001) Interaction of ammonium, glucose, and chitin regulates the expression of cell wall-degrading enzymes in *Trichoderma atroviride* strain P1. *Applied and Environmental Microbiology* 67, 5643–5647.

Duplessis, S., Tagu, D. and Martin, F. (2002) Living together underground. A molecular glimpse of the ectomycorrhizal symbiosis. In: Osiewacz H.D (ed.) *Molecular Biology of Fungal Development*. Marcel Dekker, Inc., New York, pp. 297–324.

Ebbs, S., Bushey, J., Poston, S., Kosma, D., Samiotakis, M. and Dzombak, D. (2003) Transport and metabolism of free cyanide and iron cyanide complexes by willow. *Plant Cell and Environment* 26, 1467–1478.

Enkerli, J., Felix, G. and Boller, T. (1999) The enzymatic activity of fungal xylanase is not necessary for its elicitor activity. *Plant Physiology* 121, 391–397.

Espósito, E. and da Silva, M. (1998) Systematic and environmental application of the genus *Trichoderma*. *Critical Reviews in Microbiology* 24, 89–98.

Ezzi, M.I. and Lynch, J.M. (2002) Cyanide catabolising enzymes in *Trichoderma* spp. *Enzyme and Microbial Technology* 31, 1042–1047.

Ezzi, M.I. and Lynch, J.M. (2005) Biodegradation of cyanide by *Trichoderma* spp. and *Fusarium* spp. *Enzyme and Microbial Technology* 36, 849–854.

Franken, P., Khun, G. and Gianinazzi-Pearson, V. (2002) Development and molecular biology of arbuscular mycorrhizal fungi. In: Osiewacz H.D (ed.) *Molecular Biology of Fungal Development*. Marcel Dekker, Inc., New York, pp. 325–348.

Grondona, I., Hermosa, R., Tejada, M., Gomis, M.D., Mateos, P.F., Bridge, P.D., Monte, E. and García-Acha, I. (1997) Physiological and biochemical characterization of *Trichoderma harzianum*, a biological control agent against soilborne fungal plant pathogens. *Applied and Environmental Microbiology* 63, 3189–3198.

Hananaia, U. and Avni, A. (1997) High-affinity binding site for ethylene-inducing xylanase elicitor on *Nicotiana tabacum* membranes. *Plant Journal* 12, 113–120.

Harman, G.E. (2000) Myths and dogmas of biocontrol. Changes in perceptions derived from research on *Trichoderma harzianum* T22. *Plant Disease* 84, 377–393.

Harman, G.E. (2006) Overview of mechanism and uses of *Trichoderma* spp. *Phytopathology* 96, 190–194.

Harman, G.E. and Kubicek, C.P. (eds) (1998) *Trichoderma and Gliocladium. Enzymes, Biological Control and Commercial Applications*, Vol. 2. Taylor & Francis Ltd., London.

Harman, G.E., Taylor, A.G. and Stazs, T.E. (1989) Combining effective strains of *Trichoderma harzianum* and solid matrix priming to improve biological seed treatments. *Plant Disease* 73, 631–637.

Harman, G.E., Howell, C.R., Viterbo, A., Chet, I. and Lorito, M. (2004a) *Trichoderma* species – opportunistic, avirulent plant symbionts. *Nature Reviews Microbiology* 2, 43–56.

Harman, G.E., Lorito, M. and Lynch, J.M. (2004b) Uses of *Trichoderma* spp. to alleviate or remediate soil and water pollution. *Advances in Applied Microbiology* 56, 313–330.

Hjeljord, I. and Tronsmo, A. (1998) *Trichoderma* and *Gliocladium* in biological control, an overview. In: Harman, G.E. and Kubicek, C.P. (eds) *Trichoderma & Gliocladium*, Vol. 2. Taylor & Francis Ltd, London, pp. 131–152.

Hornby, D. (1990) *Biological Control of Soilborne Plant Pathogens*. CAB International, Wallingford, UK.

Howell C.R. (1998) The role of antibiosis in biocontrol. In: Harman, G.E. and Kubicek, C.P. (eds) *Trichoderma & Gliocladium*, Vol. 2. Taylor & Francis Ltd, London, pp. 173–184.

Howell, C.R. (2003) Mechanisms employed by *Trichoderma* species in the biological control of plant disease, the history and evolution of currents concepts. *Plant Disease* 87, 4–10.

Howell, C.R. (2006) Understanding the mechanisms employed by *Trichoderma virens* to effect biological control of cotton diseases. *Phytopatology* 96, 178–180.

Howell, C.R. and Stipanovic, R.D. (1995) Mechanisms in the biocontrol of *Rhizoctonia solani*-induced cotton seedlings disease by *Gliocladium virens*, antibiosis. *Phytopathology* 85, 469–472.

Ilmen, M., Thrane, C. and Penttilä, M. (1996) The glucose repressor gene *cre1* of *Trichoderma*, isolation and expression of a full-length and a truncated mutant form. *Molecular and General Genetics* 251, 451–460.

Inbar, J. and Chet, I. (1996) The role of lectins in recognition and adhesion of the mycoparasitic fungus *Trichoderma* spp. to its host. *Advances in Experimental Medicine and Biology* 408, 229–231.

Kissi, M., Mountadar, M., Assobhei, O., Gargiulo, E., Palmieri, G., Giardina, P. and Sannia, G. (2001) Roles of two white-rot basidiomycete fungi in decolorisation and detoxification of olive mill waste water. *Applied Microbiology and Biotechnology* 57, 221–226.

Klemsdal, S.S., Clarke, J.L., and Elen, O. (2004) Chitinase gene from *Trichoderma atroviride* confers *Fusarium* resistance to GM-barley. In: McIntyre, M., Nielsen, J., Arnau, J., van der Brink, H., Hansen, K. and Madrid, S. (eds) *Proceedings of the 7th European Conference on Fungal Genetics*, DTU, Copenhagen, p. 68.

Krzysko-Lupicka, T., Strof, W., Kubs, K., Skorupa, M., Wieczorek, P., Lejczak, B. and Kafarski, P. (1997) The ability of soil-borne fungi to degrade organophosphonate carbon-to-phosphorus bonds. *Applied Microbiology and Biotechnology* 48, 549–552.

Kudla, B., Caddick, M.X., Langdon, T., Martinez-Rossi, N.M., Bennett, C.F., Sibley, S., Davies, R.W. and Arst, H.N.Jr. (1990) The regulatory gene *areA* mediating nitrogen metabolite repression in *Aspergillus nidulans*. Mutations affecting specificity of gene activation alter a loop residue of a putative zinc finger. *EMBO Journal* 9, 1355–1364.

Li, A., Antizar-Ladislao, B. and Khraisheh, M. (2007) Bioconversion of municipal solid waste to glucose for bio-ethanol production. *Bioprocess and Biosystems Engineering* 30, 189–196.

Limón, M.C. and Benítez, T. (2002) Function and regulation of fungal chitinases. *Recent Research Developments in Genetics* 2, 97–119.

Limón, M.C., Pintor-Toro, J.A. and Benítez, T. (1999) Increased antifungal activity of *Trichoderma harzianum* transformants that overexpress a 33-kDa chitinase. *Phytopathology* 89, 254–261.

Limón, M.C., Chacón, M.R., Mejías, R., Delgado-Jarana, J., Rincón, A.M., Codón, A.C. and Benítez, T. (2004) Increased antifungal and chitinase specific activities of *Trichoderma harzianum* CECT 2413 by addition of a cellulose binding domain. *Applied Microbiology and Biotechnology* 64, 675–685.

Lo, C.-T., Nelson, E.B., Hayes, C.K. and Harman, G.E. (1998) Ecological studies of transformed *Trichoderma harzianum* strain 1295-22 in the rhizosphere and on the phylloplane of creeping bentgrass. *Phytopathology* 88, 129–136.

Lorito, M. (1998) Chitinolytic enzymes and their genes. In: Harman, G.E. and Kubicek, C.P. (eds) *Trichoderma & Gliocladium*, Vol. 2. Taylor & Francis Ltd, London, pp. 73–99.

Lorito, M., Mach, R.L., Sposato, P., Strauss, J., Peterbauer, C.K. and Kubicek, C.P. (1996a) Mycoparasitic interaction relieves binding of the Cre1 carbon catabolite repressor protein to promoter sequences of the *ech42* (endochitinase-encoding) gene in *Trichoderma harzianum*. *Proceedings of the National Academy of Sciences of the United States of America* 93, 14868–14872.

Lorito, M., Woo, S.L., D'Ambrosio, M.D., Harman, G.E., Hayes, C.K., Kubicek, C.P. and Scala, F. (1996b) Synergistic interaction between cell wall degrading enzymes and membrane affecting compounds. *Molecular Plant–Microbe Interactions* 9, 206–213.

Lorito, M., Woo, S.L., García, I., Colucci, G., Harman, G.E., Pintor-Toro, J.A., Filippone, E., Muccifora, S., Lawrence, C.B., Zoina, A., Tuzun, S., Scala, F. and Fernández, I.G. (1998) Genes from mycoparasitic fungi as a source for improving plant resistance to fungal pathogens. *Proceedings of the National Academy of Sciences of the United States of America* 95, 7860–7865.

Lynch, J.M. and Moffat, A.J. (2005) Bioremediation – prospects for the future application of innovative applied biological research. *Annals of Applied Biology* 146, 217–221.

Ma, L.Q., Komar, K.M., Tu, C., Zhang, W., Cai, Y. and Kennelley, E.D. (2001) A fern that hyperaccumulates arsenic. *Nature* 409, 579.

Marra R., Ambrosino, P., Carbone, V., Vinale, F., Woo, S.L., Ruocco, M., Ciliento, R., Lanzuise, S., Ferraioli, S., Soriente, I., Gigante, S., Turra, D., Fogliano, V., Scala, F. and Lorito, M. (2006) Study of the three-way interaction between *Trichoderma atroviride*, plant and fungal pathogens by using a proteomic approach. *Current Genetics* 50, 307–321.

Martínez C., Blanc, F., Le Claire, E., Besnard, O., Nicole, M. and Baccou, J.C. (2001) Salicylic acid and ethylene pathways are differentially activated in melon cotyledons by active or heat-denatured cellulase from *Trichoderma longibrachiatum*. *Plant Physiology* 127, 334–344.

Mendoza-Mendoza, A., Pozo, M.J., Grzegorski, D., Martínez, P., García, J.M., Olmedo-Monfil, V., Cortés, C., Kenerley, C. and Herrera-Estrella, A. (2003) Enhanced biocontrol activity of *Trichoderma* through inactivation of a mitogen-activated protein kinase. *Proceedings of the National Academy of Sciences of the United States of America* 100, 15965–15970.

Metcalfe, D.A. and Wilsom, C.R. (2004) The process of antagonism of *Sclerotium cepivorum* in white rot affected onion roots by *Trichoderma koningii*. *Plant Pathology* 50, 249–257.

Migheli, Q., González-Candelas, L., Dealessi, L., Camponogara, A. and Ramón-Vidal, D. (1998) Transformants of *Trichoderma longibrachiatum* overexpressing the beta-1,4-endoglucanasa gene *egl1* show enhanced biocontrol of *Pythium ultimum* on cucumber. *Biological Control* 88, 673–677.

Monte, E. (2001) Understanding *Trichoderma*, between biotechnology and microbial ecology. *International Microbiology* 4, 1–4.

Montero, M., Sanz, L., Rey, M., Monte, E. and Llobell, A. (2005) BGN16.3, a novel acidic beta-1,6-glucanase from mycoparasitic fungus *Trichoderma harzianum* CECT 2413. *The FEBS Journal* 272, 3441–3448.

Mora, A. and Earle, E.D. (2001) Combination of *Trichoderma harzianum* endochitinase and a membrane-affecting fungicide on control of *Alternaria* leaf spot in transgenic broccoli plants. *Applied Microbiology and Biotechnology* 55, 306–310.

Moreno-Mateos, M.A. (2006) Regulación por pH ambiental en el agente de biocontrol *Trichoderma harzianum* CECT 2413. MSc thesis, Universidad de Sevilla, Spain.

Moreno-Mateos, M.A., Delgado-Jarana, J., Codón, A.C. and Benítez, T. (2007) pH and Pac1 control development and antifungal activity in *Trichoderma harzianum*. *Fungal Genetics and Biology* 44, 1355–1367.

Mukherjee, P.K., Latha, J., Hadar, R. and Horwitz, B.A. (2003) TmkA, a mitogen-activated protein kinase of *Trichoderma virens*, is involved in biocontrol properties and repression of conidiation in the dark. *Eukaryotic Cell* 2, 446–455.

Mukherjee, P.K., Latha, J., Hadar, R. and Horwitz, B.A. (2004) Role of two G-protein alpha subunits, TgaA and TgaB, in the antagonism of plant pathogens by *Trichoderma virens*. *Applied and Environmental Microbiology* 70, 542–549.

Mukherjee, M., Mukherjee, P.K. and Kale, S.P. (2007) cAMP signalling is involved in growth, germination, mycoparasitism and secondary metabolism in *Trichoderma virens*. *Microbiology* 153, 1734–1742.

Olmedo-Monfil, V., Mendoza-Mendoza, A., Gómez, I., Cortés, C. and Herrera-Estrella, A. (2002) Multiple environmental signals determine the transcriptional activation of the mycoparasitism related gene *prb1* in *Trichoderma atroviride*. *Molecular Genetics and Genomics* 267, 703–712.

Osiewacz, H.D. (2002) *Molecular Biology of Fungal Development*. Marcel Dekker, New York, USA.

Peberdy, J.F. (1990) Fungal cell walls – a review. In: Kuhn, P.J., Trinci, A.P.J., Jung, M.J., Goosey, M.W. and Copping, L.G. (eds) *Biochemistry of Cell Walls and Membranes in Fungi*. Springer-Verlag, Heidelberg, Germany, pp. 5–24.

Pentillä, M., Limón, M.C. and Nevalainen, H. (2004) Molecular biology of *Trichoderma* and biotechnological applications. In: Arora, D.K. (ed.) *Handbook of Fungal Biotechnology*. Marcel Dekker, Inc., New York, pp. 413–427.

Peterbauer, C.K., Litscher, D. and Kubicek, C.P. (2002) The *Trichoderma atroviride seb1* (stress response element binding) gene encodes an AGGGG-binding protein which is involved in the response to high osmolarity stress. *Molecular Genetics and Genomics* 268, 223–231.

Ravelet, C., Krivobok, S., Sage, L. and Steiman, R. (2000) Biodegradation of pyrene by sediment fungi. *Chemosphere* 40, 557–563.

Reithner, B., Brunner, K., Schuhmacher, R., Peissl, I., Seidl, V., Krska, R. and Zeilinger, S. (2005) The G protein alpha subunit Tga1 of *Trichoderma atroviride* is involved in chitinase formation and differential production of antifungal metabolites. *Fungal Genetics and Biology* 42, 749–760.

Reithner, B., Schuhmacher, R., Stoppacher, N., Pucher, M., Brunner, K. and Zeilinger, S. (2007) Signalling via the *Trichoderma atroviride* mitogen-activated protein kinase Tmk1 differentially affects mycoparasitism and plant protection. *Fungal Genetics and Biology* 44, 1123–1133.

Rey, M., Delgado-Jarana, J. and Benítez, T. (2001) Improved antifungal activity of a mutant of *Trichoderma harzianum* CECT 2413 which produces more extracellular proteins. *Applied Microbiology and Biotechnology* 55, 604–608.

Rincón, A.M. (2004) Mejora de la capacidad antagonista de *Trichoderma harzianum* y de la resistencia a patógenos de *Nicotiana tabacum* mediante sobreexpresión de la beta-1,3-glucanasa I. MSc thesis, Universidad de Sevilla, Spain.

Rosado, I.V., Rey, M., Codón, A.C., Govantes, J., Moreno-Mateos, M.A. and Benítez, T. (2007) QID74 Cell wall protein of *Trichoderma harzianum* is involved in cell protection and adherence to hydrophobic surfaces. *Fungal Genetics and Biology* 44, 950–964.

Sahai, A.S. and Manocha, M.S. (1993) Chitinases of fungi and plants, their involvement in morphogenesis and host–parasite interaction. *Microbiological Reviews* 11, 317–338.

Saikia, N. and Gopal, M. (2004) Biodegradation of beta-cyfluthrin by fungi. *Journal of Agricultural and Food Chemistry* 52, 1220–1223.

Samuels, G.J. (2006) *Trichoderma*, systematics, the sexual state, and ecology. *Phytopatology* 96, 195–206.

Saraswathy, A. and Hallberg, R. (2002) Degradation of pyrene by indigenous fungi from a former gasworks site. *FEMS Microbiology Letters* 210, 227–232.

Seidl, V., Seiboth, B., Karaffa, L. and Kubicek, C.P. (2004) The fungal STRE-element-binding protein Seb1 is involved but not essential for glycerol dehydrogenase (gld1) gene expression and glycerol accumulation in *Trichoderma atroviride* during osmotic stress. *Fungal Genetics and Biology* 41, 1132–1140.

Seidl, V., Schmoll, M., Scherm, B., Balmas, V., Seiboth, B., Migheli, Q. and Kubicek, C.P. (2006) Antagonism of *Pythium* blight of zucchini by *Hypocrea jecorina* does not require cellulase gene expression but is improved by carbon catabolite derepression. *FEMS Microbiology Letters* 257, 145–151.

Sharon, E., Bar-Eyal, M., Chet, I., Herrera-Estrella, A., Kleifeld, O. and Spiegel, Y. (2001) Biological control of the root-knot nematode *Meloidogyne javanica* by *Trichoderma harzianum*. *Phytopathology* 91, 687–693.

Suárez, B., Rey, M., Castillo, P., Monte, E. and Llobell, A. (2004) Isolation and characterization of PRA1, a trypsin-like protease from the biocontrol agent *Trichoderma harzianum* CECT 2413 displaying nematicidal activity. *Applied Microbiology and Biotechnology* 65, 46–55.

Tabet, J.C. and Lichtenstein, E.P. (1976) Degradation of [14C]photodieldrin by *Trichoderma viride* as affected by other insecticides. *Canadian Journal of Microbiology* 22, 1345–1356.

Van Leeuwen, J.A., Nicholson, B.C., Hayes, K.P. and Mulcahy, D.E. (1996) Resistance of bound chloroguaiacols and AOX from pulp mill effluent to degradation by *Trichoderma harzianum* isolated from Lake Bonney, south-eastern South Australia. *Marine and Freshwater Research* 47, 961–969.

Viterbo, A., Haran, S., Friesem, D., Ramot, O. and Chet, I. (2001) Antifungal activity of a novel endochitinase gene (*chit36*) from *Trichoderma harzianum* Rifai TM. *FEMS Microbiology Letters* 200, 169–174.

Viterbo, A., Montero, M., Ramot, O., Friesem, D., Monte, E., Llobell, A. and Chet, I. (2002) Expression regulation of the endochitinase chit36 from *Trichoderma asperellum* (*T. harzianum* T-203). *Current Genetics* 42, 114–122.

Viterbo, A., Harel, M., Horwitz, B.A., Chet, I. and Mukherjee, P.K. (2005) *Trichoderma* mitogen-activated protein kinase signalling is involved in induction of plant systemic resistance. *Applied and Environmental Microbiology* 71, 6241–6246.

Windham, M.T., Elad, Y. and Baker, R. (1986) A mechanism for increased plant growth induced by *Trichoderma* spp. *Phytopathology* 76, 518–521.

Woo, S.L., Donzelli, B., Scala, F., Mach, R.L., Harman, G.E., Kubicek, C.P., Del Sorbo, G. and Lorito, M. (1999) Disruption of the ech42 (endochitinase-encoding) gene affects biocontrol activity in *Trichoderma harzianum* P1. *Molecular Plant–Microbe Interactions* 12, 419–429.

Woo, S.L., Scala, F., Ruocco, M. and Lorito, M. (2006) The molecular biology of the interactions between *Trichoderma* spp., phytopathogenic fungi and plants. *Phytopatology* 96, 181–185.

Yedidia, I., Srivastva, A.K., Kapulnik, Y. and Chet, I. (2001) Effect of *Trichoderma harzianum* on microelement concentrations and increased growth of cucumber plants. *Plant and Soil* 235, 235–242.

Yedidia, I., Shoresh, M., Kerem, Z., Benhamou, N., Kapulnik, Y. and Chet, I. (2003) Concomitant induction of systemic resistance to *Pseudomonas syringae* pv. *lachrymans* in cucumber by *Trichoderma asperellum* (T-203) and accumulation of phytoalexins. *Applied and Environmental Microbiology* 69, 7343–7353.

Zayed, S.M., Mostafa, I.Y., Farghaly, M.M., Attaby, H.S., Adam, Y.M. and Mahdy, F.M. (1983) Microbial degradation of trifluralin by *Aspergillus carneus, Fusarium oxysporum* and *Trichoderma viride*. *Journal of Environmental Science and Health B* 18, 253–267.

Zeilinger, S., Reithner, B., Scala, V., Peissl, I., Lorito, M. and Mach, R.L. (2005) Signal transduction by Tga3, a novel G protein alpha subunit of *Trichoderma atroviride*. *Applied and Environmental Microbiology* 71, 1591–1597.

13 *Agrobacterium tumefaciens* as a Molecular Tool for the Study of Fungal Pathogens

Carol M. McClelland[1] and Brian L. Wickes[2]

[1]*Department of Biology, McMurry University, Abilene, Texas, USA;*
[2]*Department of Microbiology and Immunology, The University of Texas Health Science Center at San Antonio, San Antonio, Texas, USA*

Introduction

The discovery of *Agrobacterium tumefaciens* as the causative agent of crown gall disease in plants occurred a century ago (Smith and Townsend, 1907). In the 21st century, the ability of this Gram-negative soil bacterium to mediate trans-kingdom transfer of DNA is crucial to our ability to genetically manipulate many plant and non-plant species. The number of plant and non-plant species that have been stably transformed by *Agrobacterium*-mediated transformation (AMT) has steadily increased in recent years, due to improved laboratory techniques and the development of recombinant *Agrobacterium* plasmid vectors.

The non-plant species transformed via AMT include a wide variety of fungi and, for many of these, AMT offers significant advantages over other more traditional transformation techniques. Advantages include straightforward and inexpensive transformation protocols, transformation of diverse host morphologies, single-copy integration and the ability to generate both heterologous and homologous transformants. Additionally, the use of *A. tumefaciens* as a model system to study genetic transformation has increased our understanding of DNA transfer mechanisms from bacteria to eukaryotic organisms, and allowed investigation of the mechanisms governing both homologous and non-homologous recombination.

Principles of *Agrobacterium*-mediated Transformation

Genetic mechanisms of AMT

In nature, *A. tumefaciens* induces tumour formation in a plant host by transferring part of its DNA, the T-DNA, to the host cell. The T-DNA is carried on an extrachromosomal Ti plasmid, which can vary in size (De Vos *et al.*, 1981; Komari *et al.*, 1986; Gerard *et al.*, 1992). Following integration into the host genome, expression of oncogenes and opine-catabolism genes located on the T-DNA leads to neoplastic growth and the production of opines that are used as a nitrogen source by the bacterium (Gaudin *et al.*, 1994; Gelvin, 1998). The T-DNA region, which is approximately 10–30 kb long (Lemmers *et al.*, 1980), is delineated by T-DNA border sequences (see Fig. 13.1). These are 24 bp imperfect direct repeats, which act as the *cis*-acting signal for the DNA delivery system (Wang *et al.*, 1984). Since no other signal is required from the T-DNA for trans-kingdom DNA transfer, the intervening sequence between the T-DNA borders can be eliminated and replaced by other gene(s) of interest.

Processing of the T-DNA from the Ti plasmid and its subsequent export to the plant cell depends on the activity of virulence (*vir*) genes, which are also carried by the Ti plasmid. The mechanism by which the Vir proteins act has been the subject of several reviews (see Jin *et al.*, 1990; Gelvin, 2003) and is summarized in Fig. 13.2. Briefly, proteins VirA and VirG are members of a two-component regulatory signal transduction system. Upon phosphorylation, the VirG protein activates/increases the level of transcription of the *vir* genes. The transfer of T-DNA from *A. tumefaciens* to plant cells is via a type IV secretion system encoded by the *virB* operon and the *virD4* gene. However, the exact mechanism by which the T-DNA is integrated into the host genome remains unknown.

AMT of fungi

The ability of *Agrobacterium* to mediate trans-kingdom DNA transfer is dependent on host cell mechanisms common to many organisms; therefore,

A 5'-tggcaggata tattgtggtg taaac-3' 5'-tgacaggata tattggcggg taaac-3'

Left Border Right Border

B

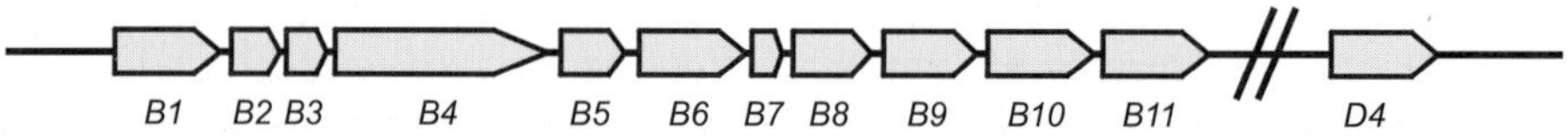

Fig. 13.1. Sequence of the 25-bp direct imperfect repeats constituting the left and right borders of the T-DNA region.

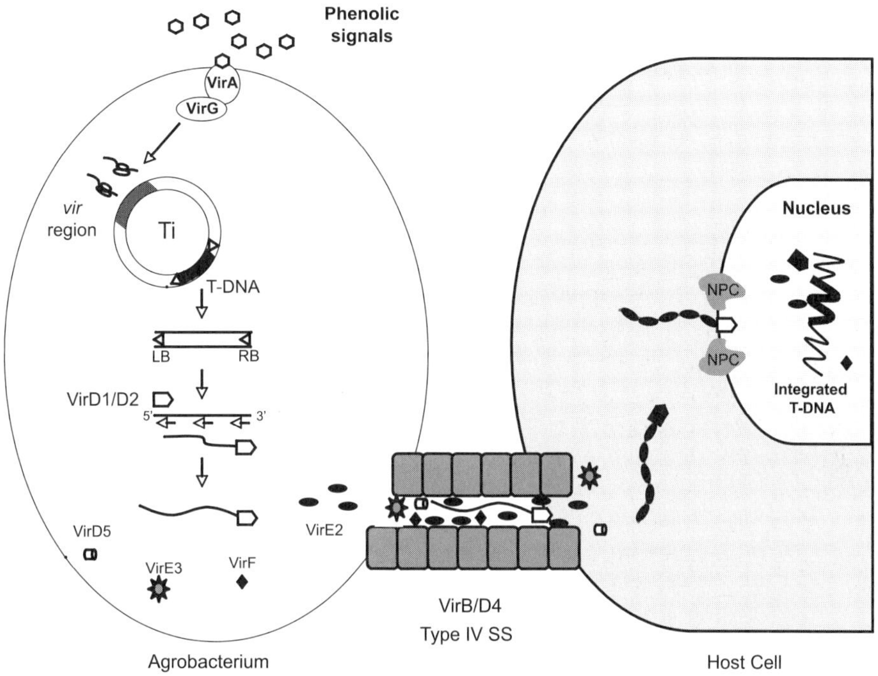

Fig. 13.2. General schematic representation of *Agrobacterium*-mediated transformation. Following activation and autophosphorylation of VirA, VirG is phosphorylated and induces *vir* gene transcription. A single-stranded T-DNA molecule is released after nicking by VirD1/VirD2 border-specific endonucleases. The T-DNA border sequence also serves as the covalent attachment site for the VirD2 protein. Export of the virD2/T–strand complex to the host cell occurs via a type 4 secretion system (SS) formed by the VirB and VirD4 proteins. Several other Vir proteins (VirD5, VirE2, VirE3 and VirF) are independently transported into the host cell cytoplasm by the same VirB/VirD4 channel. Once inside the host cell, the virD2/T-strand conjugate is coated with VirE2 molecules, imported into the host cell nucleus and integrated into the host cell genome. NPC, nuclear pore complex.

it is reasonable that *Agrobacterium* could be used to transform non-plant species. *Agrobacterium*-mediated transformation of the yeast *Saccharomyces cerevisiae* was first reported in 1995. A Ura^- yeast strain was converted to Ura^+ following introduction of the *URA3* gene by T-DNA transfer and integration (Bundock *et al.*, 1995). As with plants, the presence of the small phenolic compound acetosyingone, a powerful inducer of the *vir* genes (Stachel *et al.*, 1986), was required for *A. tumefaciens* T-DNA transfer into yeast, suggesting an active role for the *Agrobacterium* Vir proteins. Additionally, *A. tumefaciens* strains with mutations in some *vir* genes known to be essential for AMT of plant species were either avirulent or attenuated in virulence for yeast cells. Insertion of the T-DNA into the *Saccharomyces* genome was random; however, if the T-DNA contained sequences that were homologous to the yeast genome, homologous recombination was observed (Bundock *et al.*, 1995).

de Groot *et al.* (1998) transformed *Aspergillus awamori* and several other filamentous fungi by AMT. As with *S. cerevisiae*, random T-DNA integration was observed initially, but later studies proved that homologous recombination was possible if the T-DNA shared larger regions of homology with the host genome (Michielse *et al.*, 2004a). Work with *vir*⁻-defective *A. tumefaciens* strains also indicated that transformation in fungi was a result of T-DNA activation and insertion, as opposed to conjugative plasmid transfer (Michielse *et al.*, 2004a). Since these original transformation studies, over 60 species of fungi have been successfully transformed by AMT, including members of the ascomycetes, basidiomycetes, zygomycetes and some oomycetes. It is important to note that AMT has been successfully applied to those fungi not easily transformed by other more conventional methods, including fungi of medical, agricultural and economic significance, such as *Magnaporthe grisea* (Khang *et al.*, 2005), *Coccidioides immitis* (Abuodeh *et al.*, 2000) and *Agaricus bisporus* (de Groot *et al.*, 1998).

While various transformation protocols have been successfully employed for different species of fungi, AMT offers significant advantages for many species. For example, one benefit of AMT is the variety of host morphologies that can be transformed. In most studies, conidia or yeast cells are used as the starting material; however, spores, protoplasts, mycelium and even fruiting body tissue have been successfully transformed. Fungal transformation efficiency rates are also often higher with AMT than with other methods. AMT rates in *A. awamori* were 600 times more efficient than with PEG/$CaCl_2$ (de Groot *et al.*, 1998) and 140 times more efficient in *Calonectria morganii* (Malonek and Meinhardt, 2001). Finally, AMT transformation of fungi often results in primarily single-copy T-DNA integration events that are mitotically stable, even upon passage (Bundock *et al.*, 2002; Hanif *et al.*, 2002; Combier *et al.*, 2003; McClelland *et al.*, 2005; Sugui *et al.*, 2005). In contrast, conventional transformation methods often result in multiple insertions, genetic deletions, rearrangements and unstable phenotypes (Leclerque *et al.*, 2004; Michielse *et al.*, 2004c).

Genetic Tools for AMT

Vectors

Since natural Ti plasmids are large and not amenable to genetic manipulation, an *Agrobacterium* binary vector system has been developed. In this binary system, the two main components required for *Agrobacterium*-mediated gene transfer, the T-DNA and the *vir* region, reside on separate plasmids (Hoekema *et al.*, 1983). The *vir* genes reside on a disarmed Ti plasmid within the *Agrobacterium* strain, while the T-DNA is located on a separate Ti vector suitable for genetic manipulation. Some of the most commonly used Ti vectors are listed in Table 13.1. Recently, Takken *et al.* (2004) developed a one-step protocol to convert bacterial artificial chromosomes (BACs) or other

Table 13.1. Commonly used binary vectors.

Vector	Size (bp)	LacZ	Bacterial selection marker[a]	Replication origin for *A. tumefaciens*	Replication origin for *E. coli*	Mobilization	Reference/accession number/URL
pBIN19	11,777	Yes	Kan	pRK2	pRK2	Yes	Bevan, 1984/U09365
pCV001	9,200	No	Amp	pRK2	ColE1	Yes	Koncz and Schell, 1986
pGreen series	3,000–5,000	Yes	Kan	pSA	pUC	No	Hellens *et al.*, 2000/ http://www.pgreen.ac.uk/
pPZP series	≈ 6,800	Yes	Cm, Sp	pVS1	ColE1	Yes	Hajdukiewicz *et al.*, 1994
pBIBAC series	≈ 21,000	No	Kan	pRi	F factor	Yes	Hamilton, 1997; Takken *et al.*, 2004/ http://www.biotech.cornell.edu/BIBAC/BIBACHomePage.html
pCambia series	7,000–12,000	Yes	Kan, Cm	pVS1	pBR322	Yes	http://www.cambia.org/
pSB11	6,300	No	Sp	None	ColE1	Yes	Komari *et al.*, 1996/AB027256/ http://www.jti.co.jp/plantbiotech

Kan, kanamycin; Amp, ampicillin; Cm, chloramphenicol, Sp, spectinomycin.
[a]Multiple markers for a plasmid series indicates that multiple plasmids are available with individual selection markers.

circular DNA constructs into binary vectors by *in vivo* recombination between a PCR-based gene replacement cassette and a BAC cosmid. As BAC libraries already exist for a number of fungi, this method can be applied to create fungal binary BACs (BIBACs) for functional genomic and complementation studies.

Selection markers

The primary dominant selection marker used in fungal AMT protocols is the *E. coli* hygromycin resistance gene *hph*. Other drug resistance markers include (i) the phleomycin resistance gene (*BLE*); (ii) the aminoglycoside 3′-phosphotransferase type I gene (*APHI*) from the bacterial transposon Tn903, which confers resistance to the genticin analogue G418; (iii) the nourseothricin acetyltransferase gene (*NAT*), which confers resistance to nourseothricin; and (iv) the neomycin phosphotransferase gene (*nptII*), which confers resistance to neomycin. Selection using the toxic analogue 5-fluoroorotic acid allows for selection of targeted gene disruptions in the *pyrG* gene, which encodes orotidine-5-phosphate decarboxylase in *A. awamori* and *A. niger* (de Groot *et al.*, 1998; Gouka *et al.*, 1999).

Markers based on complementation of gene function are also available for some fungi. This strategy circumvents the introduction of antibiotic resistance genes or other foreign DNA, which is significant for agricultural and medical applications, but is limited to species where nutritional auxotrophs are available. Both green fluorescent protein (GFP) and *Discosoma sp.* red fluorescent protein (DsRED) have also been successfully introduced into several species of fungi via AMT (Fitzgerald *et al.*, 2004; Eckert *et al.*, 2005; Grimaldi *et al.*, 2005; Almeida *et al.*, 2007; Li *et al.*, 2007). Examples include co-inoculated cultures of either *Leptosphaeria maculans* and *L. biglobosa*, or *Oculimacula yallundae* and *O. acuformis*, transformed with the genes encoding either DsRed or GFP, where transformed mycelium could be distinguished (Eckert *et al.*, 2005). Transformants of *L. maculans* or *L. biglobosa* expressing DsRed or GFP can be observed in the leaves of *Brassica napus* (Rapeseed) and have been used to observe the colonization of leaf petioles and growth in xylem vessels (Eckert *et al.*, 2005).

Factors Influencing Fungal AMT

A number of parameters affect the success of AMT in any species. The choice of *Agrobacterium* strain, the recipient starting material and co-cultivation conditions can dramatically alter transformation efficiency. These parameters must often be determined empirically for each species, and even for different strains within the same species. Some general guidelines on how the various parameters may influence the success of AMT in fungi are discussed below.

Agrobacterium host strain

Agrobacterium strains used in AMT are defined by their chromosomal background and resident Ti plasmid (see Table 13.2). The C58 chromosomal background is the basis for several strains used in fungal AMT. Strains with increased expression or activation of the virG protein, which activates transcription of the *vir* gene cluster, often have higher transformation frequencies (Komari *et al.*, 1996). Systematic comparisons of different strains in relation to transformation frequencies have not been performed in fungi, but the choice of *Agrobacterium* strain affects transformation efficiency in several different species. Studies using *Agrobacterium* strains derived from the hypervirulent *A. tumefaciens* strain A281 resulted in higher transformation frequencies in *S. cerevisiae, Monascus purpureus* and *Phytophthora infestans* than were obtained in transformations with the LBA1100 strain (Piers *et al.*, 1996). *Agrobacterium* strain A281 and a derivative (AGL-1) were also more efficient than *Agrobacterium* strain LBA4404 in transferring T-DNA into *Cryphonectria parasitica* (Park and Kim, 2004).

Starting material

As discussed previously, conidia or yeast cells are most often used as starting material for AMT, but a variety of fungal morphologies have been successfully transformed. The presence of actively dividing cells or DNA may be important for T-DNA integration. Germination of spores and extended germination times increased transformation frequency in a study of *C. immitis* (Abuodeh *et al.*, 2000), while actively dividing cells of *Blastomyces dermatitidis* were transformed more efficiently than older cultures (Sullivan *et al.*, 2002). However, the opposite phenomenon was observed with *M. purpureus, Agaricus giganteus* and *A. bisporis* (Meyer *et al.*, 2003).

Table 13.2. Commonly used *Agrobacterium tumefaciens* strains.

Strain	Chromosomal background	Ti plasmid (marker gene)	Opine[a]	Reference
LB4404	TiAch5	pAL4404	Octopine	Ooms *et al.*, 1982
LBA1100	TiAch5	pAL1100ΔT-DNA, Δtra, Δocc	Octopine	Beijersbergen *et al.*, 1992
A348	C58	pTiAGNc	Octopine	Sciaky *et al.*, 1978
GV3101	C58	Cured	Nopaline	Holsters *et al.*, 1980
EHA101	C58	pEHA101 (pTiBo542ΔT-DNA)	Nopaline	Hood *et al.*, 1986
EHA105	C58	pEHA105 (pTiBo542ΔT-DNA)	Succinamopine	Hood *et al.*, 1993
A281	C58	pTiBo542	Succinamopine	Sciaky *et al.*, 1978
AGL-1	C58, RecA	pTiBo542ΔT-DNA	Succinamopine	Lazo *et al.*, 1991

[a]Generally accepted classification scheme of *Agrobacterium* strains. Indicates opine catabolism of the original progenitor wild-type strain and/or non-disarmed parental Ti plasmid. Does not necessarily indicate opine production.

A recent study by Roberts *et al.* (2003) demonstrated that adenine auxotrophs of *S. cerevisiae* were supersensitive to AMT. Furthermore, plant cells exposed to purine synthesis inhibitors, but not to pyrimidine synthesis inhibitors, were more sensitive to AMT. In contrast, adenine auxotrophs of *Cryptococcus neoformans* were not more sensitive to AMT, but several uracil auxotrophic *C. neoformans* strains showed higher transformation levels as compared with their wild-type counterparts (McClelland *et al.*, 2005). These increased sensitivities may be a factor in transformation frequency if complementing nutritional markers are used.

Co-cultivation conditions

Co-cultivation conditions include a number of factors, such as the ratio between *A. tumefaciens* and the fungal cells, temperature, length of the co-cultivation period and acetosyringone concentration. Each species of fungus appears to have an optimal combination of these various factors for maximal transformation frequency. Growth rates of the fungal host strain and differences in susceptibility to *A. tumefaciens* probably account for these differences. An appropriate *Agrobacterium*:fungal cell ratio is essential for efficient AMT. For example, altering the bacteria:fungal cell ratio from 1:1 to 500:1 resulted in a tenfold increase in transformation frequency for *C. immitis* (Abuodeh *et al.*, 2000). A bacteria:fungal cell ratio of 26,000:1 was needed for high-efficiency transformation of *Pseudozyma antarctica* (Marchand *et al.*, 2007), but a lower input of *A. tumefaciens* cells compared with fungal cells (1:5 and 1:10) led to higher numbers of transformants per co-cultivation in *Paracoccidioides brasiliensis* (Almeida *et al.*, 2007).

The addition of too many *A. tumefaciens* cells may decrease transformation efficiency due to nutritional or space limitations. Conversely, the presence of too many fungal cells may result in high background levels of untransformed cells. Altering the bacteria:fungal cell ratio may also distort the copy number of T-DNA insertions into a single host cell, as increasing the *A. tumefaciens* concentration during co-cultivation in *Blastomyces dermatitidis* and *Suillus bovinus* resulted in multiple T-DNA insertions (Hanif *et al.*, 2002; Sullivan *et al.*, 2002).

The length of the co-cultivation period has been examined for several different species (de Groot *et al.*, 1998; Abuodeh *et al.*, 2000; Zwiers and De Waard, 2001; Combier *et al.*, 2003; Leclerque *et al.*, 2004; Michielse *et al.*, 2004b; McClelland *et al.*, 2005). In general, longer co-cultivation periods lead to increased transformation efficiency. For a serotype D strain of *C. neoformans*, a co-cultivation time of 72 h resulted in more than a log-fold increase in transformation frequency as compared with one of 24 h (McClelland *et al.*, 2005). The co-cultivation period for most fungal AMT protocols ranges from 48 to 96 h (de Groot *et al.*, 1998; Abuodeh *et al.*, 2000; Zwiers and De Waard, 2001; Combier *et al.*, 2003; Leclerque *et al.*, 2004; Michielse *et al.*, 2004b; McClelland *et al.*, 2005). However, there is a limit to how long the co-cultivation period can be extended. Transformation periods longer than 72 h decreased transformation rates for a serotype D strain of *C. neoformans* (C.M. McClelland, unpublished), and a prolonged co-cultivation time in *A. awamori* led to irreproducible results (Michielse *et al.*, 2004a).

The co-cultivation temperature greatly influences transformation efficiency, and temperatures between 22 and 25°C have been found to be optimal (Bundock *et al.*, 1995; Abuodeh *et al.*, 2000; Zwiers and De Waard, 2001; Combier *et al.*, 2003; Idnurm *et al.*, 2004; Michielse *et al.*, 2004a, 2005b; Almeida *et al.*, 2007). This observation is similar to what has been observed for AMT in plants (Dillen *et al.*, 1997), and may be due to the lability of the T-pilus. Although *vir* gene induction is maximal at approximately 25–27°C, T-pilus formation is most stable at 18–20°C (Fullner and Nester, 1996; Fullner *et al.*, 1996).

Acetosyringone (AS) is required during co-cultivation for transformation. Leclerque *et al.* (2004) investigated the effect of AS concentration on *Agrobacterium* during co-cultivation. They found that increasing the AS concentration to 1 M increased the number of transformants, and high transformation frequencies were obtained only when sufficient amounts of AS were present in the co-cultivation plates and during pre-culture in induction media. However, the addition of AS during pre-culture of *Agrobacterium* does not appear to be an absolute requirement for transformation (Combier *et al.*, 2003).

For species where co-cultivation is performed on a filter membrane, the choice of filter can influence transformation frequency. Different filters, such as nitrocellulose, Hybond (Amersham, Piscataway, New Jersey), filter paper, cellophane sheets and polyvinylidene difluoride, have been successfully used for co-cultivation. No difference was seen in transformation rates between filter paper, Hybond N or Hybond N+ in AMT of *A. awamori* (Michielse *et al.*, 2005b). However, nylon and cellulose were found to be better substrates for co-cultivation than nitrocellulose in *A. fumigatus* (Sugui *et al.*, 2005). In summary, a unique set of parameters appears to exist for each individual fungal species that promotes efficient T-DNA transfer and integration during AMT. The variability seen in co-cultivation factors indicates that considerable optimization may be needed to perform AMT for some species.

AMT as a Tool for Mutagenesis in Fungi

Insertional mutagenesis

Insertional mutagenesis is a classic forward genetics approach for determining gene function. The genome is saturated with mutations, and collections of such mutants are screened for deficiencies in specific processes to identify mutants of interest. Genes containing the mutation are identified and correlated with the observed phenotype using classical molecular methods of cloning, disruption and complementation.

For an insertional mutagenesis methodology to be successful, a number of criteria must be met. First, large numbers of transformants must be easily generated. Second, the transforming DNA must carry a tag so that it can be re-isolated following screening of the transformant library for the desired phenotype. Finally, insertion into the host genome must be as a single-copy or in low-copy numbers in a random and non-sequence-specific manner. While AMT is now a standard tool for insertional mutagenesis in plants, there is some question as to the randomness of T-DNA integration. Bakarat *et al.* (2000) reported that T-DNA integration was random in the *Arabidopsis thaliana* genome, but integration in the rice genome occurred predominantly in gene-rich and transcriptionally active regions.

Mapping of a large collection of T-DNA insertions in *S. cerevisiae* indicated that T-DNA insertion was random and non-sequence specific, as no extensive homology between the T-DNA and integration site or between the integration sites was found (Bundock *et al.*, 2002). T-DNA integrations were located in upstream elements (24%), ORFs (26%), downstream elements (6%) and intergenic regions (41%). Additionally, analysis of isolated flanking fungal chromosomal DNA and T-DNA borders has shown that truncation of both the right and left borders often occurs following T-DNA integration in fungi (de Groot *et al.*, 1998; Michielse *et al.*, 2004c), but to a much lesser extent than in plants (Gheysen *et al.*, 1991). A recent analysis of 1110 T-DNA-tagged mutants of the rice blast fungus *Magnaporthe oryzae* found that integration was biased towards chromosomes 1 and 2, and showed a twofold preference for the promoter region of genes (Choi *et al.*, 2007). Chromosomal rearrangements and read-through of plasmid vectors were also observed; however, the overall stability of the junction regions between the T-DNA insertion and the flanking chromosomal DNA was still relatively conservative (Choi *et al.*, 2007).

Insertional mutagenesis studies with AMT have been performed in several fungi, including *C. neoformans* (Idnurm *et al.*, 2004; Idnurm and Heitman, 2005), *Coniothyrium minitans* (Rogers *et al.*, 2004), *Colletotrichum lagenaria* (current name *C. orbiculare*) (Tsuji *et al.*, 2003) and *L. maculans* (Blaise *et al.*, 2007), leading to the identification of a variety of mutants involved in sporulation, pathogenicity, pigmentation and antibiotic production. Several genes involved in melanization, including a gene encoding a voltage-gated chloride channel (*CLC1*) that appears to be involved in pigmentation and a gene involved in controlling light responses, (*BWC2*) have all been identified in *C. neoformans* as a result of *Agrobacterium*-mediated insertional mutagenesis (Idnurm *et al.*, 2004; Idnurm and Heitman, 2005). In the *C. neoformans* study only 50% of the mutant phenotypes could be directly correlated to a T-DNA insertion. In a recent study with *Leptosphaeria maculans,* a total of 53 transformants out of 1388 displayed reproducible pathogenicity defects. Further analysis of 12 mutants showed that the phenotype could be definitively linked to the T-DNA insertion upon passage in 50% of the isolates (Blaise *et al.*, 2007).

The first large-scale functional genomics study using AMT in fungi was recently completed with *M. oryzae* (Jeon *et al.*, 2007). A total of 21,080 mutants were generated by AMT and catalogued in multi-well plates. The mutants were mitotically and meiotically stable, and over 80% of the mutants generated were estimated to have only a single T-DNA insertion site. Genome saturation was estimated to be 61%. A high-throughput phenotype screening was developed to assay alterations in seven life cycle traits, including sporulation, appressoria formation and the ability to cause disease on rice leaves. Thus far, 202 new pathogenicity loci have been identified. This study indicates that AMT is an effective tool for large-scale random mutagenesis studies in fungi, and is likely to be applied to many more species in the near future.

Targeted mutagenesis

The most direct routine for assessing gene function is to disrupt or replace the gene of interest and observe the resulting phenotype. This technique is effective in species where gene targeting and homologous recombination (HR) rates are high. Specific gene targeting in *S. cerevisiae* is extremely efficient, and homologous flanking sequences as short as 50 bp can efficiently target the gene of interest to the chromosome.

However, for most other species of fungi, large-scale disruption of the predicted genes is seriously limited by two considerations. First, large (1000 bp or more) homologous flanks are often needed to obtain a reasonable HR frequency (Asch and Kinsey, 1990). The efficiency of HR is determined primarily by the dominant pathway of the host for repairing double-stranded DNA breaks (Schaefer, 2001). The length and G/C content of the flanking homologous DNA, the transcriptional status of the targeted gene and the location of the targeted gene on the chromosome have also been shown to influence HR efficiency (Hua *et al.*, 1997; Gray and Honigberg, 2001). Several methods have been developed to increase gene-targeting efficiency and/or identify those transformants that have undergone recombination in fungi (reviewed in Wendland, 2003).

AMT is important for targeted gene disruption because it appears to promote HR in several fungal species. In the yeast *Kluyveromyces lactis*, introduction of a gene-disruption cassette by AMT resulted in a ten- to 71-fold increase in the gene targeting frequency, as compared with electroporation (Bundock *et al.*, 1999). Introducing gene replacement cassettes by AMT into *A. awamori* also led to increased frequencies as compared with gene replacement cassette introduction by $CaCl_2$/polyethylene glycol-mediated transformation (Michielse *et al.*, 2005a). Furthermore, introduction of the gene replacement cassette by AMT allowed for the use of shorter homologous DNA flanking sequence in order to obtain a reasonable gene-targeting frequency (Michielse *et al.*, 2005a).

AMT has been used to disrupt genes involved in various processes (cell wall biosynthesis, melanin synthesis and pathogenicity) with high efficiency in many fungal species, including *A. awamori* (de Groot *et al.*, 1998; Gouka *et al.*, 1999; Michielse *et al.*, 2004a, 2005a; Almeida *et al.*, 2007), *A. fumigatus*

(Sugui *et al.*, 2005), *Ceratocystis resinifera* (Loppnau *et al.*, 2004), *Coccidioides posadasii* (Kellner *et al.*, 2005; Li *et al.*, 2007), *Fusarium graminearum* (Malz *et al.*, 2005), *Fusarium oxysporum* (Khang *et al.*, 2005, 2006), *Glarea lozoyensis* (Zhang *et al.*, 2003; Lu *et al.*, 2005), *Histoplasma capsulatum* (Sullivan *et al.*, 2002; Marion *et al.*, 2006), *Kluyveromyces lactis* (Bundock *et al.*, 1999), *Leptosphaeria maculans* (Gardiner and Howlett, 2004; Eckert *et al.*, 2005; Gardiner *et al.*, 2005; Blaise *et al.*, 2007), *Maganaporthe grisea* (Khang *et al.*, 2005, 2006), *Mycosphaerella graminicola* (Zwiers and De Waard, 2001), *Ophiostoma piliferum* (Hoffman and Breuil, 2004), *Trichoderma atroviride* (Cardoza *et al.*, 2006), *Verticillium dahliae* (Chen and Fukuhara, 1988; Klimes *et al.*, 2006) and *V. fungicola* (Amey *et al.*, 2002). Gene-targeting efficiencies ranged from 14 to 90%, which are higher than those normally achieved by traditional transformation methods.

Targeted gene disruption with AMT has been combined with double-marker enrichment in *A. awamori* and *Leptosphaeria maculans* (Gardiner and Howlett, 2004; Michielse *et al.*, 2005a). In this strategy, a second selectable marker is added to the gene disruption cassette to distinguish between homologous and non-homologous recombination. If the gene disruption cassette is integrated by HR, the second selection marker is lost. Both markers are retained if integration is mediated by NHR (see Fig. 13.3).

The addition of the dominant *A. nidulans amdS* selection marker to the gene replacement cassette led to twofold enrichment in putative gene replacement transformants in *A. awamori* (Michielse *et al.*, 2005a). The negative selection marker, thymidine kinase, was used as a second selection marker in *L. maculans*, and this combination of double-marker enrichment

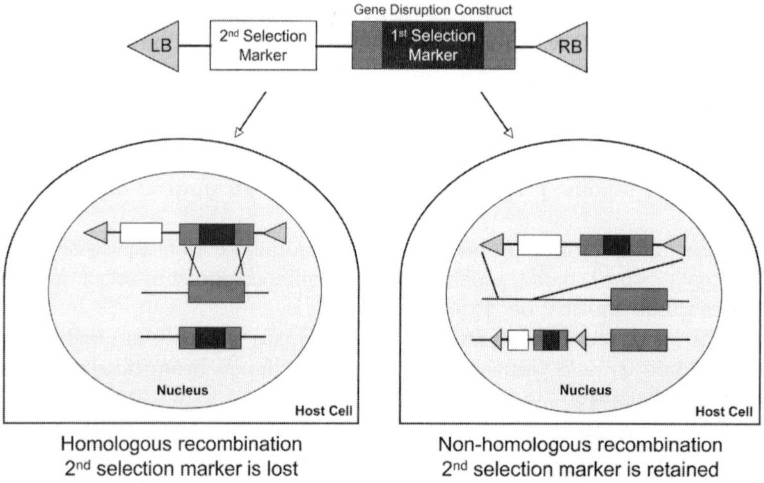

Fig. 13.3. Targeted gene disruption using two-marker selection. A second selection marker (2nd SM) is incorporated into the T-DNA gene-disruption cassette, illustrated at the left border, but it can be placed on either side of the disruption cassette. Following AMT, transformants are screened using both markers. If recombination occurred via the homologous recombination pathway, the second selection marker gene is lost. If the T-DNA insertion was via non-homologous recombination, the second selection marker gene is retained.

and AMT led to a 17% homologous integration rate (Gardiner and Howlett, 2004). Homologous integration was unsuccessful in *L. maculans* without the presence of the second selection marker. The thymidine kinase gene has also been successfully used as a negative selection marker in AMT-targeted gene disruption experiments with *M. grisea* and *F. oxysporum* (Khang *et al.*, 2005), however, this method was unsuccessful in targeted gene disruption experiments in *C. neoformans* (McClelland *et al.*, 2005). The reason for this outcome was unclear.

It is important to note that a number of false-positive gene disruptants were obtained in all these studies, even with the use of the second selection marker. Higher numbers of false-positive transformants were observed when the second selection marker was located at the left T-DNA border repeat. Analysis showed that the false positives were primarily a result of ectopic integrations of the T-DNA truncated at the left border, leading to the loss of the second selection marker (Khang *et al.*, 2005; Michielse *et al.*, 2005a). This is consistent with plants, where T-DNA truncation is more profound at the left T-DNA border than at the right (Tzfira *et al.*, 2004).

In species where gene disruption and replacement is unfeasible, RNA interference (RNAi) has proved a useful alternative. Recently, RNAi was combined with AMT in *Venturia inaequalis* to silence a green fluorescent protein (*GFP*) transgene and the endogenous trihydroxynaphthalene reductase (*THN*) gene, which is involved in melanin production. High-frequency gene silencing was achieved using hairpin constructs for the *GFP* or the *THN* genes, which were transferred by AMT (Fitzgerald *et al.*, 2004). AMT and RNAi have also been combined to study the role of the α-(1,4)-amylase gene in α-(1,3)-glucan production and virulence in *Histoplasma capsulatum* (Marion *et al.*, 2006). The high homologous recombination frequency obtained in fungi using AMT technology should provide a good basis for the development of large-scale targeted gene-disruption approaches, either alone or in combination with existing methods.

Conclusions

As sequence information for a wide variety of fungi becomes available, effective genetic transformation protocols will be essential for performing large-scale genomic studies. These efforts are important, not only scientifically but economically. For example, rice blast caused by *M. grisea* destroys between 11 and 30% of the crop annually. This percentage represents a loss of 157 million tons of rice (http://www.riceblast.org). Incidents of human infection with both primary and opportunistic fungal pathogens continue to rise, especially as the population of immunocompromised patients grows.

AMT has proved to be an important tool for the transformation of a diverse number of yeasts and fungi. The lack of a need for special equipment, the simplicity of transformation protocols and the variety of starting materials that can be transformed make AMT a preferable alternative to other more laborious and time-consuming transformation protocols. The success of a large-scale functional genomics study using AMT in *M. oryzae* indicates that this method may be particularly useful as a method for high-throughput functionality studies for many of the more than 30 species of fungi that have been sequenced. Additionally, AMT can be used to mediate targeted-gene disruption in several medically and agriculturally important species that are recalcitrant to other transformation techniques (e.g. *H. capsulatum*, *C. immitis*, *P. brasiliensis*, *F. graminearum*, *M. oryzae*). Both targeted and insertional mutagenesis studies in these species will contribute to our understanding of the pathogenicity of these organisms.

References

Abuodeh, R.O., Orbach, M.J., Mandel, M.A., Das, A. and Galgiani, J.N. (2000) Genetic transformation of *Coccidioides immitis* facilitated by *Agrobacterium tumefaciens*. *Journal of Infectious Diseases* 181(6), 2106–2110.

Almeida, A.J., Carmona, J.A., Cunha, C., Carvalho, A., Rappleye, C.A., Goldman, W.E., Hooykaas, P.J., Leao, C., Ludovico, P. and Rodrigues, F. (2007) Towards a molecular genetic system for the pathogenic fungus *Paracoccidioides brasiliensis*. *Fungal Genetics and Biology* 44(12), 1387–1398.

Amey, R.C., Athey-Pollard, A., Burns C., Mills, P.R., Bailey, A. and Foster, G.D. (2002) PEG-mediated and *Agrobacterium*-mediated transformation in the mycopathogen *Verticillium fungicola*. *Mycological Research* 106(1), 4–11.

Asch, D.K. and Kinsey, J.A. (1990) Relationship of vector insert size to homologous integration during transformation of *Neurospora crassa* with the cloned am (GDH) gene. *Molecular and General Genetics* 221(1), 37–43.

Barakat, A., Gallois, P., Raynal, M., Meatre-Ortega, D., Sallaud, C., Guiderdoni, E., Delseny, M. and Bernardi, G. (2000) The distribution of T-DNA in the genomes of transgenic *Arabdopsis* and rice. *FEBS Letters* 471, 161–164.

Beijersbergen, A., Dulk-Ras, A.D., Schilperoort, R.A. and Hooykaas, P.J. (1992) Conjugative transfer by the virulence system of *Agrobacterium tumefaciens*. *Science* 256(5061), 1324-–1327.

Bevan, M. (1984) Binary *Agrobacterium* vectors for plant transformation. *Nucleic Acids Research* 12(22), 8711–8721.

Blaise, F., Remy, E., Meyer, M., Zhou, L., Narcy, J.P., Roux, J., Balesden, M.H. and Rouxel, T. (2007) A critical assessment of *Agrobacterium tumefaciens*-mediated transformation as a tool for pathogenicity gene discovery in the phytopathogenic fungus *Leptosphaeria maculans*. *Fungal Genetics and Biology* 44(2), 123–138.

Bundock, P., den Dulk-Ras, A., Beijersbergen, A. and Hooykaas, P.J. (1995) Trans-kingdom T-DNA transfer from *Agrobacterium tumefaciens* to *Saccharomyces cerevisiae*. *The EMBO Journal* 14(13), 3206–3214.

Bundock, P., Mroczek, K., Winkler, A.A., Steensma, H.Y. and Hooykaas, P.J. (1999) T-DNA from *Agrobacterium tumefaciens* as an efficient tool for gene targeting in *Kluyveromyces lactis*. *Molecular and General Genetics* 261(1), 115–121.

Bundock, P., van Attikum, H., den Dulk-Ras, A. and Hooykaas, P.J. (2002) Insertional mutagenesis in yeasts using T-DNA from *Agrobacterium tumefaciens*. *Yeast* 19(6), 529–536.

Cardoza, R.E., Vizcaino, J.A., Hermosa, M.R., Monte, E. and Gutierrez, S. (2006) A comparison of the phenotypic and genetic stability of recombinant *Trichoderma* spp. generated by protoplast- and *Agrobacterium*-mediated transformation. *Journal of Microbiology* 44(4), 383–395.

Chen, X.J. and Fukuhara, H. (1988) A gene fusion system using the aminoglycoside 3′-phosphotransferase gene of the kanamycin-resistance transposon Tn903: use in the yeast *Kluyveromyces lactis* and *Saccharomyces cerevisiae*. *Gene* 69(2), 181–192.

Choi, J., Park, J., Jeon, J., Chi, M.H., Goh, J., Yoo, S.Y., Jung, K., Kim, H., Park, S.Y., Rho, H. S., Kim, S., Kim, B.R., Han, S.S., Kang, S. and Lee, Y.H. (2007) Genome-wide analysis of T-DNA integration into the chromosomes of *Magnaporthe oryzae*. *Molecular Microbiology* 66(2), 371–382.

Combier, J.P., Melayah, D., Raffier, C., Gay, G. and Marmeisse, R. (2003) *Agrobacterium tumefaciens*-mediated transformation as a tool for insertional mutagenesis in the symbiotic ectomycorrhizal fungus *Hebeloma cylindrosporum*. *FEMS Microbiology Letters* 220(1), 141–148.

de Groot, M.J., Bundock, P., Hooykaas, P.J. and Beijersbergen, A.G. (1998) *Agrobacterium tumefaciens*-mediated transformation of filamentous fungi. *Nature Biotechnology* 16(9), 839–842.

De Vos, G., De Beuckeleer, M., Van Montagu, M. and Schell, J. (1981) Restriction endonuclease mapping of the octopine tumor-inducing plasmid pTiAch5 of *Agrobacterium tumefaciens*. *Plasmid* 6(2), 249–253.

Dillen, W., Clercq, J., Kapila, J., Zambre, M., Montagu, M. and Angenon, G. (1997) The effect of temperature on *Agrobacterium tumefaciens*-mediated gene transfer to plants. *The Plant Journal* 12, 1459–1463.

Eckert, M., Maguire, K., Urban, M., Foster, S., Fitt, B., Lucas, J. and Hammond- Kosack, K. (2005) *Agrobacterium tumefaciens*-mediated transformation of *Leptosphaeria* spp. and *Oculimacula* spp. with the reef coral gene DsRed and the jellyfish gene gfp. *FEMS Microbiology Letters* 253(1), 67–74.

Fitzgerald, A., Van Kan, J.A and Plummer, K.M. (2004) Simultaneous silencing of multiple genes in the apple scab fungus, *Venturia inaequalis*, by expression of RNA with chimeric inverted repeats. *Fungal Genetics and Biology* 41(10), 963–971.

Fullner, K.J. and Nester, E.W. (1996) Temperature affects the T-DNA transfer machinery of *Agrobacterium tumefaciens*. *Journal of Bacteriology* 178(6), 1498–1504.

Fullner, K.J., Lara, J.C. and Nester, E.W. (1996) Pilus assembly by *Agrobacterium* T-DNA transfer genes. *Science* 273(5278), 1107–1109.

Gardiner, D.M. and Howlett, B.J. (2004) Negative selection using thymidine kinase increases the efficiency of recovery of transformants with targeted genes in the filamentous fungus *Leptosphaeria maculans*. *Current Genetics* 45(4), 249–255.

Gardiner, D.M., Jarvis, R.S. and Howlett, B.J. (2005) The ABC transporter gene in the sirodesmin biosynthetic gene cluster of *Leptosphaeria maculans* is not essential for sirodesmin production but facilitates self-protection. *Fungal Genetics and Biology* 42(3), 257–263.

Gaudin, V., Vrain, T. and Jouanin, L. (1994) Bacterial genes modifying hormonal balances in plants. *Plant Physiology and Biochemistry* 32, 11–29.

Gelvin, S.B. (1998) The introduction and expression of transgenes in plants. *Current Opinions in Biotechnology* 9(2), 227–232.

Gelvin, S.B. (2003) *Agrobacterium*-mediated plant transformation: the biology behind the 'gene-jockeying' tool. *Microbiology and Molecular Biology Reviews* 67(1), 16–37, table of contents.

Gerard, J.C., Canaday J., Szegedi, E., de la Salle, H. and Otten, L. (1992) Physical map of the vitopine Ti plasmid pTiS4. *Plasmid* 28(2), 146–156.

Gheysen, G., Villarroel, R. and Van Montagu, M. (1991) Illegitimate recombination in plants: a model for T-DNA integration. *Genes and Development* 5(2), 287–297.

Gouka, R.J., Gerk, C., Hooykaas, P.J., Bundock, P., Musters, W., Verrips, C.T. and de Groot, M.J. (1999) Transformation of *Aspergillus awamori* by *Agrobacterium tumefaciens*-mediated homologous recombination. *Nature Biotechnology* 17(6), 598–601.

Gray, M. and Honigberg, S.M. (2001) Effect of chromosomal locus, GC content and length of homology on PCR-mediated targeted gene replacement in *Saccharomyces*. *Nucleic Acids Research* 29(24), 5156–5162.

Grimaldi, B., de Raaf, M.A., Filetici, P., Ottonello, S. and Ballario, P. (2005) *Agrobacterium*-mediated gene transfer and enhanced green fluorescent protein visualization in the mycorrhizal ascomycete *Tuber borchii*: a first step towards truffle genetics. *Current Genetics* 48(1), 69–74.

Hajdukiewicz, P., Svab, Z. and Maliga, P. (1994) The small, versatile pPZP family of *Agrobacterium* binary vectors for plant transformation. *Plant Molecular Biology* 25(6), 989–994.

Hamilton, C.M. (1997) A binary-BAC system for plant transformation with high-molecular weight DNA. *Gene* 200(1–2), 107–116.

Hanif, M., Pardo, A.G., Gorfer, M. and Raudaskoski, M. (2002) T-DNA transfer and integration in the ectomycorrhizal fungus *Suillus bovinus* using hygromycin B as a selectable marker. *Current Genetics* 41(3), 183–188.

Hellens, R.P., Edwards, E.A., Leyland, N.R., Bean, S. and Mullineaux, P.M. (2000) pGreen: a versatile and flexible binary Ti vector for *Agrobacterium*-mediated plant transformation. *Plant Molecular Biology* 42(6), 819–832.

Hoekema, A., Hirsch, P.R., Hooykaas, P.J.J. and Schilperoort, R.A. (1983) A binary plant vector strategy based on separation of vir- and T-region of the *Agrobacterium tumefaciens* Ti-plasmid. *Nature* 303(5913), 179–180.

Hoffman, B. and Breuil, C. (2004) Disruption of the subtilase gene, albin1, in *Ophiostoma piliferum*. *Applied and Environmental Microbiology* 70(7), 3898–3903.

Holsters, M., Silva, B., Van Vliet, F., Genetello, C., De Block, M., Dhaese, P., Depicker, A., Inzé, D., Engler, G., Villarroel, R., Van Montagu, M. and Schell, J. (1980) The functional organization of the nopaline *A. tumefaciens* plasmid pTiC58. *Plasmid* 3(2), 212–230.

Hood, E.E., Helmer, G.L., Fraley, R.T. and Chilton, M.D. (1986) The hypervirulence of *Agrobacterium tumefaciens* A281 is encoded in a region of pTiBo542 outside of T-DNA. *Journal of Bacteriology* 168(3), 1291–1301.

Hood, E.E., Gelvin, S.B., Melchers, L.S. and Hoekema, A. (1993) New *Agrobacterium* helper plasmids for gene transfer to plants. *Transgenic Research* 2(4), 208–218.

Hua, S.B., Qiu, M., Chan, E., Zhu, L. and Luo, Y. (1997) Minimum length of sequence homology required for *in vivo* cloning by homologous recombination in yeast. *Plasmid* 38(2), 91–96.

Idnurm, A. and Heitman, J. (2005) Light controls growth and development via a conserved pathway in the fungal kingdom. *PLoS Biology* 3(4), e95.

Idnurm, A., Reedy, J.L., Nussbaum, J.C. and Heitman, J. (2004) *Cryptococcus neoformans* virulence gene discovery through insertional mutagenesis. *Eukaryotic Cell* 3(2), 420–429.

Jeon, J., Park, S.Y., Chi, M.H., Choi, J., Park, J., Rho, H.S. *et al.* (2007) Genome-wide functional analysis of pathogenicity genes in the rice blast fungus. *Nature Genetics* 39(4), 561–565.

Jin, S.G., Roitsch, T., Christie, P.J. and Nester, E.W. (1990) The regulatory VirG protein specifically binds to a cis-acting regulatory sequence involved in transcriptional activation of *Agrobacterium tumefaciens* virulence genes. *Journal of Bacteriology* 172(2), 531–537.

Kellner, E.M., Orsborn, K.I., Siegel, E.M., Mandel, M.A., Orbach, M.J. and Galgiani, J.N. (2005) *Coccidioides posadasii* contains a single 1,3-beta-glucan synthase gene that appears to be essential for growth. *Eukaryotic Cell* 4(1), 111–120.

Khang, C.H., Park, S.Y., Lee, Y.H. and Kang, S. (2005) A dual selection based, targeted gene replacement tool for *Magnaporthe grisea* and *Fusarium oxysporum*. *Fungal Genetics and Biology* 42(6), 483–492.

Khang, C.H., Park, S.Y., Rho, H.S., Lee, Y.H. and Kang, S. (2006). Filamentous fungi

(*Magnaporthe grisea* and *Fusarium oxysporum*). *Methods in Molecular Biology* 344, 403–420.

Klimes, A., Neumann, M.J., Grant, S.J. and Dobinson, K.F. (2006) Characterization of the glyoxalase I gene from the vascular wilt fungus *Verticillium dahlia*. *Canadian Journal of Microbiology* 52(9), 816–822.

Komari, T., Halperin, W. and Nester, E.W. (1986) Physical and functional map of supervirulent *Agrobacterium tumefaciens* tumor-inducing plasmid pTiBo542. *Journal of Bacteriology* 166(1), 88–94.

Komari, T., Hiei, Y., Saito, Y., Murai, N. and Kumashiro, T. (1996) Vectors carrying two separate T-DNAs for co-transformation of higher plants mediated by *Agrobacterium tumefaciens* and segregation of transformants free from selection markers. *The Plant Journal* 10(1), 165–174.

Koncz, C. and Schell, J. (1986) The promoter of TL-DNA gene 5 controls the tissue-specific expression of chimaeric genes carried by a novel type of *Agrobacterium* binary vector. *Molecular and General Genetics* 204(3), 383–396.

Lazo, G.R., Stein, P.A. and Ludwig, R.A. (1991) A DNA transformation-competent *Arabidopsis* genomic library in *Agrobacterium*. *Biotechnology* 9(10), 963–967.

Leclerque, A., Wan, H., Abschutz, A., Chen, S., Mitina, G.V., Zimmermann, G. and Schairer, H.U. (2004) *Agrobacterium*-mediated insertional mutagenesis (AIM) of the entomopathogenic fungus *Beauveria bassiana*. *Current Genetics* 45(2), 111–119.

Lemmers, M., De Beuckeleer, M., Holsters, M., Zambryski, P., Depicker, A., Hernalsteens, J.P., Van Montagu, M. and Schell, J. (1980) Internal organization, boundaries and integration of Ti-plasmid DNA in nopaline grown gall tumours. *Journal of Molecular Biology* 144(3), 353–376.

Li, L., Schmelz, M., Kellner, E.M., Galgiani, J.N. and Orbach, M.J. (2007). Nuclear labeling of *Coccidioides posadasii* with green fluorescent protein. *Annals of the New York Academy of Sciences* 1111, 198–207.

Loppnau, P., Tanguay, P. and Breuil, C. (2004) Isolation and disruption of the melanin pathway polyketide synthase gene of the softwood deep stain fungus *Ceratocystis resinifera*. *Fungal Genetics and Biology* 41(1), 33–41.

Lu, P., Zhang, A., Dennis, L.M., Dahl-Roshak, A.M., Xia, Y.Q., Arison, B., An, Z. and Tkacz, J.S. (2005) A gene (*pks2*) encoding a putative 6-methylsalicylic acid synthase from *Glarea lozoyensis*. *Molecular Genetics and Genomics* 273(2), 207–216.

Malonek, S. and Meinhardt, F. (2001) *Agrobacterium tumefaciens*-mediated genetic transformation of the phytopathogenic ascomycete *Calonectria morganii*. *Current Genetics* 40(2), 152–155.

Malz, S., Grell, M.N., Thrane, C., Maier, F.J., Rosager, P., Felk, A., Albertsen, K.S., Salomon, S., Bohn, L., Schafer, W. and Giese, H. (2005) Identification of a gene cluster responsible for the biosynthesis of aurofusarin in the *Fusarium graminearum* species complex. *Fungal Genetics and Biology* 42(5), 420–433.

Marchand, G., Fortier, E., Neveu, B., Bolduc, S., Belzile, F. and Belanger, R.R. (2007) Alternative methods for genetic transformation of *Pseudozyma antarctica*, a basidiomycetous yeast-like fungus. *Journal of Microbiological Methods* 70(3), 519–527.

Marion, C.L., Rappleye, C.A., Engle, J.T. and Goldman, W.E. (2006) An alpha-(1,4)-amylase is essential for alpha-(1,3)-glucan production and virulence in *Histoplasma capsulatum*. *Molecular Microbiology* 62(4), 970–983.

McClelland, C.M., Chang, Y.C. and Kwon-Chung, K.J. (2005) High frequency transformation of *Cryptococcus neoformans* and *Cryptococcus gattii* by *Agrobacterium tumefaciens*. *Fungal Genetics and Biology* 42(11), 904–913.

Meyer, V., Mueller D., Strowig, T. and Stahl, U. (2003) Comparison of different transformation methods for *Aspergillus giganteus*. *Current Genetics* 43(5), 371–377.

Michielse, C.B., Ram, A.F., Hooykaas, P.J. and Hondel, C.A. (2004a) Role of bacterial virulence proteins in *Agrobacterium*-mediated transformation of *Aspergillus awamori. Fungal Genetics and Biology* 41(5), 571–578.

Michielse, C.B., Ram A.F., Hooykaas, P.J. and van den Hondel, C.A. (2004b) *Agrobacterium*-mediated transformation of *Aspergillus awamori* in the absence of full-length VirD2, VirC2, or VirE2 leads to insertion of aberrant T-DNA structures. *Journal of Bacteriology* 186(7), 2038–2045.

Michielse, C.B., Salim, K., Ragas, P., Ram, A.F., Kudla, B., Jarry, B., Punt, P.J. and van den Hondel, C.A. (2004c) Development of a system for integrative and stable transformation of the zygomycete *Rhizopus oryzae* by *Agrobacterium*-mediated DNA transfer. *Molecular Genetics and Genomics* 271(4), 499–510.

Michielse, C.B., Arentshorst, M., Ram, A.F. and van den Hondel, C.A. (2005a) *Agrobacterium*-mediated transformation leads to improved gene replacement efficiency in *Aspergillus awamori. Fungal Genetics and Biology* 42(1), 9–19.

Michielse, C.B., Hooykaas, P.J., van den Hondel, C.A. and Ram, A.F. (2005b) *Agrobacterium*-mediated transformation as a tool for functional genomics in fungi. *Current Genetics* 48(1), 1–17.

Ooms, G., Hooykaas, P.J., Van Veen, R.J., Van Beelen, P., Regensburg-Tuink, T.J. and Schilperoort, R.A. (1982) Octopine Ti-plasmid deletion mutants of *Agrobacterium tumefaciens* with emphasis on the right side of the T-region. *Plasmid* 7(1), 15–29.

Park, S.M. and Kim, D.H. (2004) Transformation of a filamentous fungus *Cryphonectria parasitica* using *Agrobacterium tumefaciens. Biotechnology and Bioprocess Engineering* 9, 217–222.

Piers, K.L., Heath J.D., Liang, X., Stephens, K.M. and Nester, E.W. (1996) *Agrobacterium tumefaciens*-mediated transformation of yeast. *Proceedings of the National Academy of Sciences of the United States of America* 93(4), 1613–1618.

Roberts, R.L., Metz, M., Monks, D.E., Mullaney, M.L., Hall, T. and Nester, E.W. (2003) Purine synthesis and increased *Agrobacterium tumefaciens* transformation of yeast and plants. *Proceedings of the National Academy of Sciences* 100, 6634–6639.

Rogers, C.W., Challen, M.P., Green, J.R. and Whipps, J.M. (2004) Use of REMI and *Agrobacterium*-mediated transformation to identify pathogenicity mutants of the biocontrol fungus *Coniothyrium minitans. FEMS Microbiology Letters* 241(2), 207–214.

Schaefer, D.G. (2001) Gene targeting in *Physcomitrella patens. Current Opinion in Plant Biology* 4(2), 143–150.

Sciaky, D., Montoya, A.L. and Chilton, M.D. (1978) Fingerprints of *Agrobacterium* Ti plasmids. *Plasmid* 1(2), 238–253.

Smith, E.F. and Townsend, C.O. (1907) A plant tumor of bacterial origin. *Science* 25(643), 671–673.

Stachel, S.E., Nester, E.W. and Zambryski, P.C. (1986) A plant cell factor induces *Agrobacterium tumefaciens vir* gene expression. *Proceedings of the National Academy of Sciences of the United States of America* 83(2), 379–383.

Sugui, J.A., Chang, Y.C. and Kwon-Chung, K.J. (2005) *Agrobacterium tumefaciens*-mediated transformation of *Aspergillus fumigatus*: an efficient tool for insertional mutagenesis and targeted gene disruption. *Applied and Environmental Microbiology* 71(4), 1798–1802.

Sullivan, T.D., Rooney, P.J. and Klein, B.S. (2002) *Agrobacterium tumefaciens* integrates transfer DNA into single chromosomal sites of dimorphic fungi and yields homokaryotic progeny from multinucleate yeast. *Eukaryotic Cell* 1(6), 895–905.

Takken, F.L., Van Wijk, R., Michielse, C.B., Houterman, P.M., Ram, A.F. and Cornelissen, B.J. (2004) A one-step method to convert vectors into binary vectors suited for *Agrobacterium*-mediated transformation. *Current Genetics* 45(4), 242–248.

Tsuji, G., Fujii S., Fujihara N., Hirose, C., Tsuge, S., Shiraishi, T. and Kubo, Y. (2003) *Agrobacterium tumefaciens*-mediated transformation for random insertional mutagenesis in *Colletotrichum lagenarium*. *Journal of General Plant Pathology* 69(4), 230–239.

Tzfira, T., Li, J., Lacroix, B. and Citovsky, V. (2004) *Agrobacterium* T-DNA integration: molecules and models. *Trends in Genetics* 20(8), 375–383.

Wang, K., Herrera-Estrella, L., Van Montagu, M. and Zambryski, P. (1984). Right 25 bp terminus sequence of the nopaline T-DNA is essential for and determines direction of DNA transfer from *Agrobacterium* to the plant genome. *Cell* 38(2), 455–462.

Wendland, J. (2003) PCR-based methods facilitate targeted gene manipulations and cloning procedures. *Current Genetics* 44(3), 115–123.

Zhang, A., Lu, P., Dahl-Roshak, A.M., Paress, P.S., Kennedy, S., Tkacz, J.S. and An, Z. (2003) Efficient disruption of a polyketide synthase gene (*pks1*) required for melanin synthesis through *Agrobacterium*-mediated transformation of *Glarea lozoyensis*. *Molecular Genetics and Genomics* 268(5), 645–655.

Zwiers, L.H. and De Waard, M.A. (2001) Efficient *Agrobacterium tumefaciens*-mediated gene disruption in the phytopathogen *Mycosphaerella graminicola*. *Current Genetics* 39(5–6), 388–393.

14 Myconanotechnology: a New and Emerging Science

Mahendra Rai[1], Alka Yadav[1], Paul Bridge[2] and Aniket Gade[1]

[1]*Department of Biotechnology, SGB Amravati University, Amravati, Maharashtra, India;* [2]*British Antarctic Survey, NERC, Cambridge, UK*

Introduction

The term nanotechnology was defined by the Tokyo Science University Professor Norio Taniguchi in 1974 (Taniguchi, 1974) as the creation and exploitation of materials in the size range of 1–100 nm. Nanotechnology is multidisciplinary across science and includes aspects of research and technology development in many areas of physics, chemistry and biology (McNeil, 2005; Uskokovic, 2008). Nanoparticles are metal particles smaller than 100 nm that can be synthesized in numerous shapes (e.g. spherical, triangular, rods) from various metal ions. Nanoparticles have many different applications and are used in a number of fields, including medicine, pharmacology, environmental monitoring and electronics (Liu, 2006).

Myconanotechnology (myco = fungi, nanotechnology = the creation and exploitation of materials in the size range of 1–100 nm) is a new term that is proposed here for the first time. It is defined as the fabrication of nanoparticles by fungi and their subsequent application, particularly in medicine. Myconanotechnology is the interface between mycology and nanotechnology, and is an exciting new applied interdisciplinary science that may have considerable potential, partly due to the wide range and diversity of fungi.

The current interest in metallic nanoparticles is due to their variable chemical, physical and optical properties, and recent developments in the field of nanostructures have ensured that nanotechnology will play a crucial role in the future development of many scientific applications. The use of nanotechnology for the synthesis of nanomaterials is a rapidly developing and emerging field. Numerous protocols have been developed to synthesize nanoparticles of different shapes and sizes by physical and chemical methods (e.g. Jana *et al.*, 2000; Rao and Cheetham, 2001; Kohler *et al.*, 2007).

Physical and chemical synthesis can have limitations, and in recent years there has been an increased interest in the use of biological agents as tools for synthesizing nanoparticles (Kowshik *et al.*, 2003). One of the earliest reports of the synthesis of metallic nanoparticles was on the formation of CdS (cadmium sulfide) microcrystals in yeasts (Dameron *et al.*, 1989) and, to date, the synthesis of nanoparticles has been investigated in various biological agents, including actinomycetes and other bacteria, fungi and plants (Nair and Pradeep, 2002; Sastry *et al.*, 2003; Senapati *et al.*, 2005; Chandran *et al.*, 2006). Various plant species have been used to produce metal nanoparticles, including *Medicago sativa* (Gardea-Torresdey *et al.*, 2003), *Geranium graveolans* (Shankar *et al.*, 2003), *Azadirachta indica* (Shankar *et al.*, 2004), *Triticum* sp. (Armendariz *et al.*, 2004), *Cinnamomum camphora* (Huang *et al.*, 2006), *Capsicum annum* (Li *et al.*, 2007) and *Pinus* sp. (Song and Kim, 2008).

Among the microbes metallic nanoparticles have been reported from bacteria and yeasts. The bacterial species used have included *Bacillus subtilis* (Southam and Beveridge, 1994), *Pseudomonas stutzeri* (Klaus *et al.*, 1999), *Rhodococus* sp., (Ahmad *et al.*, 2003a), *Thermonospora* sp. (Ahmad *et al.*, 2003b), *Shewanella algae* (Konishi *et al.*, 2004) and *Lactobacillus* strains (Prasad *et al.*, 2007), while yeast species have included *Candida glabrata* (Dameron *et al.*, 1989) and *Schizosaccaharomyces pombe* (Kowshik *et al.*, 2002).

Fungi have a number of advantages for nanoparticle synthesis compared with other organisms, particularly as they are relatively easy to isolate and culture, and secrete large amounts of extracellular enzymes (Mandal *et al.*, 2006). Many of the proteins secreted by fungi are capable of hydrolyzing metal ions quickly and through non-hazardous processes. As a result, fungi have the potential to produce nanoparticles faster than some chemical synthesis methods. In addition, nanoparticles of high monodispersity and dimensions can be obtained from fungi (Bhattacharya and Gupta, 2005; Senapati *et al.*, 2005; Mohanpuria *et al.*, 2007; Mukherjee *et al.*, 2008), and the downstream processing and the handling of fungal biomass can be much simpler than the processes needed for chemical synthesis (Mandal *et al.*, 2006).

This chapter will focus on the emergence of myconanotechnology as a new science for the synthesis of nanoparticles by using fungi, and will consider some of the potential future applications and challenges in this area.

Current State of the Art

There has been considerable interest in the potential use of fungi and fungal systems as nanofactories for the synthesis of metal nanoparticles (Mandal *et al.*, 2006; Ingle *et al.*, 2008). These particles are typically produced as a result of reactions between fungal biomass and aqueous metallic salt solutions, and current research has considered both extracellular and intracellular methods of synthesis (see Table 14.1).

Table 14.1. Examples of metal nanoparticles produced from fungi.

Fungal species	Method of synthesis	Nanoparticles	Reference
Fusarium oxysporum	Extracellular	CdS	Ahmad *et al.*, 2002
Usnea longissima	Extracellular	Usnic acid	Shahi and Patra, 2003
Fusarium oxysporum	Extracellular	ZrO_2	Bansal *et al.*, 2004
Fusarium oxysporum	Extracellular	SiO_2, TiO_2	Bansal *et al.*, 2005
Fusarium oxysporum Verticillium sp.	Extracellular	Magnetite	Bharde *et al.*, 2006
Aspergillus fumigatus	Extracellular	Ag	Bhainsa and D'Souza, 2006
Fusarium oxysporum	Extracellular	CdSe quantum dots	Kumar *et al.*, 2007a
Fusarium semitectum	Extracellular	Ag	Basavaraja *et al.*, 2007
Fusarium oxysporum	Extracellular	Ag	Kumar *et al.*, 2007b
Aspergillus niger	Extracellular	Ag	Gade *et al.*, 2008
Fusarium accuminatum	Extracellular	Ag	Ingle *et al.*, 2008
Trichoderma asperellum	Extracellular	Ag	Mukherjee *et al.*, 2008
Penicillium sp.	Extracellular	Ag	Sadowski *et al.*, 2008
Verticillium sp.	Intracellular	Au	Mukherjee *et al.*, 2001
Phoma sp.	Intracellular	Ag	Chen *et al.*, 2003
Fusarium oxysporum f. sp. lycopersici	Intra- and extracellular	Pt	Riddin *et al.*, 2006

CdS, cadmium sulfide; ZrO_2, zirconia; SiO_2, silica; TiO_2, titania; Ag, silver; CdSe, cadmium selenide; Au, gold; Pt, platinum.

Extracellular synthesis of nanoparticles

Shahi and Patra (2003) synthesized bioactive nanoparticles of usnic acid from the lichen-forming fungus *Usnea longissima* in a defined medium. The nanoparticles they obtained were uniform in shape, with a diameter of 50–200 nm, and their bioactivity was investigated both *in vitro* and *in vivo* in a nanoemulsion. The *in vitro* results showed that the nanoemulsion had a minimum inhibitory concentration (MIC) of 0.1 μl/ml against the dermatophytes *Epidermophyton floccosum, Microsporum audouinii, M. canis, M. gypseum, M. nanum, Trichophyton mentagrophytes, T. rubrum, T. tonsurans* and *T. violaceum*. The *in vivo* results showed that the nanoemulsion was safe and effective in controlling dermatophyte infections in humans. Shankar *et al.* (2003) have reported the production of spherical nanoparticles of gold by using *Geranium* leaves and an unidentified endophytic fungus *Colletotrichum*. In this study the reducing agents were found to be fungal polypeptides and enzymes.

Bansal *et al.* (2004) used an isolate of *Fusarium oxysporum* to produce zirconia (ZrO_2) nanoparticles by reacting fungal cell filtrate with an aqueous solution of zirconium hexafluoride at room temperature for 24 h. The regular quasi-spherical particles were between 3 and 11 nm in diameter and fourier transform infrared (FTIR) analysis showed that they also contained proteins. Bansal *et al.* (2005) also used *Fusarium oxysporum* to produce nanoparticles of silica (SiO_2) and titania (TiO_2). As previously, a culture filtrate was treated with aqueous solutions, in this case with di-potassium silicon and titanium hexafluorides, for 24 h. The silica nanoparticles produced were quasispherical and of 5–15 nm in diameter, and the titania nanoparticles were of a similar size (6–13 nm) and spherical. FTIR analysis showed SiF_6, Si-O-Si and amide bonds in the silica nanoparticles, and Ti-O-Ti and amide bonds in the titania particles. The energy dispersion x-ray analysis (EDAX) spectra indicated that, in addition to silicon and titanium, the nanoparticles also contained oxygen, potassium and fluorine.

Bharde *et al.* (2006) used isolates of *Fusarium oxysporum* and an unidentified *Verticillium* species to produce magnetite nanoparticles. The culture filtrates were reacted for 24 h with 2:1 molar mixtures of $K_3[Fe(Cn)_6]$ and $K_4[Fe(Cn)_6]$. Irregular quasi-spherical particles of 20–50 nm were detected by transmission electron microscopy (TEM), and the crystalline nature of these was confirmed by selected area electron diffraction (SAED).

Much of the research on fungal extracellular synthesis of nanoparticles has been undertaken with silver (see Table 14.1). Silver nanoparticles have been produced by extracellular synthesis from 1 mM silver nitrate and the culture filtrate of *Aspergillus fumigatus* (Bhainsa and D'Souza, 2006). The formation of the silver nanoparticles was monitored during the 24 h reaction by UV-visible spectrophotometry, and particle formation was confirmed by TEM. The nanoparticles varied in size (5–25 nm) and shape, with most being spherical although some were triangular. The crystalline nature of the particles was confirmed by X-ray diffraction studies.

Extracellular biosynthesis of silver nanoparticles has also been achieved with *Fusarium semitectum* by a similar methodology (Basavraja *et al.*, 2007). The spherical nanoparticles were between 10 and 60 nm in diameter and were stable for 6–8 weeks. FTIR was used to identify any capping agents, and amide and carbonyl groups were detected. Capping agents prevent agglomeration of nanoparticles and contribute to their stability in the medium. Kumar *et al.* (2007b) have reported a nitrate reductase-mediated method for the synthesis of silver nanoparticles. Nitrate reductase was purified from a fungal cell filtrate by column chromatography, and this was then used in an anaerobic reaction with silver nitrate, 4-hydroxyquinoline and NADPH. This resulted in both individual silver particles of 10–25 nm and aggregates. The particles were well dispersed, crystalline and consisted predominantly of Ag, C, O, N and Na.

Silver nanoparticles have also been reported from *Trichoderma asperellum* and *Penicillium* sp. Mukherjee *et al.* (2008) obtained nanoparticles in the range of 13–18 nm from the agriculturally important fungus *Trichoderma asperellum.* These were found to be stable for up to six months, and FTIR spectra suggested that nitrogen, amide and amine groups were responsible for the bioreduction of silver nanoparticles. Sadowski *et al.* (2008) obtained silver nanoparticles from an unnamed isolate of *Penicillium*. In this case the particles ranged from hundreds of nanometres to micrometres in size.

The potential antibacterial activity of fungus-derived silver nanoparticles has been investigated in some studies. Gade *et al.* (2008) considered the potential bioactivity of silver nanoparticles of around 20 nm produced by *Aspergillus niger.* They tested the antibacterial effect of the spherical silver particles against a range of bacteria and reported maximum activity against *S. aureus* and minimum activity against *Escherichia coli*. Ingle *et al.* (2008) also investigated the antibacterial properties of 5–40 nm spherical silver nanoparticles from a *Fusarium acuminatum* biosynthesis. These nanoparticles were found to have effective antibacterial activity against a number of common human pathogenic bacteria (e.g. *Staphlyococcus epidermis, S. aureus, Salmonella typhi* and *E. coli*).

Intracellular synthesis of silver nanoparticles

In addition to the various studies examining the extracellular synthesis of nanoparticles by culture filtrates, there have also been a limited number of studies that have investigated the intracellular formation of metallic nanoparticles within fungal cells. Mukherjee *et al.* (2001) reported the bioreduction of auric chloride by an isolate of *Verticillium* sp. They found that when fungal cells grown in a defined medium were transferred into aqueous auric chloride solution, the pale yellow colour of the fungal cells changed to vivid purple over 24 h. The UV-visible spectra of the fungal cells showed an absorption peak at 540 nm, which is characteristic for gold nanoparticles, and TEM micrographs of the cells showed gold nanoparticles with an average size of 20–28 nm on the cell wall as well, as in the cytoplasm. Higher magnification TEM showed large numbers of largely spherical gold nanoparticles, although some were triangular or hexagonal.

The intracellular synthesis of platinum nanoparticles has been described in *Fusarium oxysporum* f. sp. *lycopersici* by Riddin *et al.* (2006). The formation of the nanoparticles was evident by a change in colour of the fungal biomass from yellow to dark brown. The particles produced were of various shapes, ranging from spherical, hexagonal and pentagonal to square, rectangular, etc. Riddin *et al.* (2006) also screened the same fungus for the extracellular biosynthesis of platinum nanoparticles. The fungal filtrate was treated with hydrogen hexacholoroplatinate and the nanoparticles produced were between 10 and 100 nm and of varied shapes, including spherical, hexagonal, pentagonal, square and rectangular.

Other methods

As well as being produced through fungal growth and activity, nanoparticles have also been produced in dried fungal material. In such an experiment Chen *et al.* (2003) obtained freeze-dried mycelium from an isolate of *Phoma* and treated this with silver nitrate solution for 50 h. Adsorption assays indicated that the mycelium had absorbed some 13 mg of silver, and TEM micrographs showed the presence of a large number of silver nanoparticles of around 70 nm within the fungal mycelium.

Mechanisms involved in the mycosynthesis of nanoparticles

A number of possible mechanisms have been proposed for the formation of metallic nanoparticles, but to date no single generalized mechanism has been identified.

Cell walls and cell wall sugars are likely to play an important role in the reduction of metal ions (Mukherjee *et al.*, 2001). Nanoparticles are formed on cell wall surfaces, and the first step in bioreduction is the trapping of the metal ions on this surface. This probably occurs due to the electrostatic interaction between the metal ions and positively charged groups in enzymes present at the cell wall. This may be followed by enzymatic reduction of the metal ions, leading to their aggregation and the formation of nanoparticles.

Proteins have been implicated in nanoparticle formation in a number of different studies. Bansal *et al.* (2004) found that during the formation of zirconia nanoparticles, the fungus secreted proteins capable of extracellularly hydrolyzing compounds with zirconium ions, and this was confirmed in subsequent studies with silica and titania (Bansal *et al.*, 2005). They found that their fungus was also capable of hydrolysing metal halide precursors under acidic conditions. Their studies indicated that the proteins involved in the reduction of metal nanoparticles were cationic proteins with molecular weights of around 21–24 kDa. Bharde *et al.* (2006) also suggested that a cationic protein of around 55 kDa found in extracellular extracts of *Verticillium* sp. might be responsible for the hydrolysis of $[Fe(CN)_6]^3$ and $[Fe(CN)_6]^4$.

Bhainsa and D'Souza (2006) concluded that the absorption peaks they obtained at 220 and 280 nm corresponded to amide bonds and to tryptophan and tyrosine residues present in the proteins, respectively, and that this indicated proteinaceous compounds in fungal biomass which had a major role in the bioreduction of metal nanoparticles.

Gade *et al.* (2008) found proteins bound to the nanoparticle surface, and the presence of S atoms around the silver nanoparticles was taken to suggest an association between nanoparticles and fungal proteins. Ingle *et al.* (2008) showed that NADH transfer and NADH-dependent enzymes were involved in the reduction of metal ions, and they further demonstrated the presence of nitrate reductase in the fungal cell filtrate.

Applications

There are many possible uses for biologically produced nanoparticles in diverse applications in different fields (Salata, 2004; Kvishek and Prucek, 2005; Uskokovic, 2008). Zirconium is a particularly hard metal, and zirconium nanoparticles may be useful for providing a hard, abrasive surface for micro-cutting tools, or for high-temperature coatings in engine components (Bansal *et al.*, 2004). Silver nanoparticles have uses in optical–electronic applications, as coatings for solar energy absorption materials, in electric batteries and as catalysts for chemical processing (Bhainsa and D'Souza, 2006; Kumar *et al.*, 2007b).

The broad-spectrum antibacterial properties of silver nanoparticles may also lead to a future role in pharmaceutical development (Ingle *et al.*, 2008). Similarly, the bioactive nanoparticles synthesized by the lichen-fungi *Usnea longissima* may have potential as a future natural nanomedicine against microbial infections (Shahi and Patra, 2003).

Titanium dioxide nanoparticles have practical applications in many forms of transport such as automobiles, aircraft, ships, etc. Titania nanoparticles may also be particularly useful for the production of pigments, paints and cosmetics (Bansal *et al.*, 2005; Prasad *et al.*, 2007). Some of the nanoparticles that can be produced by fungi may be of particular relevance to new and emerging technologies.

The use of silver coatings in solar absorption systems has already been mentioned. Other applications in such areas include the use of gold nanoparticles as precursors to coatings for electronic applications (Mukherjee *et al.*, 2001) and the use of platinum nanoparticles in the production of fuel cells (Riddin *et al.*, 2006). Magnetite nanoparticles have ferromagnetic properties and can be employed in remediation processes (Bharde *et al.*, 2005).

Future Challenges

The production of metallic nanoparticles by fungi appears to be a relatively simple biotechnological process, predominantly involving only the reaction of fungal culture filtrates with solutions of metal salts. There are, however, a number of issues where further research is required. The work undertaken to date indicates that there may be a number of different kinds of reducing agents involved in the mechanism of synthesis of metal nanoparticles. These may also have effects on the final shapes and sizes of the nanoparticles. There is a need for studies to elucidate the particular mechanisms involved in nanoparticle formation, and for comparative studies to determine whether the same or different pathways are used by different fungi for different metals. Once the basic mechanisms have been determined there will be a need to consider methods for optimizing the concentration, sizes and shapes of the nanoparticles. In order for fungal synthesis of nanoparticles to become commercially feasible, it will also be necessary to develop low-cost recovery techniques to separate the particles from the fungal mycelium that can be used routinely in industrial processes.

Conclusion

Nanotechnology and the capability of synthesizing nanoscale materials with specific physical, chemical and optoelectronic properties have increasingly important roles in the development of future technology. Microorganisms have been shown to have potential for the biosynthesis of metallic nanoparticles by reduction of the metal ions from aqueous salt solutions. There is an increasing interest in the use of fungi for these processes, and fungi may have the potential to provide relatively quick and ecologically 'clean' biofactories for metallic nanoparticles. Simple protocols have been used for the production of metallic nanoparticles of gold, silver, platinum, silica, titania and zirconia from fungi, and these processes have been observed in both intra- and extracellular preparations.

Many industrial areas, including food, enzyme and pharmaceutical production and processing, currently use fungal biomass, and so downstream handling and processing procedures for fungal material are already well established. There are a number of challenges that need to be undertaken before the potential of myconanotechnology can be fully evaluated, and these include determining the detailed chemical processes involved in the formation of the particles, and optimizing these to provide sufficient quantities of a consistent product that can be easily extracted or separated. In relation to many other uses of fungi, the production of metallic nanoparticles is a relatively new development. While much of the basic information on the chemistry and practicalities of the process is still unknown, there would appear to be considerable potential for this technology to develop in the future.

References

Ahmad, A., Mukherjee, P., Mandal, D., Senapati, S., Khan,. M.I., Kumar, R. and Sastry, M. (2002) Enzyme-mediated extracellular biosynthesis of CdS nanoparticles by the fungus *Fusarium oxysporum*. *Journal of the American Chemical Society* 124, 12108–12109.

Ahmad, A., Senapati, S., Khan, M.I., Kumar, R., Ramani, R., Shrinivas, V. and Sastry, M. (2003a) Intracellular synthesis of gold nanoparticles by a novel alkalotolerant actinomycete, *Rhodococcus* species. *Nanotechnology* 14, 824–828.

Ahmad, A., Senapati, S., Khan, M.I., Kumar, R. and Sastry, M. (2003b) Extracellular biosynthesis of monodisperse gold nanoparticles by a novel extremophilic actinomycete, *Thermonospora* sp. *Langmuir* 19, 3550–3553.

Armendariz, V., Gardea- Torresdey, J.L., Herrera, I., Moller, A.D., Jose-Yacaman, M., Peralta-Videa, J.R. and Troiani, H. (2004) HRTEM characterization of gold nanoparticles produced by wheat biomass. *Revista Mexicana de Fisica* 50, 7–11.

Bansal, V., Rautray, D., Ahamd, A. and Sastry, M. (2004) Biosynthesis of zirconia nanoparticles using the fungus *Fusarium oxysporum*. *Journal of Materials Chemistry* 14, 3303–3305.

Bansal, V., Rautray, D., Bharde, A., Ahire, K., Sanyal, A., Ahmad, A. and Sastry, M. (2005) Fungus-mediated biosynthesis of silica and titania particles. *Journal of Materials Chemistry* 15, 2583–2589.

Basavaraja, S., Balaji, S.D., Lagashetty, A., Rajasab, A.H. and Venkataraman, A. (2007) Extracellular biosynthesis of silver nanoparticles using the fungus *Fusarium semitectum*. *Materials Research Bulletin* 43(5), 1164–1170.

Bhainsa, K.C. and D'Souza, S.F. (2006) Extracellular biosynthesis of silver nanoparticles using the fungus *Aspergillus fumigatus*. *Colloids and Surfaces B: Biointerfaces* 47, 160–164.

Bharde, A., Rautray, D., Bansal, V., Ahmad, A., Sarkar, I., Yusuf, S.M., Sanyal, M. and Sastry, M. (2006) Extracellular biosynthesis of magnetite using fungi. *Small* 2(1), 135–141.

Bhattacharya, D. and Gupta, R.K. (2005) Nanotechnology and potential of microorganisms. *Critical Reviews in Biotechnology* 24(4), 199–204.

Chandran, S.P., Ahmad, A., Chaudhary, M., Pasricha, R. and Sastry, M. (2006) Synthesis of gold nanotriangles and silver nanoparticles using *Aloe vera* plant extract. *Biotechnology Program* 22(2), 577–583.

Chen, J.C., Lin, Z.H. and Ma, X.X. (2003) Evidence of the production of silver nanoparticles via pretreatment of *Phoma* sp. 3.2883 with silver nitrate. *Letters in Applied Microbiology* 37, 105–108.

Dameron, C.T., Reese, R.N., Mehra, R.K., Korton, A.R., Caroll, P.J., Steigerwald, M.L., Brus, L.E. and Winge, D.R. (1989) Biosynthesis of cadmium sulfide quantum semiconductor crystallites. *Nature* 338, 596–597.

Gade, A.K., Bonde, P., Ingle, A.P., Marcato, P.D., Duran, N. and Rai, M.K. (2008) Exploitation of *Aspergillus niger* for synthesis of silver nanoparticles. *Journal of Biobased Materials and Bioenergy* 2, 243–247.

Gardea-Torresedey, J.L., Armendariz, V., Herreira, I., Parsons, J.G., Peralta-Videa, J.R., Teimann, K.J. and Torresday, K.J. (2003) Binding of silver (I) ions by alfalfa biomass (*Medicago sativa*): batch $_{p}$H, time, temperature, and ionic strength studies. *Journal of Hazardous Substance Research* 4, 1–15.

Huang, J., Chen, C., He, N., Hong, J., Lu, Y., Qingbiao, L., Shao, W., Sun, D., Wang, X.H., Wang, Y. and Yiang, X. (2006) Biosynthesis of silver and gold nanoparticles by novel sundried *Cinnamomum camphora* leaf. *Nanotechnology* 18, 105–106.

Ingle, A., Gade, A. Pierrat, S., Sonnichsen, C and Rai, M.K. (2008) Mycosynthesis of silver nanoparticles using the fungus *Fusarium acuminatum* and its activity against some human pathogenic bacteria. *Current Nanoscience,* 4, 141–144.

Jana, N.R., Pal, T., Sau, T.K. and Wang, Z.L. (2000) Seed-mediated growth method to prepare cubic copper nanoparticles. *Current Science* 79(9), 1367–1370.

Klaus, T., Granqvist, C.G., Joerger, R. and Olsson, E. (1999) Silver-based crystalline nanoparticles, microbially fabricated. *Proceedings of the National Academy of Sciences* 96(24), 13611–13614.

Kohler, J.M., Hubner, U., Romanus, H. and Wagner, J. (2007) Formation of star-like and core-shell AuAg nanoparticles during two and three step preparation in batch and microfluidic systems. *Journal of Nanomaterials* 1155, 98134–98141.

Konishi, Y., Nomura, T., Tsukiyama, T. and Saioth, N. (2004) Microbial preparation of gold nanoparticles by anaerobic bacterium. *Transactions of the Materials Research Society of Japan* 29, 2341–2343.

Kowshik, M., Deshmukh, N., Kulkarni, S.K., Paknikar, K.M., Vogel, W. and Urban, J. (2002) Microbial synthesis of semiconductor CdS nanoparticles, their characterization, and their use in fabrication of an ideal diode. *Biotechnology and Bioengineering* 78(5), 583–588.

Kowshik, M., Ashataputre, S., Kharrazi, S., Kulkarni, S.K., Paknikari, K.M., Vogel, W. and Urban, J. (2003) Extracellular synthesis of silver nanoparticles by a silver-tolerant yeast strain MKY3. *Nanotechnology* 14, 95–100.

Kumar, S.A., Ayoobul, A.A., Absar, A. and Khan, M.I. (2007a) Extracellular biosynthesis of CdSe quantum dots by the fungus, *Fusarium oxysporum. Journal of Biomedical Nanotechnology* 3, 190–194.

Kumar, S.A., Abyaneh, M.K., Gosavi, S.W., Kulkarni, S.K., Pasricha, R., Ahmad, A. and Khan, M.I. (2007b) Nitrate reductase-mediated synthesis of silver nanoparticles from $AgNO_3$. *Biotechnology Letters* 29, 439–445.

Kvishek, L. and Prucek, R. (2005) The preparation and application of silver nanoparticles. *Journal of Materials Science* 22, 2461–2473.

Li, S., Qui, L., Shen, Y., Xie, A., Yu, X., Zhang, L. and Zhang, Q. (2007) Green synthesis of silver nanoparticles using *Capsicum annum* L. extract. *Green Chemistry* 9, 852–858.

Liu, W.T. (2006) Nanoparticles and their biological and environmental applications. *Journal of Bioscience and Bioengineering* 102(1), 1–7.

Mandal, D., Bolander, M.E., Mukhopadhyay, D., Sarkar, G. and Mukherjee, P. (2006) The use of microorganisms for the formation of metal nanoparticles and their application. *Applied Microbiology and Biotechnology* 69, 485–492.

McNeil, S.E. (2005) Nanotechnology for the biologist. *Journal of Leukocyte Biology* 78, 585–593.

Mohanpuria, P., Rana, N.K. and Yadav, S.K. (2007) Biosynthesis of nanoparticles: technological concepts and future applications. *Journal of Nanoparticle Research* 7, 9275–9280.

Mukherjee, P., Ahmad, A., Mandal, D., Senapati, S., Sainkar, S.R., Khan, M.I., Ramani, R., Parischa, R., Ajayakumar, P.V., Alam, M., Sastry, M. and Kumar, R. (2001) Bioreduction of $AuCl_4^-$ ions by the fungus, *Verticillium* sp. and surface trapping of the gold nanoparticles formed. *Angewandte Chemie International Edition* 40(19), 3585–3588.

Mukherjee, P., Roy, M., Mandal, B.P., Dey, G.K., Mukherjee, P.K., Ghatak, J., Tyagi, A.K. and Kale, S.P. (2008) Green synthesis of highly stabilized nanocrystalline silver particles by a non-pathogenic and agriculturally important fungus *T. asperellum*. *Nanotechnology* 19, 103–110.

Nair, B. and Pradeep, T. (2002) Coalescence of nanoclusters and formation of submicron crytallites assisted by *Lactobacillus* strains. *Crystal Growth and Design* 2, 293–298.

Prasad, K., Jha, A.K. and Kulkarni, A.R. (2007) *Lactobacillus*-assisted synthesis of titanium nanoparticles. *Nanoscale Research Letters* 2, 248–250.

Rao, C.N. and Cheetham, A.K. (2001) Science and technology of nanomaterials: current status and future prospects. *Journal of Materials Chemistry* 11, 2887–2894.

Riddin, T.L., Gericke, M. and Whiteley, C.G. (2006). Analysis of the inter- and extracellular formation of platinum nanoparticles by *Fusarium oxysporum* f. sp. *lycopersici* using response surface methodology. *Nanotechnology* 17, 3482–3489.

Salata, O.V. (2004). Applications of nanoparticles in biology and medicine. *Journal of Nanobiotechnology* 2, 1–6.

Sadowski, Z., Maliszewska, I.H., Grochowalska, B., Polowczyk, I. and Kozlecki, T. (2008). Synthesis of silver nanoparticles using microorganisms. *Materials Science – Poland* 26(2), 419–425.

Sastry, M., Ahmad, A., Khan, M.I. and Kumar, R. (2003) Biosynthesis of metal nanoparticles using fungi and actinomycete. *Current Science* 85(2), 162–170.

Senapati, S., Ahmed, A., Khan, M.I., Kumar, R. and Sastry, M. (2005) Extracellular biosynthesis of bimetallic Au–Ag alloy nanoparticles. *Small* 1, 517–520.

Shahi, S.K. and Patra, M. (2003) Microbially synthesized bioactive nanoparticles and their formulation active against human pathogenic fungi. *Reviews on Advanced Material Science* 5, 501–509.

Shankar, S.S., Ahmad, A., Pasricha, R. and Sastry, M. (2003) Bioreduction of chloroaurate ions by *Geranium* leaves and its endophytic fungus yields gold nanoparticles of different shapes. *Journal of Material Chemistry* 13, 1822–1826.

Shankar, S.S., Ahmad, A., Rai, A. and Sastry, M. (2004) Rapid synthesis of Au, Ag and bimetallic Au core-Ag shell nanoparticles by using neem (*Azardirachta indica*) leaf broth. *Journal of Colloid and Interface Science* 275 (5), 496–502.

Song, Y.S. and Kim, B.S. (2008) Rapid biological synthesis of silver nanoparticles using plant leaf extracts. *Bioprocess and Biosystem Engineering* 449 (8), 224–230.

Southam, G. and Beveridge, T.J. (1994) The *in vitro* formation of placer gold by bacteria. *Geochimica et Cosmochimicha Acta* 58, 4527–4530.

Taniguchi N. (1974) On the basic concept of nano-technology. In: *Proceedings of the International Conference on Production Engineering*, Tokyo, Part II. Japan Society of Precision Engineering, Tokyo.

Uskokovic, V. (2008) Nanomaterials and nanotechnologies: approaching the crest of this big wave. *Current Nanoscience* 4, 119–129.

15 Current Advances in Fungal Chitinases

Duochuan Li and Anna Li

Department of Plant Pathology, Shandong Agricultural University, Taian, China

Introduction

Chitin, an unbranched homopolymer of 1,4-β-linked *N*-acetyl-D-glucosamine (GlcNAc), is widely distributed in nature. It is believed to be the second most abundant renewable polymer on earth, next to cellulose. Chitinases are chitin-degrading enzymes that hydrolyze the β-(1,4) linkages of chitin. Chitinases occur in a wide range of organisms including viruses, bacteria, fungi, insects, plants and animals. The roles of chitinases in these organisms are diverse and, in fungi, chitinases are thought to have autolytic, nutritional and morphogenetic roles (Adams, 2004). Chitinases in mycoparasitic fungi are most commonly thought to be involved in mycoparasitism (Haran *et al.*, 1996). Chitinases in bacteria have been shown to play a role in the utilization of chitin as a carbon and energy source and in the recycling of chitin in nature (Park *et al.*, 1997).

In insects, chitinases are associated with post-embryonic development and degradation of old cuticle (Merzendorfer and Zimoch, 2003). Plant chitinases are involved in defence and development (Graham and Sticklen, 1994), viral chitinases have roles in pathogenesis (Patil *et al.*, 2000) and mammalian chitinases are suggested as playing a role in defence against chitinous mammalian pathogens (Van Eijk *et al.*, 2005). There is considerable potential for the application of chitinases in commercial agricultural, biological and environmental processes.

There has been a large amount of research on fungal chitinases in recent years, and this chapter will focus on the nomenclature, structure, expression and function, and application of fungal chitinases.

Classification and Assays of Chitinase

Chitinases can be divided into two major categories: endochitinases and exochitinases (Graham and Sticklen, 1994). Endochitinases (EC3.2.1.14) cleave chitin randomly at internal sites, generating soluble, low-molecular mass multimers of GlcNAc, such as chitotetraose, chitotriose and diacetylchitobiose. Exochitinases can be divided into two subcategories: chitobiosidases and β-(1,4)-*N*-acetyl-glucosaminidases. Chitobiosidases (EC3.2.1.29) or chitin-1,4-β-chitobiosidases catalyze the progressive release of diacetylchitobiose starting at the non-reducing end of chitin chains. They produce diacetylchitobioses, and no monosaccharides or oligosaccharides are formed. β-(1,4)-*N*-acetyl-glucosaminidases (GlcNAcase, EC3.2.1.30) or chitobiases split diacetylchitobiose and higher chitin polymers, including chitotriose and chitotetraose, into GlcNAc monomers in an exo-type fashion. The enzyme can also cleave the chromogenic substrates *p*-nitrophenyl-GlcNAc and pNP-*N*-acetyl-β-D-galactosaminide to release pNP, and 4-MU-GlcNAc to release 4-MU, providing simple assay procedures. As the enzyme has broad substrate specificity, it is also called β-(1,4)-*N*-acetylhexosaminidase (HexNAcase, EC3.2.1.52) (Cannon *et al.*, 1994).

Techniques and methods for detecting chitinases have been developed, including chromogenic assays involving *p*-nitrophenyl-labelled substrates, such as *p*NP-GlcNAc, *p*NP-$(GlcNAc)_2$ and *p*NP-$(GlcNAc)_3$. Glucosaminidase, chitobiosidase and endochitinase activities can be determined by measuring the release of *p*-nitrophenyl from *p*NP-GlcNAc, *p*NP-$(GlcNAc)_2$ and *p*NP-$(GlcNAc)_3$, respectively (Robert and Selitrennikoff, 1988; Harman *et al.*, 1993). Endochitinase activity has also been measured by the reduction of turbidity of a suspension of colloidal chitin (Tronsmo and Harman, 1993). In addition, chitinase activity is often determined by measuring the amount of reducing sugars liberated from colloidal chitin by enzyme activity (Reissig *et al.*, 1955).

To detect chitinases after SDS-PAGE, proteins are prepared in Laemmli buffer. The proteins are separated by SDS-PAGE. Enzymes are reactivated in the gels by removing SDS by the casein–EDTA procedure (McGrew and Green, 1990). Enzyme activity is detected in gels by using fluorescent substrates (Tronsmo and Harman, 1993), where fluoresence under UV light is due to the enzymatic hydrolysis of 4-methylumbelliferone from GlcNAc mono- and oligosaccharides, such as 4-MU-GlcNAc, 4-MU-$(GlcNAc)_2$ and 4-MU-$(GlcNAc)_3$. These substrates allow for detection of three different chitinase types by acting as dimeric, trimeric and tetrameric substrates, respectively. The dimer is the preferred substrate for GlcNAcases. Chitobiosidases release fluorescent product from only the trimeric substrate, and endochitinases are identified by digestion of the tetrameric substrate (Haran *et al.*, 1995).

Purification and Characterization of Fungal Chitinases

Although there have been many reports on the production of chitinases by fungi, data on the purification and characterization of these enzymes are limited to relatively few species. In mycoparasitic fungi most work has been undertaken on the chitinolytic system of *Trichoderma harzianum*. Several chitinases of *T. harzianum* have been purified, including endochitinases, chitobiosidases and *N*-acetyl-glucosaminidases (De La Cruz *et al.*, 1992; Ulhoa and Peberdy, 1991, 1992; Harman *et al.*, 1993; Lorito *et al.*, 1994; Haran *et al.*, 1995; Deane *et al.*, 1998; Lisboa *et al.*, 2004). The purification of chitinases from other species of mycoparasitic fungi has also been reported, including from *Aphanocladium album* (Kunz *et al.*, 1992b), *Gliocladium virens* (Di Pietro *et al.*, 1993), *Fusarium chlamydosporum* (Mathivanan *et al.*, 1998), *Trichothecium roseum* (Li *et al.*, 2004), *Stachybotrys elegans* (Taylor *et al.*, 2002) and *Talaromyces flavus* (Li *et al.*, 2005). In entomopathogenic fungi, *N*-acetyl-D-glucosaminidase, endochitinase and exochitinase have been purified and characterized from *Beauveria bassiana* (Bidochka *et al.*, 1993; Peng *et al.*, 1996; Fang *et al.*, 2005) and *Metarhizium anisopliae* (Leger *et al.*, 1996; Pinto *et al.*, 1997; Kang *et al.*, 1999).

In comparison, in the nematophagous fungi only two chitinases from *Pochonia chlamydosporia* var. *chlamydosporia* and *Pochonia suchlasporia* var. *suchlasporia* (as *Verticillium chlamydosporium* and *V. suchlasporium*) have been purified and characterized to date (Tikhonov *et al.*, 2002). Recently, chitinases from thermophilic fungi have received attention due to their thermostability, and a thermostable chitinase has been purified from *Thermomyces lanuginosus* (Guo *et al.*, 2005).

Fungal chitinases can be purified to homogeneity by various methods, including fractional ammonium sulfate precipitation, ion exchange chromatography, hydrophobic interaction chromatography, gel filtration chromatography, chitin-affinity chromatography and isoelectric focusing (IEF). Fungal chitinases are often only isolated at low purity levels (De La Cruz *et al.*, 1992; Kunz *et al.*, 1992b; Ulhoa and Peberdy, 1992; Di Pietro *et al.*, 1993; Harman *et al.*, 1993; Li *et al.*, 2004, 2005), and this may be due to both a synergistic action of the different isoforms for chitin degradation present in the crude supernatant and the loss of chitinase activity during purification progress (De La Cruz *et al.*, 1992).

Most chitinolytic fungi have been found to produce more than one kind of chitinase. *T. harzianum* can produce seven individual chitinases: two *N*-acetyl-glucosaminidases (102 and 73 kDa) (Haran *et al.*, 1995), four endochitinases (52, 42, 33 and 31 kDa) (De La Cruz *et al.*, 1992; Harman *et al.*, 1993; Haran *et al.*, 1995) and one chitobiosidase (40 kDa) (Harman *et al.*, 1993). *Talaromyces flavus* produces at least two kinds of chitinases (Li *et al.*, 2005), *Stachybotrys elegans* produces two exochitinases and one endochitinase (Taylor *et al.*, 2002) and *Metarhizium anisopliae* produces at least six different chitinases (Leger *et al.*, 1991, 1996; Pinto *et al.*, 1997; Kang *et al.*, 1999). These multiple chitinases have been shown to have mutually synergistic and complementary effects and, for example, the chitinases CHIT33, CHIT37 and CHIT42 of *T. harzianum* have a synergistic action in cell wall degradation (De La Cruz *et al.*, 1992). A synergistic increase in antifungal activity has been reported from the combined activities of the endochitinase and chitobiosidase from *T. harzianum* (Lorito *et al.*, 1993).

Most purified fungal chitinases have been characterized to ascertain molecular weight, pI, optimal pH, optimal temperature, thermostability, inhibitors and antifungal activity. Fungal endochitinases and chitobiosidases are usually single polypeptides, whereas *N*-acetyl-glucosaminidases – such as the *N*-acetyl-glucosaminidases of *T. harzianum*, *S. elegans* and *B. bassiana* – are dimeric (Ulhoa and Peberdy, 1991; Bidochka *et al.*, 1993; Taylor *et al.*, 2002). Molecular masses of fungal chitinases have a wide range, from 27 to 190 kDa and with a pI of about 3–8. Optimal pH and temperature are similar for the majority of purified fungal chitinases, and they are active in the range pH 4.0–7.0.

The optimum temperature for most fungal chitinase activity is at 20–40°C (De La Cruz *et al.*, 1992; Kunz *et al.*, 1992a ; Ulhoa and Peberdy, 1992; Di Pietro *et al.*, 1993; Harman *et al.*, 1993; Li *et al.*, 2004, 2005). However, chitinases from two thermophilic fungi – *Thermomyces lanuginosus* and *Talaromyces emersonii* – have high optimum temperatures and thermostability. The *Tal. emersonii* chitinase has maximum activity at 65°C and a half-life of 20 min at 70°C (McCormack *et al.*, 1991). The *Therm. lanuginosus* chitinase exhibits optimum catalytic activity at 55°C, with a half-life time of 25 min at 65°C (Guo *et al.*, 2005).

Fungal chitinases have a specific inhibitor – either allosamidin or demethylallosamidin – that is produced by *Streptomyces* sp. This can specifically inhibit family 18 chitinases from yeast, fungi and insects, and has been used for both functional research and the identification of chitinases. There have been many reports showing that fungal chitinases can degrade fungal cell walls and inhibit fungal growth *in vitro*, and purified endochitinase and chitobiosidase from *T. harzianum* can also inhibit spore germination and germ tube elongation in different fungal species (Lorito *et al.*, 1993).

It should be emphasized that chitinases of *T. harzianum* are substantially more active and effective against a wide range of fungi than chitinolytic enzymes from plants and other microorganisms (Lorito *et al.*, 1993). Hence chitinases of *T. harzianum* are particularly attractive as a potential source of chitinolytic enzymes for biocontrol. In addition, chitinases from thermophilic fungi are also promising for biocontrol and enzymatic conversion of chitin because of their thermostability (Guo *et al.*, 2005).

Cloning and Expression of Fungal Chitinase Genes

Many chitinase-encoding genes have been isolated and analysed from a wide range of filamentous fungi and yeasts (see Table 15.1).

In recent years complete genomes have been sequenced for several model fungi, including *Saccharomyces cerevisiae*, *Candida albicans*, *Coccidioides immitis*, *Neurospora crassa*, *Gibberella zeae*, *Magnaporthe grisea*, *Aspergillus nidulans*, *Aspergillus fumigatus* and *Trichoderma reesei*. Genomic analysis has allowed further chitinase genes to be identified. Such methods have shown the presence of at least 5, 9, 17, 11, 21, 16 and 18 ORFs encoding putative

Table 15.1. Chitinase genes isolated from fungi and yeasts.

Origin	Chitinase gene(s)	Reference(s)
Saccharomyces cerevisiae	*cts1*, *cts2*	Kuranda and Robbins, 1987; Kuranda and Robbins, 1991
Candida albicans	*cht1*, *cht2*, *cht3*, *cht4*, *chs1A*, *hex1*	Cannon *et al.*, 1994; McCreath *et al.*, 1995, 1996; Sudoh *et al.*, 1995
Kluyveromyces lactis	*KLCTS1*, killer toxin gene	Stark and Boyd, 1996; Colussi *et al.*, 2005
Coccidioides immitis	*cts1* to *cts7*, fixation-chitinase gene	Pishko *et al.*, 1995; Zimmermann *et al.*, 1996
Paracoccidioides brasiliensis	*Pbcts1*	Bonfim *et al.*, 2006
Trichoderma harzianum	*exc1y*, *exc2y*, *th-en42*, *nag1*, *chit33*, *chit42*, *chit36*, *ech42*, *exc1*, *exc2*	Carsolio *et al.*, 1994; Garcia *et al.*, 1994; Hayes *et al.*, 1994; Draborg *et al.*, 1995; Limon *et al.*, 1995; Peterbauer *et al.*, 1996; Viterbo *et al.*, 2001, 2002; Brunner *et al.*, 2003a,b; Ramot *et al.*, 2004
Trichoderma reesei	*chit18-1 to chit18-18*, *chi46*	Ike *et al.*, 2005; Seidl *et al.*, 2005
Trichoderma atroviride	*ech42*, *ech30*	Kulling *et al.*, 2000; Hoell *et al.*, 2005
Trichoderma virens	*cht42*, *ech1*, *ech2*	Baek *et al.*, 1999; Kim *et al.*, 2002a
Trichoderma aureoviride	*ech42*	Song *et al.*, 2005
Rhizopus oligosporus	*chi1*, *chi2*, *chi3*	Yanai *et al.*, 1992; Takaya *et al.*, 1998b
Aspergillus nidulans	*chiA*, *chiB*, *chiC*, *nagA*	Takaya *et al.*, 1998a; Kim *et al.*, 2002b
Aspergillus fumigatus	*chiA1*, *chiB1*	Jaques *et al.*, 2003
Metarhizium anisopilae	*chit1*, *chit2*, *chit3*, *chit11*, *chit*, *nag1*	Valadares-Inglis *et al.*, 1997; Bogo *et al.*, 1998; Kang *et al.*, 1998; Screen *et al.*, 2001; Da Silva *et al.*, 2005
Beauveria bassiana	*Bbchit1*	Fang *et al.*, 2005
Aphanocladium album	*chi1*	Blaiseau and Lafay, 1992
Nomuraea rileyi	*Chi*	Wattanalai *et al.*, 2004
Stachybotrys elegans	*sechi44*	Morissette *et al.*, 2003
Puccinia triticina	*Chi*	Zhang *et al.*, 2003
Penicillium chrysogenum	*nagA*	Diez *et al.*, 2005
Thermomyces lanuginosus	a chitinase gene	Guo and Li, 2006

chitinases in *C. albicans, N. crassa, G. zeae, M. grisea, A. fumigatus, A. nidulans* and *T. reesei*, respectively (Jones *et al.*, 2004; Seidl *et al.*, 2005).

Chitinase genes of fungi can be functionally expressed in various hosts, such as *E. coli*, yeasts, filamentous fungi and plants. Expression of ech42-encoding endochitinase of *T. harzianum* in *E. coli* results in high chitinase activity (Carsolio *et al.*, 1994). Transformation of *Schizosaccharomyces pombe* with *cts1* from *Saccharomyces cerevisiae* results in the appearance of about a five- to 13-fold increase in chitinase activity (Kuranda and Robbins, 1987). Introduction of an *Aphanocladium album* chitinase gene into *Fusarium oxysporum* results in high chitinase levels in the host (Blaiseau *et al.*, 1992).

Expression of a gene-encoding endochitinase from *T. harzianum* in *T. reesei* results in a transformant that produces 20 times more chitinase than *T.*

reesei (Margolles-Clark *et al.*, 1996). When *ech42* from *T. harzianum* is transferred to another *T. harzianum* strain, the transgenic strains carrying multiple copies of *ech42* can show increases up to 42-fold in their level of chitinase activity as compared with the original wild-type strain (Carsolio *et al.*, 1999). The *ThEn42* gene of *T. harzianum* endochitinase is highly expressed in different plant tissues, and has been transferred to tobacco, potato and apple (Lorito *et al.*, 1998; Bolar *et al.*, 2000). Overexpression of a gene (Bbchit1)-encoding endochitinase of the entomopathogenic fungus *Beauveria bassiana* also results in high chitinase activity (Fang *et al.*, 2005).

Classification and Structure of Fungal Chitinases

Families of fungal chitinases

Chitinases are classified as glycosyl hydrolases. The glycosyl hydrolase classification system is based on the amino acid sequence similarity of the catalytic domains, and chitinases have been placed in families 18, 19 and 20 (Henrissat and Bairoch, 1993). Family 18 chitinases are found in bacteria, fungi, yeasts, viruses, plants and animals and hence the family is diverse in evolutionary terms. Family 19 members are almost exclusively found in plants, while a single family 19 chitinase has been identified in *Streptomyces griseus* (Ohno *et al.*, 1996). Family 18 and family 19 chitinases do not share amino acid sequence similarity and have completely different 3-D structures and molecular mechanisms, suggesting that they have arisen from different ancestors (Hamel *et al.*, 1997; Suzuki *et al.*, 1999). Family 20 consists of the β-*N*-acetylhexosaminidases or β-*N*-acetylglucosaminidases from bacteria, fungi and mammals.

Subfamilies of fungal chitinases

Chitinases are separated into different subfamilies, classes or groups based on their structure and their amino acid sequence. Bacterial chitinases are clearly separated into three major subfamilies, A, B and C (Watanabe *et al.*, 1993). Plant chitinases are divided in five different classes. Class I and Class II chitinases are similar in their catalytic domains, but Class I chitinases have a chitin-binding domain. This domain is separated from the catalytic domain by a hinge region. Class II chitinases lack both the chitin-binding domain and the hinge region. Class III chitinases have higher homology to fungal chitinases than to other plant chitinase classes. Class IV chitinases are similar to Class I chitinases but they are smaller in size due to certain deletions. Class V chitinases show some homology with bacterial exochitinases (Neuhaus *et al.*, 1996).

Fungal chitinases are not as well classified as bacterial and plant chitinases. Within family 18 two distinct classes of fungal chitinase may be identified based on their similarities to family 18 chitinases from plants or bacteria (Pishko *et al.*, 1995; Takaya *et al.*, 1998b). Currently, fungal chitinases are divided into fungal/bacterial chitinases (corresponding to Class V), which are similar to those found in bacteria, and fungal/plant chitinases (corresponding to Class III), which are similar to those from plants (Takaya *et al.*, 1998b; Jaques *et al.*, 2003; Adams, 2004). Recently, phylogenetic analysis has been used to suggest that family 18 fungal chitinases can be divided into three groups: group A and group B (corresponding to classes V and III of plant chitinases) and a novel group C, consisting of high-molecular weight chitinases with a domain structure similar to that of the *K. lactis* killer toxins (Seidl *et al.*, 2005).

Structure of fungal chitinases

Most fungal chitinases belong to family 18 of the glycohydrolase superfamily (Henrissat, 1999). One characteristic of the family 18 chitinases is their multi-domain structure. Typically, the basic structure of family 18 fungal chitinases is composed of five domains or regions: an N-terminal signal peptide region, a catalytic domain, a serine/theronine-rich region, a chitin-binding domain and a C-terminal extension region. However, most fungal chitinases lack the last three domains, which do not seem to be necessary for chitinase activity as naturally occurring chitinases that lack these regions are still enzymatically active.

The derived amino acid sequence of a *Saccharomyces cerevisiae* endo-chitinase, CTS1, shows four distinct domains: a signal sequence, a catalytic domain, a serine/theronine-rich region and a chitin-binding domain (Kuranda and Robbins, 1991). Two chitinases (chitinase I and chitinase II) from *Rhizopus oligosporus* have the five distinct domains: a signal sequence, a catalytic domain, a serine/theronine-rich region, a chitin-binding domain and an additional C-terminal domain (Yanai *et al.*, 1992). The *T. harzianum* 33 kDa chitinase includes a putative signal peptide region and a catalytic domain, but lacks the serine/theronine-rich region, the chitin-binding region and the C-terminal region (Limon *et al.*, 1995).

Fungal chitinase structural domains have different biochemical functions. In the fungal chitinases sequenced so far, all except *Rhizopus oligosporus* chitinase III (Takaya *et al.*, 1998a) and *Trichoderma harzianum* Chi18-2, Chi18-3 and Chi18-7 (Seidl *et al.*, 2005) are predicted to have a signal peptide preceding the N-terminal region of the mature protein. The signal peptide presumably mediates secretion of the enzyme and it is cleaved off by signal peptidases after the protein has been transported across the membrane. The chitinases that lack the secretory signal sequence have been shown to be intracellular chitinases, and they may have a role during morphogenesis (Takaya *et al.*, 1998b; Seidl *et al.*, 2005). The fungal chitinase catalytic domain, responsible for the hydrolysis of the substrate, comprises the N-terminal half of the enzyme. Sequence alignments have shown two highly conserved regions within the catalytic domain. In family 18 chitinases the two consensus regions or motifs, SxGG and DxxDxDxE, correspond to a substrate-binding site and a catalytic domain, respectively (Henrissat, 1991).

Apart from the N-terminal signal peptide region and catalytic domain some fungal chitinases, such as the *S. cerevisiae* endochitinase CTS1, the *K. lactis* endochitinase KlCts1p, the *T. reesei* Chi18-13 and Chi18-16 and the *Rhizopus ologosporus* chitinase I and II, have been reported to have the serine/theronine-rich region and chitin-binding domain (ChBD) (Kuranda and Robbins, 1991; Yanai *et al.*, 1992; Colussi *et al.*, 2005). The serine/theronine-rich region of fungal chitinases is usually glycosylated with sugar chains post-translationally to yield the mature protein. The glycosylation sites may be necessary for the secretion of the protein and maintenance of its stability. Fungal chitinases are thought to be anchored to the cell wall or their substrate through the chitin-binding domain (Kuranda and Robbins, 1991).

A conserved six-cysteine motif, which is thought to mediate the tertiary structure or protein–protein interaction through the formation of disulfide bridges, is found in the chitin-binding domain of CTS1 and KlCts1p (Kuranda and Robbins, 1991; Colussi *et al.*, 2005). The chitin-binding domains of the *T. reesei* Chi18-13 and Chi18-16 have high similarity to a fungal cellulose-binding domain (CBD), consisting of four strictly conserved aromatic amino acid residues that are implicated in the interaction with cellulose (Kraulis *et al.*, 1989; Seidl *et al.*, 2005). It should be noted that the chitin-binding domain differs from the substrate-binding site in the catalytic domain (Henrissat, 1999). The chitin-binding domain of the *K. lactis* KlCts1p functions independently of the catalytic domain (Colussi *et al.*, 2005). It has been demonstrated that the presence of the chitin-binding domain in *S. cerevisiae* CTS1 does not increase the rate of chitin hydrolysis (Kuranda and Robbins, 1991), but addition of ChBD from *Nicotiana tabacum* ChiA chitinase to *Trichoderma harzianum* chitinase Chit42, which lacks a ChBD, increases the chitin-binding capacity of the native Chit42 (Limon *et al.*, 2001).

It is not clear how the C-terminal region of fungal chitinase functions. Interestingly, it has been found that there are putative glycosyl-phosphatidylinositol (GPI)-anchor and cleavage sites at the C-terminus of a *Candida albicans* chitinase, CHT2, and two *Aspergillus fumigatus* chitinases, ChiA1 and ChiA. The GPI region may anchor CHT2 and ChiA1 to the cell wall or cell membrane, and have some role during growth and morphogenesis (Takaya *et al.*, 1998b; Iranzo *et al.*, 2002; Jaques *et al.*, 2003).

The crystal structures of only two fungal chitinases are available so far, and these have provided three-dimensional models of a *Coccidioides immitis* chitinase, CiX1, and an *Aspergillus fumigatus* chitinase, ChiB1 (Hollis *et al.*, 2000; Rao *et al.*, 2005). The active site of CiX1 is composed of conserved amino acid residues (W47, 131, 315, 378, Y239, 293, R52, 295). The amino acid residue E171, as in other class 18 glycohydrolases, is thought to be the catalytic amino acid, and site-directed mutagenesis of E171 eliminates any detectable enzyme activity. The structure model showed that CiX1 is an eight-stranded β/α-barrel, with eight parallel β-strands forming the core, surrounded by eight α-helices. It has been suggested that the chitinase barrel structure forms a groove on the enzyme's surface. This groove is considered as the active centre, which binds sugar units of chitin – possibly $(GlcNAc)_6$ moieties, which are subsequently cleaved (Drouillard *et al.*, 1997).

It should be noted that studies on glycohydrolase family 20 fungal β-*N*-acetyl-glucosaminidases, have recently been undertaken. Genes encoding β-*N*-acetyl-glucosaminidases have been cloned from several fungi, including *T. harzianum*, *C. albicans* and *A. nidulans* (Peterbauer *et al.*, 1996; Kim *et al.*, 2002a,b; Van Eijk *et al.*, 2005). Cloning and sequencing of fungal β-*N*-acetyl-glucosaminidases has shown the enzymes to be composed of a signal peptide, a propeptide, a zincin-like domain, a catalytic domain and a C-terminal segment (Plihal *et al.*, 2004).

Regulation of Fungal Chitinase Genes

Regulation in culture

Chitinases produced from fungi are inducible in nature, and activity is detected when fungi are grown in a chitin-containing medium (De La Cruz *et al.*, 1992; Di Pietro *et al.*, 1993; Harman *et al.*, 1993; Haran *et al.*, 1995; Taylor *et al.*, 2002; Li *et al.*, 2004, 2005; Guo *et al.*, 2005). Chitinase gene expression in fungi has been reported as being controlled by a repressor/inducer system, in which chitin or other products of degradation (such as *N*-acetyl-β-D-glucosamine and glucasamine) act as inducers, and glucose or easily metabolizable carbon sources act as repressors (Blaiseau *et al.*, 1992; Sahai and Manocha, 1993).

Recently, the regulation of the *T. harzianum* chitinase genes *ech42* encoding a 42 kDa endochitinase, *chit36* encoding a 36 kDa endochitinase, *chit33* encoding a 33 kDa endochitinase and *nag1* encoding *N*-acetyl-β-glucosaminidase has been studied in detail. The four genes were induced by fungal cell walls, colloidal chitin or by carbon starvation (Limon *et al.*, 1995; Margolles-Clark *et al.*, 1996; Dana *et al.*, 2001; Viterbo *et al.*, 2002), whereas the presence of high concentrations of glucose or glycerol inhibited their expression (Carsolio *et al.*, 1994; Schirmbock *et al.*, 1994; Lorito *et al.*, 1996; Mach *et al.*, 1999; Dana *et al.*, 2001; Viterbo *et al.*, 2002). In addition, *chit33* and *nag1* were also induced by GlcNAc, but GlcNAc failed to induce *ech42* and *chit36* expression (Peterbauer *et al.*, 1996; Dana *et al.*, 2001; Ramot *et al.*, 2004).

Northern blot analysis of *T. harzianum* showed that the CHIT42 mRNA was specifically induced by chitin provided as a carbon source or by chitin-containing cell walls (Carsolio *et al.*, 1994; Garcia *et al.*, 1994; Irene *et al.*, 1994). Further research has shown that *ech42* expression appears not to be directly induced by chitin but by carbon starvation (nutrient depletion). Furthermore, *ech42* gene transcription is also induced by other physiological stresses, such as low temperature and high osmotic pressure. Four copies of a putative stress response element (CCCCT) were found in the *ech42* promoter (Mach *et al.*, 1999), and a similar stress response element has also been identified in *S. cerevisiae*. The stress response element in *S. cerevisiae* has been shown to mediate various stress responses, including nutrient depletion, low temperature and high osmotic pressure (Treger *et al*., 1998). Potentially, the stress response element of *T. harzianum* is likely to have similar functions. Recently, a new BrlA-like cis-acting element, different from the CCCCT element, has been identified in the *ech42* promoter under nutrient depletion. This is possibly also involved in *ech42* regulation, as expression of *ech42* under carbon starvation was antagonized via the BrlA-like cis-acting element (Brunner *et al*., 2003a).

The mechanisms for the regulation of *nag1* and *ech42* expression in *T. harzianum* are different. Expression of *nag1* is induced by low-molecular weight chito-oligosaccharides and by its own catabolic products (GlcNAc), whereas *ech42* expression is not induced by GlcNAc and chito-oligosaccharides (Mach *et al*., 1999). Induction of a hydrolase gene by its own products has also been reported for *T. reesei* genes such as α-galactosidase, β-xylosidase and α-arabinosidase (Margolles-Clark *et al*., 1997). Conversely, an *N*-acetyl-glucosaminidase of *T. harzianum* is produced at a low constitutive level when the fungus is grown on glucose as the sole carbon source (Carsolio *et al*., 1994; Haran *et al*., 1995; Inbar and Chet, 1995).

Therefore, it is possible that a low-constitutive *N*-acetyl-glucosaminidase, which may be secreted extracellularly, may be sufficient to initiate chitin degradation and to release GlcNAc which, in turn, further induces *nag1* expression. Recently two regulating elements, a CCCCT motif and a $CCAGN_{13}CTGG$ motif, have been identified in the *nag1* promoter. Similar CCAGNCTGG motifs have also been identified in the promoters of two *T. asperellum* GlcNAcase genes, *exc1y* and *exc2y* (Ramot *et al*., 2004). Further work has shown that mutation of the two regulating elements in the *nag1* promoter results in loss of inducibility of *nag1* expression by GlcNAc (Peterbauer *et al*., 2002).

The *chit33* gene from *T. harzianum* is reported to be strongly repressed by glucose, and de-repressed under starvation conditions (Limon *et al*., 1995). Northern blot analysis has shown that the *chit33* mRNA is strongly expressed under starvation conditions, but very little chitinase activity is observed, suggesting post-transcriptional regulation of *chit33* or a high rate of proteolysis of CHIT33. Conversely the mRNA and protein levels of *chit42* are similar, implying that no post-transcriptional regulation is involved in the expression of that gene (Irene *et al*., 1994). In addition, *chit33* expression is also induced by heating (Dana *et al*., 2001), whereas *ech42* expression is induced by low temperature (Mach *et al*., 1999).

It should be noted that there are other regulation mechanisms for fungal chitinases. Two mRNA species (spliced and unspliced mRNAs) of *chi18-3* and *chi18-13* are present in *T. harzianum*, and the ratio of the two transcripts and their relative abundance seems to depend on growth conditions (Seidl *et al*., 2005). Such differences in mRNA levels suggest that this may be post-transcriptional regulation. Fungal chitinase activity may also be regulated by secretion. In *T. harzianum N*-acetyl-glucosaminidase activity is divided into secreted activity and activity that is bound to the mycelium, and the relative proportions of these activities vary for different carbon sources

(Peterbauer *et al.*, 1996). Similar results have also been reported for *N*-acetyl-glucosaminidase activity in *T. asperellum* (Ramot *et al.*, 2004).

In addition, there is regulation between individual chitinases: for example, the *N*-acetyl-glucosaminidase of *T. atroviride* is essential for the induction of *ech42* by chitin, as *chit42* is not induced by chitin in a mutant that does not express the gene *nag1* (Brunner *et al.*, 2003b). This is possibly not a universal occurrence, as the absence of expression of *exc2y* (a homologue of *nag1*) in *T. asperellum* does not affect the transcription of *chit42* (Ramot *et al.*, 2004). This differential regulation of *chit42* may reflect differences in *nag1* and *exc2y* enzymatic activities (Ramot *et al.*, 2004).

Glucose repression is a common phenomenon in the regulation of fungal chitinase genes, and a better understanding of the molecular basis of this is currently being developed. It has been shown that the URS (upstream regulatory sequence) in the fungal chitinase promoters plays a key role in the regulation of glucose repression. The protein product of the regulatory gene *creA*/*cre1* is a negatively acting transcription factor that binds to DNA sequence motifs with the consensus sequence SYGGRG in the URS (Ilmen *et al.*, 1996; Stapleton and Dobson, 2003). In the presence of glucose, the activated CreA/Cre1 protein binds to the consensus motif, and represses transcription of chitinase genes.

Regulation during mycoparasitism

Regulation of chitinase genes during mycoparasitism has also been investigated in the induction of chitinases CHIT102 and CHIT73 in *T. harzianum* during parasitism on *Sclerotium rolfsii* (Inbar and Chet, 1995). *T. harzianum* CHIT42 chitinase expression was strongly enhanced during interactions of *T. harzianum* with both *Rhizoctonia solani* (Carsolio *et al.*, 1994) and *Botrytis cinerea* (Lorito *et al.*, 1996). The transcripts of *ech42* and *chit36* were found in cultures of *T. harzianum* during the pre-contact stage of confrontation with the hosts (Kulling *et al.*, 2000; Viterbo *et al.*, 2002). Recently, *ech42* expression has been shown to be triggered by diffusible factors whose formation does not require contact between *T. harzianum* and *R. solani* (Cortes *et al.*, 1998; Zeilinger *et al.*, 1999). One of the diffusible factors is believed to be a macromolecule produced by *T. harzianum*.

Further research has shown the macromolecule to be an enzyme and it is most likely that it is a constitutive chitinase, as its action can be inhibited by allosamidin. The action of the enzyme on *R. solani* is crucial for *ech42* expression (Zeilinger *et al.*, 1999). This work has been used to suggest a model whereby the action of the constitutive chitinase on *R. solani* causes *T. harzianum* to release a low-molecular weight inducer of *ech42* expression (Zeilinger *et al.*, 1999; Kulling *et al.*, 2000). Interestingly, it is also believed that mycoparasitic interaction relieves the binding of the Cre1 carbon catabolite repressor protein to promoter sequences (SYGGRG) of the *ech42* gene (Lorito *et al.*, 1996). However, the mechanism of this is currently unknown.

Regulation of GlcNAcase during mycoparasitism is simpler than *ech42*, and is similar to the regulation of *nag1* in culture. Two GlcNAcases (EXC1Y and EXC2Y) of *T. asperellum* are up-regulated just before the physical interaction between the parasite and its host. GlcNAc is a diffusible element that up-regulates the two GlcNAcases, and it may be generated by the action of the constitutive activity of GlcNAcases starting positive feedback (Ramot *et al.*, 2004).

Unlike *ech42*, *chit33* is expressed after direct contact between *T. harzianum* and *R. solani* (Mach *et al.*, 1999; Zeilinger *et al.*, 1999; Kulling *et al.*, 2000; Dana *et al.*, 2001), indicating that triggering of *chit33* is likely to be different from that of *ech42*. It has been suggested the induction of *T. harzianum* chitinases is an early event which is elicited by a recognition signal (i.e. lectin–carbohydrate interactions in mycoparasitism) (Inbar and Chet, 1995). A lectin from the cell walls of the host *Sclerotium rolfsii* has been shown to act as a recognition signal (Inbar and Chet, 1992, 1994, 1995). This implies that *T. harzianum* chitinases should be elicited after direct contact between *T. harzianum* and *R. solani*, namely after the interaction and recognition between the lectin in the cell surface of the host and the carbohydrate in the cell surface of *T. harzianum*. Whether *chit33* expression is involved in the recognition signal is not clear, and this requires further study.

Functions and Applications of Fungal Chitinases

Chitinases are biologically and physiologically important in fungi, where they have autolytic, nutritional, morphogenetic and parasitic roles. In *S. cerevisiae*, disruption of the chitinase gene *CTS*, results in cell clumping and failure of the cells to separate after division (Kuranda and Robbins, 1991). This complements evidence that cell separation of *S. cerevisiae* is inhibited by the potent chitinase inhibitor demethylallosamidin (Sakuda *et al.*, 1990). Disruption of *CTS1* can also promote pseudohyphal growth (King and Butler, 1998). A second chitinase gene, *CTS2*, has recently been identified in *S. cerevisiae*. The gene product is involved in sporulation, as disruption of *CTS2* results in abnormal spore wall biosynthesis and a failure to form mature asci (Giaver *et al.*, 2002). In *C. albicans*, results from disruption of its chitinase gene *CHT2* suggest a role for the gene product in cell separation (McCreath *et al.*, 1995). Deletion of *CHT3* generates chains of unseparated cells in the yeast growth phase (Dunkler *et al.*, 2005). In *Benjaminiella poitrasii* the membrane-bound endochitinase and *N*-acetyl-β-glucosaminidase significantly contribute to the morphological changes during the yeast-mycelium transition (Ghormade *et al.*, 2000).

Kluyveromyces lactis secretes a killer toxin, which has an essential chitinase activity (Butler *et al.*, 1991). The killer toxin is a heterotrimeric glycoprotein composed of α, β and γ subunits encoded by the killer plasmid k1. Its toxicity to competitive yeast is thought to be related to the exochitinase activity of the α subunit (Schaffrath and Breunig, 2000). Recently a second chitinase (KlCts1p) has been identified and characterized from *K. lactis* (Colussi *et al.*, 2005). This is a nuclear-encoded endochitinase, and disruption of the *KlCts1* gene results in cell separation defects, suggesting that the KlCts1p protein is required for normal cytokinesis.

In filamentous fungi, chitinases may have roles during sporulation, spore germination, hyphal growth and hyphal autolysis. The *chi3* chitinase gene of *Rhizopus oligosporus* is transcribed during hyphal growth and the deduced amino acid sequence of *chi3* has no potential secretary signal sequence in its amino terminus. It has been suggested that the *chi3* chitinase may possibly function in loosening the cell wall at the hyphal tip to enable turgor pressure to extend the hypha at the apex (Takaya *et al.*, 1998b).

Disruption of a gene that encodes for the *Aspergillus nidulans* chitinase, *ChiA*, leads to a decrease in the frequency of spore germination and to a lower hyphal growth rate (Takaya *et al.*, 1998a). In *Acremonium chrysogenum* the specific chitinase inhibitors allosamidin or demethylallosamidin inhibit fragmentation of hyphae into arthroconidia (Sandor *et al.*, 1998). Disruption of the *nagA* gene of *Aspergillus nidulans* results in poor growth on a medium containing chitiobiose as carbon source, but no phenotypic differences in growth compared with wild-type strains on media containing glucose as carbon source. This suggests that NagA is not essential for the growth of *Aspergillus nidulans* on media containg easily metabolizable carbon sources, and that its main role is in the assimilation of chitin degradation products (Kim *et al.*, 2002b).

In *T. atroviride* a strain with a *nag1* disruption gave delayed autolysis (Brunner *et al.*, 2003b). Conversely disruption of the gene encoding the ChiB1p chitinase of *Aspergillus fumigatus* had no effect on growth or morphogenesis (Jaques *et al.*, 2003). One possible reason is that related enzymes in *A. fumigatus* may compensate for the loss of ChiB1p, as the species contains large numbers of chitinase-encoding genes. On the other hand, it has also been suggested that these secreted chitinases may not have morphogenetic roles, and they may have a principal role in the digestion and utilization of exogenous chitin for energy, biosynthesis and recycling during autolysis (Jaques *et al.*, 2003; Adams, 2004).

On contact with the host, the mycoparasitic fungus *T. harzianum* mycelium coils around the host hyphae, forms hook-like structures that aid in penetration of the host's cell wall and finally absorbs nutrients from host cells. It is suggested that chitinases play an important role in this process, especially by degrading the cell walls to allow cell wall penetration and nutrient utilization (Elad *et al.*, 1982, 1983; Inbar and Chet, 1995). Chitinase genes (such as *ech42*, *chi33*, *nag1* and *chi18-13*) from *T. harzianum* have been shown to have important roles in mycoparasitism, and disruption of the *ech42* gene compromises mycoparasitic ability (Woo *et al.*, 1998; Zeilinger *et al.*, 1999; Kulling *et al.*, 2000; Seidl *et al.*, 2005).

In addition to having important biophysiological functions, there is considerable potential for the application of fungal chitinases in various areas of biotechnology. One promising area is the possibility of improving plant resistance using genetic manipulation with chitinase genes. The *chit42* gene of *T. harzianum* encodes a powerful endochitinase with strong antifungal activity against a wide range of phytopathogenic fungi. This gene can be constitutively expressed in tobacco, apple and potato, and such transgenic plants show a high level of resistance against phytopathogenic fungi (Lorito *et al.*, 1998; Bolar *et al.*, 2000). Transgenic tobacco plants expressing the *S. cerevisiae* Cts1 chitinase can inhibit spore germination and hyphal growth of *Botrytis cinerea* (Carstens *et al.*, 2003).

Fungal chitinases have also been used for the control of insect pests, and a transgenic *B. bassiana* that overexpressess the *Bbchit1* gene has shown significantly enhanced virulence against aphids. Insect bioassay results gave significantly lower 50% lethal concentrations and 50% lethal times for the transformants overexpressing the *Bbchit1* gene compared with the wild-type strain, suggesting that overproduction of endochitinase Bbchit1 increased the infection efficiency of *B. bassiana* in aphids and accelerated infection (Treger *et al.*, 1998). Other commercial uses of fungal chitinases include the generation of fungal protoplasts in biological research, (Kitomoto *et al.*, 1988), the bioconversion of shellfish and the production of single-cell protein (SCP) from chitinous waste (Vyas and Deshpande, 1991). The enzymatic conversion of chitin to ethanol has been suggested (Cody *et al.*, 1990).

Future Perspectives

Fungal chitinase research has progressed quickly at a molecular level. Future research should concentrate on the following topics: (i) the discovery of novel chitinases (especially from fungi from extreme environments); (ii) determining the biological functions of chitinases in growth and development and mycoparasitism; (iii) signalling in chitinase expression; (iv) regulation mechanisms between different chitinases; (v) gene evolution; and (vi) generating new chitinases through protein engineering.

Acknowledgement

This work was supported by the Natural Foundation for Science of China (30571246) and the Hi-Tech Research and Development Program of China (2006AA10Z304).

References

Adams, D.J. (2004) Fungal cell wall chitinases and glucanases. *Microbiology* 150, 2029–2035.

Baek, J.M., Howell, C.R. and Kenerley, C.M. (1999) The role of an extracellular chitinase from *Trichoderma virens* Gv29-8 in the biocontrol of *Rhizoctonia solani*. *Current Genet*ics 35, 41–50.

Bidochka, M.J., Tong, K.I. and Khachatourians, G.G. (1993) Partial purification and characterization of two extracellular N-acetyl-D-glucosaminidases produced by the entomopathogenic fungus *Beauveria bassiana*. *Canadian Journal of Microbiology* 39, 41–45.

Blaiseau, P.L. and Lafay, J. (1992) Primary structure of a chitinase-encoding gene (*chi1*) from the filamentous fungus *Aphanocladium album*: similarity to bacterial chitinases. *Gene* 120, 243–248.

Blaiseau, P.L., Kunz, C., Grison, R., Bertheau, Y, and Brygoo, Y. (1992) Cloning and expression of chitinase gene from the hyperparasitic fungus *Aphanocladium album*. *Current Genetics* 21, 61–66.

Bogo, M.R., Rota, C.A., Pinto, H. J.R., Ocampos, M., Correa, C.T., Vainstein, M.H. and Schrank, A. (1998) A chitinase-encoding gene (*chit1* gene) from the entomopathogen *Metarhizium anisopliae*: isolation and characterization of genomic and full-length cDNA. *Current Microbiology* 37, 221–225.

Bolar, J.P., Norelli, J.L., Wong, K.W., Hayes, C.K,, Harman, G.E. and Aldwinckle, H.S. (2000) Expression of endochitinase from *Trichoderma harzianum* in transgenic apple increases resistance to apple scab and reduces vigor. *Phytopathology* 90, 72–77.

Bonfim, S.M.R.C., Cruz, A.H.S., Jesuino, R.S.A., Ulhoa, C.J., Molinari-Madlum, E.E.W.I., Soares, C.M.A. and Pereira, M. (2006) Chitinase from *Paracoccidioides brasiliensis*: molecular cloning, structural, phylogenitic, expression and activity analysis. *FEMS Immunology and Medical Microbiology* 46, 269–283.

Brunner, K., Montero, M., Mach, R.L., Peterbauer, C.K. and Kubicek, C.P. (2003a) Expression of the *ech42* (endochitinase) gene of *Trichoderma atroviride* under carbon starvation is antagonized via a BrlA-like cis-acting element. *FEMS Microbiology Letters* 218, 258–264.

Brunner, K., Peterbauer, C.K., Mach, R.L., Lorito, M., Zeilinger, S. and Kubicek, C.P. (2003b) The *nag1* *N*-acetylglucosaminidase of *Trichoderma atroviride* is essential for chitinase induction by chitin and of major relevance to biocontrol. *Current Genetics* 43, 289–295.

Butler, A.R., O'Donnell, R.W., Martin, V.J., Gooday, G.W. and Stark, M.J.R. (1991) *Kluyveromyces lactis* toxin has an essential chitinase activity. *European Journal of Biochemistry* 199, 483–488.

Cannon, R.D., Niimi, K., Jenkinson, H.F. and Shepherd, M.G. (1994) Molecular cloning and expression of the *Candida albicans* beta-N-acetylglucosaminidase (HEX1) gene. *Journal of Bacteriology* 176, 2640–2647.

Carsolio, C., Gutierrez, A., Jimenez, B., Van Montagu, M. and Herrera-Estrella, A. (1994) Characterization of *ech42*, a *Trichoderma harzianum* endochitinase gene expressed during mycoparasitism. *Proceedings of the National Academy of Science USA* 91, 10903–10907.

Carsolio, C., Benhamou, N., Haran, S., Cortes, C., Gutierrez, A., Chet, I. and Herrera-Estrella, A. (1999) Role of the *Trichoderma harzianum* endochitinase gene, ech42, in mycoparasitism. *Applied and Environmental Microbiology* 65, 929–935.

Carstens, M., Vivier, M.A. and Pretorius, I.S. (2003) The *Saccharomyces cerevisiae* chitinase, encoded by the *CTS1-2* gene, confers antifungal activity against *Botrytis cinerea* to transgenic tobacco. *Transgenic Research* 12, 497–508.

Cody, R.M., Davis, N.D., Lin, J. and Shaw, D. (1990) Screening microorganisms for chitin hydrolysis and production of ethanol from aminosugars. *Biomass* 21, 285–295.

Colussi, P.A., Specht, C.A. and Taron, C.H. (2005) Characterization of a nucleus-encoded chitinase from the yeast *Kluyveromyces lactis*. *Applied and Environmental Microbiology* 71, 2862–2869.

Cortes, C., Gutierrez, A., Olmedo, V., Inbar, J., Chet, I. and Herrera-Estrella, A. (1998) The expression of genes involved in parasitism by *Trichoderma harzianum* is triggered by a diffusible factor. *Molecular and General Genetics* 260, 218–225.

Dana, M.M., Limon, M.C., Mejias, R., Mach, R.L., Benitez, T., Pintor-Toro, J.A. and Kubicek, C. (2001) Regulation of chitinase 33 (*chit33*) gene expression in *Trichoderma harzianum*. *Current Genetics* 38, 335–342.

Da Silva, M.V., Santi, L., Staats, C.C., Da Costa, A.M., Colodel, E.M., Driemeier, D., Vainstein, M.H. and Schrank, A. (2005) Cuticle-induced endo/exoacting chitinase CHIT30 from

Metarhizium anisopliae is encoded by an ortholog of the *chi3* gene. *Research in Microbiology* 156, 382–392.

Deane, E.E., Whipps, J.M., Lynch, J.M. and Peberdy, J.F. (1998) The purification and characterization of a *Trichoderma harzianum* exochitinase. *Biochimica et Biophysica Acta* 1383, 101–110.

De La Cruz, J., Rey, M., Lora, J.M., Hidalgo-Gallego, A., Dominguez, F., Pintor-Toro, J.A., Llobell, A. and Benitez, T. (1992) Isolation and characterization of three chitinases from *Trichoderma harzianum*. *European Journal of Biochemistry* 206, 859–867.

Diez, B., Rodriguez-Saiz, M., de la Fuente, J.L., Moreno, M.A. and Barredo, J.L. (2005) The *nagA* gene of *Penicillium chrysogenum* encoding beta-N-acetylglucosaminidase. *FEMS Microbiology Letters* 242, 257–264.

Di Pietro, A., Lorito, M., Hayes, C.K., Broadway, R.M. and Harman, G.E. (1993) Endochitinase from *Gliocladium virens*: isolation, characterization, and synergistic antifungal activity in combination with gliotoxin. *Phytopathology* 83, 308–313.

Draborg, H., Kauppinen, S., Dalboge, H. and Christgau, S. (1995) Molecular cloning and expression in *S. cerevisiae* of two exochitinases from *Trichoderma harzianum*. *Biochemistry and Molecular Biology International* 36, 781–791.

Drouillard, S., Armand, S., Davies, G.J., Vorgias, C.E. and Henrissat, B. (1997) *Serratia marcescens* chitobiase is a retaining glycosidase utilizing substrate acetamido group participation. *Biochemistry Journal* 328, 945–949.

Dunkler, A., Walther, A., Specht, C.A. and Wendland, J. (2005) *Candida albicans CHT3* encodes the functional homolog of the Cts1 chitinase of *Saccharomyces cerevisiae*. *Fungal Genetics and Biology* 42, 935–947.

Elad, Y., Chet, I. and Henis, Y. (1982) Degradation of plant pathogenic fungi by *Trichoderma harzianum*. *Canadian Journal of Microbiology* 28, 719–725.

Elad, Y., Chet, I., Boyle, P. and Henis, Y. (1983) Parasitism of *Trichoderma* spp. on *Rhizoctonia solani* and *Sclerotium rolfsii*-scanning electron microcopy studies and fluorescence microscopy. *Phytopathology* 73, 85–88.

Fang, W., Leng, B., Xiao, Y., Jin, K., Ma, J., Fan, Y., Feng, J., Yang, X., Zhang, Y. and Pei, Y. (2005) Cloning of *Beauveria bassiana* chitinase gene *Bbchit1* and its application to improve fungal strain virulence. *Applied and Environmental Microbiology* 71, 363–370.

Garcia, I., Lora, J.M., De La Cruz, J., Benitez, T., Llobell, A. and Pintor-Toro, J.A. (1994) Cloning and characterization of a chitinase (CHIT42) cDNA from the mycoparasitic fungus *Trichoderma harzianum*. *Current Genetics* 27, 83–89.

Ghormade, V.S., Lachke, S.A. and Deshpande, M.V. (2000) Dimorphism in *Benjaminiella poitrasii*: involvement of intracellular endochitinase and N-acetylglucosaminidase activities in the yeast-mycelium transition. *Folia Microbiologia* 45, 231–238.

Giaver, G., Chu, A.M., Ni, L. *et al.* (2002) Functional profiling of the *Saccharomyces cerevisiae* genome. *Nature* 418, 387–391.

Graham, L.S. and Sticklen, M.B. (1994) Plant chitinases. *Canadian Journal of Botany* 72, 1057–1083.

Guo, R.F. and Li, D.C. (2006) Molecular cloning and expression of a chitinase gene from the thermophilic fungus *Thermomyces lanuginosus*. *Acta Microbiologica Sinica* 46, 99–103 [in Chinese].

Guo, R.F., Li, D.C. and Wang, R. (2005) Purification and properties of a thermostable chitinase from thermophilic fungus *Thermomyces lanuginosus*. *Acta Microbiologica Sinica* 45, 270–274 [in Chinese].

Hamel, F., Boivin, R., Tremblay, C. and Bellemare, G. (1997) Structural and evolutionary relationships among chitinases of flowering plants. *Journal of Molecular Evolution* 44, 614–624.

Haran, S., Schickler, H., Oppenheim, A. and Chet, I. (1995) New components of the chitinolytic system of *Trichoderma harzianum*. *Mycological Research* 99, 441–446.

Haran, S., Schickler, H. and Chet, I. (1996) Molecular mechanisms of lytic enzymes involved in the biocontrol activity of *Trichoderma harzianum*. *Microbiology* 142, 2321–2331.

Harman, G.E., Hayes, C.K., Lorito, M., Broadway, R.M., Di Pietro, A., Peterbauer, C. and Tronsmo, A. (1993) Chitinolytic enzymes of *Trichoderma harzianum*: purification of chitobiosidase and endochitinase. *Phytopathology* 83, 313–318.

Hayes, C.K., Klemsdal, S., Lorito, M., Di Pietro, A., Peterbauer, C., Nakas, J.P., Tronsmo, A. and Harman, G.E. (1994) Isolation and sequence of an endochitinase gene from a cDNA library of *Trichoderma harzianum*. *Gene* 138, 143–148.

Henrissat, B. (1991) A classification of glycosyl hydrolases based on amino acid sequence similarities. *Biochemistry Journal* 280, 309–316.

Henrissat, B. (1999) Classification of chitinase modules. In: Jolles, P. and Muzzarelli, R.A.A. (eds) *Chitin and Chitinases*. Burkhauser, Basel, Switzerland, pp. 137–156.

Henrissat, B. and Bairoch, A. (1993) New families in the classification of glycosylhydrolases based on amino acid sequence similarities. *Biochemistry Journal* 293, 781–788.

Hoell, I.A., Klemsdal, S.S., Vaaje-Kolstad, G., Hora, S.J. and Eijsink, V.G. (2005) Overexpression and characterization of a novel chitinase from *Trichoderma atroviride* strain P1. *Biochimica et Biophysica Acta* 1748, 180–190.

Hollis, T., Monzingo, A.F., Bortone, K., Ernst, S., Cox, R. and Robertus, J.D. (2000) The X-ray structure of a chitinase from the pathogenic fungus *Coccidioides immitis*. *Protein Science* 9, 544–551.

Ike, M., Nagamatsu, K., Shioya, A., Nogawa, M., Ogasawara, W., Okada, H. and Morikawa, Y. (2005) Purification, characterization, and gene cloning of 46 kDa chitinase (Chi46) from *Trichoderma reesei* PC-3-7 and its expression in *Escherichia coli*. *Applied Microbiology and Biotechnology* 71, 294–303.

Ilmen, M., Thrane, C. and Penttila, M. (1996) The glucose repressor gene cre1 of *Trichoderma*: isolation and expression of a full-length and a truncated mutant form. *Molecular and General Genetics* 251, 451–460.

Inbar, J. and Chet, I. (1992) Biomimics of fungal cell–cell recognition by use of lectin-coated nylon fibers. *Journal of Bacteriology* 174, 1055–1059.

Inbar, J. and Chet, I. (1994) A newly isolated lectin from the plant pathogenic fungus *Sclerotium rolfsii*: purification, characterization and its role in mycoparasitism. *Microbiology* 140, 651–657.

Inbar, J. and Chet, I. (1995) The role of recognition in the induction of specific chitinases during mycoparasitism by *Trichoderma harzianum*. *Microbiology* 141, 2823–2829.

Iranzo, M., Aguado, C., Pallotti, C., Canizares, J.T. and Mormeneo, S. (2002) The use of trypsin to solubilize wall proteins from *Candida albicans* led to the identification of chitinase 2 as an enzyme covalently linked to the yeast wall structure. *Research in Microbiology* 153, 227–232.

Irene, G., Jose, M.L., De La Cruz, J., Tahia, B., Antonio, L., Jose, A. and Pintor, T. (1994) Cloning and characterization of chitinase (CHIT 42) cDNA from mycoparasitic fungus *Trichoderma harzianum*. *Current Genetics* 27, 83–89.

Jaques, A.K., Fukamizo, T., Hall, D., Barton, R.C., Escott, G.M., Parkinson, T., Hitchcock, C.A. and Adams, D.J. (2003) Disruption of the gene encoding the ChiB1 chitinase of *Aspergillus fumigatus* and characterization of a recombinant gene product. *Microbiology* 149, 2931–2939.

Jones, T., Federspiel, N.A., Chibana, H., Dungan, J., Kalman, S., Magee, B.B., Newport, G., Thorstenson, Y.R., Agabian, N., Magee, P.T., Davis, R.W. and Scherer, S. (2004) The diploid genome sequence of *Candida albicans*. *Proceedings of the National Academy of Science USA* 101, 7329–7334.

Kang, S.C., Park, S. and Lee, D.G. (1998) Isolation and characterization of a chitinase cDNA from the entomopathogenic fungus, *Metarhizium anisopliae*. *FEMS Microbiology Letters* 165, 267–271.

Kang, S.C., Park, S. and Lee, D.G. (1999) Purification and characterization of a novel chitinase from the entomopathogenic fungus *Metarhizium anisopliae*. *Journal of Invertebrate Pathology* 73, 276–281.

Kim, D.J., Baek, J.M., Uribe, P., Kenerley, C.M. and Cook, D.R. (2002a) Cloning and characterization of multiple glycosyl hydrolase genes from *Trichoderma rirens*. *Current Genetics* 40, 374–384.

Kim, S., Matsuo, I., Ajisaka, K., Nakajima, H. and Kitamoto, K. (2002b) Cloning and characterization of the *nagA* gene that encoded β-*N*-acetylglucosaminidase from *Aspergillus nidulans* and its expression in *Aspergillus oryzae*. *Bioscience Biotechnology and Biochemistry* 66, 2168–2175.

King, L. and Butler, G. (1998) Ace2p, a regulator of CTS1 (chitinase) expression, affects pseudohyphal production in *Saccharomyces cerevisiae*. *Current Genetics* 34, 183–191.

Kitomoto, Y., Mori, N., Yamamoto, M., Ohiwa, T. and Ichiwaka, Y. (1988) A simple method of protoplast formation and protoplast regeneration from various fungi using an enzyme from *Trichoderma harzianum*. *Applied Microbiology and Biotechnology* 28, 445–450.

Kraulis, J., Clore, G.M., Nilges, M., Jones, T.A., Pettersson, G., Knowles, J. and Gronenborn, A.M. (1989) Determination of the three-dimensional solution structure of the C-terminal domain of cellobiohydrolase I from *Trichoderma reesei*. A study using nuclear magnetic resonance and hybrid distance geometry-dynamical simulated annealing. *Biochemistry* 28, 7241–7257.

Kulling, C.M., Mach, R.L., Lorito, M. and Kubicek, C.P. (2000) Enzyme diffusion from *Trichoderma atroviride* (= *T. harzianum* P1) to *Rhizoctonia solani* is a prerequisite for triggering of *Trichoderma ech42* gene expression before mycoparasitic contact. *Applied and Environmental Microbiology* 66, 2232–2234.

Kunz, C., Ludwig, A., Bertheau, Y. and Boller, T. (1992a) Evaluation of the antifungal activity of the purified chitinase 1 from the filamentus fungus *Aphanocladium album*. *FEMS Microbiology Letters* 90, 105–110.

Kunz, C., Sellam, O. and Bertheau, Y. (1992b) Purification and characterization of a chitinase from the hyperparasitic fungus *Aphanocladium album*. *Physiological and Molecular Plant Pathology* 40, 117–131.

Kuranda, M.J. and Robbins, P.W. (1987) Cloning and heterologous expression of glycosidase genes from *Saccharomyces cerevisiae*. *Proceedings of the National Academy of Science USA* 84, 2585–2589.

Kuranda, M.J. and Robbins, P.W. (1991) Chitinase is required for cell separation during growth of *Saccharomyces cerevisiae*. *Journal of Biological Chemistry* 266, 19758–19767.

Leger, R.J., Cooper, R.M. and Charnley, A.K. (1991) Characterization of chitinase and chitobiase produced by the entomopathogenic fungus *Metarhizium anisopliae*. *Journal of Invertebrate Pathology* 58, 415–426.

Leger, R.J., Joshi, L., Bidochka, M.J., Rizzo, N.W. and Roberts, D.W. (1996) Characterization and ultrastructural localization of chitinase from *Metarhizium anisopliae*, *M. flavoviride* and *Beauveria bassiana* during fungal invasion of host (*Manduca sexta*) cuticle. *Applied and Environmental Microbiology* 62, 907–912.

Li, D.C., Zhang, S.H., Liu, K.Q. and Lu, J. (2004) Purification and characterization of a chitinase from the mycoparasitic fungus *Trichothecium roseum*. *Journal of General and Applied Microbiology* 50, 35–39.

Li, D.C., Chen, S. and Lu, J. (2005) Purification and partial characterization of two chitinase from the mycoparasitic fungus *Talaromyces flavus*. *Mycopathologia* 159, 223–229.

Limon, M.C., Lora, J.M., Garcia, I., De La Cruz, J., Llobell, A., Benitez, T. and Pintor-Toro, J.A. (1995) Primary structure and expression pattern of the 33-kDa chitinase gene from the mycoparasitic fungus *Trichoderma harzianum*. *Current Genetics* 28, 478–483.

Limon, M.C., Margolles-Clark, E., Benitez, T. and Penttila, M. (2001) Addition of substrate-binding domains increases substrate-binding capacity and specific activity of a chitinase from *Trichoderma harzianum*. *FEMS Microbiology Letters* 198, 57–63.

Lisboa, J., Valadares-Inglis, M.C. and Felix, C.R. (2004) Purification and characterization of an *N*-acetylglucosaminidase produced by a *Trichoderma harzianum* strain which controls *Crinipellis perniciosa*. *Applied Microbiology and Biotechnology* 64, 70–75.

Lorito, M., Harman, G.E., Hayes, C.K., Broadway, R.M., Tronsmo, A., Woo, S.L. and Di Pietro, A. (1993) Chitinolytic enzymes produced by *Trichoderma harzianum*: antifungal activity of purified endochitinase and chitobiosidase. *Phytopathology* 83, 302–307.

Lorito, M., Hayes, C.K., Di Pietro, A., Woo, S.L. and Harman, G.E. (1994) Purification, characterization, and synergistic activity of a glucan 1,3-β-glucosidase and an *N*-acetylglucosaminidase from *Trichoderma harzianum*. *Phytopathology* 84, 398–405.

Lorito, M., Mach, R.L., Sposato, P., Strauss, J., Peterbauer, C.K. and Kubicek, C.P. (1996) Mycoparasitic interaction relieves binding of Cre1 carbon catabolite repressor protein to promoter sequence of *ech-42* (endochitinase-encoding) gene of *Trichoderma harzianum*. *Proceedings of the National Academy of Science USA* 93, 14868–14872.

Lorito, M., Woo, S.L., Fernandez, I.G., Colucci, G., Harman, G.E., Pintor-Toro, J.A., Filippone, E., Muccifora, S., Lawrence, C.B., Zoina, A., Tuzun, S. and Scala, F. (1998) Genes from mycoparasitic fungi as a source for improving plant resistance to fungal pathogens. *Proceedings of the National Academy of Science USA* 95, 7860–7865.

Mach, R.L., Peterbauer, C.K., Payer, K., Jaksits, S., Woo, S.L., Zeilinger, S., Kullnig, C.M., Lorito, M. and Kubicek, C.P. (1999) Expression of two major chitinase genes of *Trichoderma atroviride* (*T. harzianum* P1) is triggered by different regulatory signals. *Applied and Environmental Microbiology* 65, 1858–1863.

Margolles-Clark, E., Hayes, C.K., Harman, G.E. and Penttila, M.E. (1996) Improved production of *Trichoderma harzianum* endochitinase by expression in *Trichoderma reesei*. *Applied and Environmental Microbiology* 62, 2145–2151.

Margolles-Clark, E., Ilmen, E.M. and Penttila, M.E. (1997) Expression patterns of ten hemicellulase genes in the filamentous fungus *Trichoderma reesei* on various carbon sources. *Journal of Biotechnology* 57, 167–179.

Mathivanan, N., Kabilan, V. and Murugesan, K. (1998) Purification, characterization, and antifungal activity of chitinase from *Fusarium chlamydosporum*, a mycoparasite to groundnut rust, *Puccinia arachidis*. *Canadian Journal of Microbiology* 44, 646–651.

McCormack, J., Hackett, T.J., Tuohy, M.G. and Coughlan, M.P. (1991) Chitinase production by *Talaromyces emersonii*. *Biotechnology Letters* 13, 677–682.

McCreath, K.J., Specht, C.A. and Robbins, P.W. (1995) Molecular cloning and characterization of chitinase genes from *Candida albicans*. *Proceedings of the National Academy of Science USA* 92, 2544–2548.

McCreath, K.J, Specht, C.A., Liu, Y. and Robbins, P.W. (1996) Molecular cloning of a third chitinase gene (*CHT1*) from *Candida albicans*. *Yeast* 12, 501–504.

McGrew, B.R. and Green, D.M. (1990) Enhanced removal of detergent and recovery of enzymatic activity following sodium dodecyl sulfate-polyacrylamide gel electrophoresis: use of casein in gel wash buffer. *Analytical Biochemistry* 189, 68–74.

Merzendorfer, H. and Zimoch, L. (2003) Chitin metabolism in insects: structure, function and regulation of chitin synthases and chitinases. *Journal of Experimental Biology* 206, 4393–4412.

Morissette, D.C., Driscoll, B.T. and Jabaji-Hare, S. (2003) Molecular cloning, characterization, and expression of a cDNA encoding an endochitinase gene from the mycoparasite *Stachybotrys elegans*. *Fungal Genetics and Biology* 39, 276–285.

Neuhaus, J.M., Fritig, B., Linthorst, H.J.M. and Meins, F.J. (1996). A revised nomenclature for chitinase genes. *Plant Molecular Biology Reporter* 14, 102–104.

Ohno, T., Armand, S., Hata, T., Nikaidou, N., Henrissat, B., Mitsutomi, M, and Watanabe, T. (1996) A modular family 19 chitinase found in the prokaryotic organism *Streptomyces griseus* HUT 6037. *Journal of Bacteriology* 178, 5065–5070.

Park, J.K., Morita, K., Fukumoto. I., Yamasaki, Y., Nakagawa, T., Kawamukai, M. and Matsuda, H. (1997) Purification and characterization of the chitinase (ChiA) from *Enterobacter* sp. G-1. *Bioscience Biotechnology and Biochemistry* 61, 684–689.

Patil, S.R., Ghormade, V. and Deshpande, M.V. (2000) Chitinolytic enzymes: an exploration. *Enzyme and Microbial Technology* 26, 473–483.

Peng, R.W., Guan, K.M. and Huang, X.L. (1996) Induction and purification of two extracellular chitinases from *Beauveria bassiana*. *Acta Microbiologica Sinica* 36, 103–108 [in Chinese].

Peterbauer, C.K., Lorito, M., Hayes, C.K., Harman, G.E. and Kubicek, C.P. (1996) Molecular cloning and expression of the *nag1* gene (N-acetyl-β-*D*-glucosaminidase-encoding gene) from *Trichoderma harzianum* P1. *Current Genetics* 30, 325–331.

Peterbauer, C.K., Brunner, K., Mach, R.L. and Kubicek, C.P. (2002) Identification of the N-acetyl-D-glucosamine-inducible element in the promoter of the *Trichoderma atroviride* nag1 gene encoding N-acetyl-glucosaminidase. *Molecular Genetics and Genomics* 267, 162–170.

Pinto, D.S., Barreto, C.C., Schrank, A., Ulhoa, C.J. and Henning V.M. (1997). Purification and characterization of an extracellular chitinase from the entomopathogen *Metarhizium anisopliae*. *Canadian Journal of Microbiology* 43, 322–327.

Pishko, E.J., Kirkland, T.N. and Cole, G.T. (1995) Isolation and characterization of two chitinase-endoding genes (*cts1*, *cts2*) from the fungus *Coccidioides immitis*. *Gene* 167, 173–177.

Plihal, O., Sklenar, J., Kmonickova, J., Man, P., Pompach, P., Havlicek, V., Kren, V. and Bezouska, K. (2004) N-glycosylated catalytic unit meets O-glycosylated propeptide: complex protein architecture in a fungal hexosaminidase. *Biochemical Society Transactions* 32, 764–765.

Ramot, O., Viterbo, A., Friesem, D., Oppenheim, A. and Chet, I. (2004) Regulation of two homodimer hexosaminidases in the mycoparasitic fungus *Trichoderma asperellum* by glucosamine. *Current Genetics* 45, 205–213.

Rao, F.V., Anderson, O.A., Vora, K.A., Demartino, J.A. and Van Aalten, D.M. (2005) Methylxanthine drugs are chitinase inhibits: investigation of inhibition and binding model. *Chemical Biology* 12, 973–980.

Reissig, J.L., Strominger, J.L. and Leloir, L.F. (1955) A modified colorimetri method for the estimation of N-acetyllamino sugars. *Journal of Biological Chemistry* 217, 959–967.

Robert, W.K. and Selitrennikoff, C.P. (1988) Plant and bacterial chitinases differ in antifungal activity. *Journal of General Microbiology* 134, 169–176.

Sahai, A.S. and Manocha, M.S. (1993) Chitinases of fungi and plants: their involvement in morphogenesis and host–parasite interaction. *FEMS Microbiology Reviews* 11, 317–338.

Sakuda, S., Nishimoto, Y., Ohi, M., Watanabe, M., Takayama, S., Isogai, A. and Yamada, Y. (1990) Effects of demethylallosamidin, a potent yeast chitinase inhibitor on the cell division of yeast. *Agricultural and Biological Chemistry* 54, 1333–1335.

Sandor, E., Pusztahelyi, T., Karaffa, L., Karanyi, Z., Pocsi, I., Biro, S., Szentirmai, A. and Pocsi, I. (1998) Allosamidin inhibits the fragmentation of *Acremonium chrysogenum* but does not influence the cephalosporin-C production of the fungus. *FEMS Microbiology Letters* 164, 231–236.

Schaffrath, R. and Breunig, K.D. (2000) Genetics and molecular physiology of the yeast *Kluyveromyces lactis*. *Fungal Genetics and Biology* 30, 173–190.

Schirmbock, M., Lorito, M., Wang, Y.L., Hayes, C.K., Arisan-Atac, I., Scala, F., Harman, G.E. and Kubicek, C.P. (1994) Parallel formation and synergism of hydrolytic enzymes and peptaibol antibiotics: molecular mechanisms involved in the antagonistic action of *Trichoderma harzianum* against phytopathogenic fungi. *Applied and Environmental Microbiology* 60, 4364–4370.

Screen, S.E., Hu, G. and Leger, R.T. (2001) Transformants of *Metarhizium anisopliae* sf. *anisopliae* overexpressing chitinase from *Metarhizium anisopliae* sf. *acridum* show early induction of native chitinase but are not altered in pathogenicity to *Manduca sexta*. *Journal of Invertebrate Pathology* 78, 260–266.

Seidl, V., Huemer, B., Seiboth, B. and Kubicek, C.P. (2005) A complete survey of *Trichoderma* chitinases reveals three distinct subgroups of family 18 chitinases. *FEBS Journal* 272, 5923–5939.

Song, J., Yang, Q., Liu, B. and Chen, D. (2005) Expression of the chitinase gene from *Trichoderma aureoviride* in *Saccharomyces cerevisiae*. *Applied Microbiology and Biotechnology* 69, 39–43.

Stapleton, P.C. and Dobson, A.D.W. (2003) Carbon repression of cellobiose dehydrogenase production in the white rot fungus *Trametes versicolor* is mediated at the level of gene transcription. *FEMS Microbiology Letters* 221, 167–172.

Stark, M.J.R. and Boyd, A. (1996) The killer toxin of *Kluyveromyces lactis*: characterization of the toxin subunits and identification of the genes that encode them. *EMBO Journal* 3, 1995–2002.

Sudoh, M., Watanabe, M., Mio, T., Nagahashi, S., Yamada-Okabe, H., Takagi, M. and Arisawa, M. (1995) Isolation of canCHS1A, a variant gene of *Candida albicans* chitin synthase. *Microbiology* 141, 2673–2679.

Suzuki, K., Taiyoji, M., Sugawara, N., Nikaidou, N., Henrissat, B. and Watanabe, T. (1999) The third chitinase gene (*chiC*) of *Serratia marcescens* 2170 and the relationship of its product to other bacterial chitinases. *Biochemistry Journal* 343, 587–596.

Takaya, N., Yamazaki, D., Horiuchi, H., Ohta, A. and Takagi, M. (1998a) Cloning and characterization of a chitinase-encoding gene (*chiA*) from *Aspergillus nidulans*, disruption of which decreases germination frequency and hyphal growth. *Biosciences Biotechnology and Biochemistry* 62, 60–65.

Takaya, N., Yamazaki, D., Horiuchi, H., Ohta, A. and Takagi, M. (1998b) Intracellular chitinase gene from *Rhizopus oligosporus*: molecular cloning and characterization. *Microbiology* 144, 2647–2654.

Taylor, G., Jabaji-Hare, S., Charest, P.M. and Khan, W. (2002) Purification and characterization of an extracellular exochitinase, β-*N*-acetylhexosaminidase, from fungal mycoparasite *Stachybotrys elegans*. *Canadian Journal of Microbiology* 48, 311–319.

Tikhonov, V.E., Lopez-Llorca, L.V., Salinas, J. and Jansson, H.B. (2002) Purification and characterization of chitinases from the nematophagous fungi *Verticillium chamydosporium* and *V. suchlasporium*. *Fungal Genetics and Biology* 35, 67–78.

Treger, J.M., Magee, T.R. and McEntee, K. (1998) Functional analysis of the stress response element and its role in the multistress response of *Saccharomyces cerevisiae*. *Biochemistry and Biophysics Research Communications* 243, 13–19.

Tronsmo, A. and Harman, G. (1993) Detection and quantification of *N*-acetyl-β-D-glucosaminidase, chitobiosidase, and endochitinase in solutions and on gels. *Annals of Biochemistry* 208, 74–79.

Ulhoa, C.J. and Peberdy, J.F. (1991) Purification and characterization of an extracellular chitobiase from *Trichoderma harzianum*. *Current Microbiology* 23, 285–289.

Ulhoa, C.J. and Peberdy, J.F. (1992) Purification and some properties of the extracellular chitinase produced by *Trichoderma harzianum*. *Enzyme and Microbial Technology* 14, 236–241.

Valadares-Inglis, M.C., Inglis, P.W. and Peberdy, J.F. (1997) Sequence analysis of catalytic domain of a *Metarhizium anisopliae* chitinase. *Brazilian Journal of Genetics* 20, 161–164.

Van Eijk, M., Van Roomen, C.P.A.A., Renkema, G.H., Bussink, A.P., Andrews, L., Blommaart, E.F.C., Sugar, A., Verhoeven, A.J., Boot, R.G. and Aerts, J.M.F.G. (2005) Characterization of human phagocyte-derived chitotriosidase, a component of innate immunity. *International Immunology* 17, 1505–1512.

Viterbo, A., Haran, S., Friesem, D., Ramot, O. and Chet, I. (2001) Antifungal activity of a novel endochitinase gene (chit36) from *Trichoderma harzianum* Rifai TM. *FEMS Microbiology Letters* 200, 169–174.

Viterbo, A., Montero, M., Ramot, O., Friesem, D., Monte, E., Llodell, A. and Chet, I. (2002) Expression regulation of the endochitinase *vhit36* from *Trichoderma asperellum* (*T. harzianum* T-203). *Current Genetics* 42, 114–122.

Vyas, P.R. and Deshpande, M.V. (1991) Enzymatic hydrolysis of chitin by *Myrothecium verrucaria* chitinase complex and its utilization to produce SCP. *Journal of General and Applied Microbiology* 37, 267–275.

Watanabe, T., Kobori, K., Miyashita, K., Fujii, T., Sakai, H., Usshida, M. and Tanaka, H. (1993) Identification of glutamic acid 204 and aspartic acid 200 in chitinase A1 of *Bacillus circulans* WL-12 as essential residues for chitinase activity. *Journal of Biological Chemistry* 268, 18567–18572.

Wattanalai, R., Wiwat, C., Boucias, D.G. and Tartar, A. (2004) Chitinase gene of the dimorphic mycopathogen, *Nomuraea rileyi*. *Journal of Invertebrate Pathology* 85, 54–57.

Woo, S.L., Donzelli, B., Scala, F., Mach, R.L., Harman, G.E., Kubicek, C.P., Del Sorbo, G. and Lorito, M. (1998) Disruption of *ech42* (endochitinase-encoding) gene affects biocontrol activity in *Trichoderma harzianum* strain P1. *Molecular Plant–Microbe Interactions* 12, 419–429.

Yanai, K., Takaya, N., Kojima, N., Horiuchi, H., Ohta, A. and Takagi, M. (1992) Purification of two chitinases from *Rhizopus oligosporus* and isolation and sequencing of the encoding genes. *Journal of Bacteriology* 174, 7398–7406.

Zeilinger, S., Galhaup, C., Payer, K., Woo, S.L., Mach, R.L., Fekete, C., Lorito, M. and Kubicek, C.P. (1999) Chitinase gene expression during mycoparasitic interaction of *Trichoderma harzianum* with its host. *Fungal Genetics and Biology* 26, 131–140.

Zhang, L., Meakin, H. and Dickinson, M. (2003) Isolation of genes expressed during compatible interactions between leaf rust (*Puccinia triticina*) and wheat using cDNA-AFLP. *Molecular Plant Pathology* 4, 469–477.

Zimmermann, C.R., Johnson, S.M., Martens, G.W., White, A.G. and Pappagianis, D. (1996) Cloning and expression of the complement fixation antigen-chitinase of *Coccidioides immitis*. *Infection and Immunity* 64, 4967–4975.

16 Extracellular Proteases of Mycoparasitic and Nematophagous Fungi

László Kredics, Sándor Kocsubé, Zsuzsanna Antal, Lóránt Hatvani, László Manczinger and Csaba Vágvölgyi

Department of Microbiology, Faculty of Science and Informatics, University of Szeged, Hungary

Introduction

Some filamentous fungi are able to restrict the growth and/or survival of fungal plant pathogens and nematodes, thereby contributing to biological control (Chet and Inbar, 1997). After mycoinsecticides, the most frequently applied and thoroughly studied fungal biocontrol agents are the myco-fungicides and myconematicides.

Only a few of the mechanisms involved in the suppression of the target pathogen have been studied in detail for some of the numerous mycofungicides proposed. Antagonistic interactions among fungi in nature include: (i) competition for limiting nutritional factors; (ii) antibiosis, defined as the inhibition or destruction of a fungus by the metabolic production of another fungus; and (iii) mycoparasitism, defined as a direct attack on a fungal thallus followed by nutrient utilization by the parasite. The use of beneficial, naturally antagonistic, fungi such as *Trichoderma* species as biocontrol agents to control fungal plant pathogens could be a suitable environmentally friendly alternative to chemical fungicides.

Trichoderma species are asexual, soil-inhabiting filamentous fungi with teleomorphs belonging to the genus *Hypocrea* (Ascomycota: Hypocreales). The taxonomy of the genus *Trichoderma* is currently under revision, and molecular techniques have already confirmed some 100 phylogenetically different *Trichoderma* species (Druzhinina *et al.*, 2006). Certain *Trichoderma* species are well known as cellulase producers of biotechnological importance (Kubicek *et al.*, 1990), while others are emerging as opportunistic pathogens of humans (Kredics *et al.*, 2003) or as the causative agents of green mould disease of cultivated mushrooms (Komoń-Zelazowska *et al.*, 2007).

Furthermore, well-characterized representatives of the genus are effective antagonists of plant pathogenic fungi, and so may be potential candidates for biological control of agriculturally important plant diseases.

The most effective *Trichoderma* species from a biocontrol point of view include *T. asperellum, T. atroviride, T. hamatum, T. harzianum, T. viride* and *T. virens*. *Trichoderma* species possess good antagonistic abilities against plant-pathogenic fungi, including *Alternaria, Botrytis, Colletotrichum, Crinipellis, Fusarium, Phoma, Pythium, Rhizoctonia, Sclerotinia* and *Verticillium* species. Mechanisms of antagonism involve the competition for nutrients, production of antifungal compounds, facilitation of seed germination, plant growth promotion and induction of plant defence responses, as well as mycoparasitism by the action of cell wall-degrading enzymes (Benítez *et al.*, 2004). During the process of mycoparasitism, *Trichoderma* species recognize the host hyphae, coil around them and subsequently penetrate the host cell wall by enzymatic degradation that occurs via the secretion of extracellular chitinases, β-1,3-glucanases and proteases. Several commercial biocontrol products based on *Trichoderma* species are available in many countries (Verma *et al.*, 2007).

Although biological control of plant-parasitic nematodes has been central to many biocontrol studies since the 1940s, successful biological nematode control using myconematicides has only been achieved in a few cases. Nematophagous fungi include nematode-trapping fungi that use modified, adhesive mycelia to capture nematodes. One example is *Arthrobotrys oligospora*, which is capable of germinating directly into conidial traps. Other control mechanisms include endoparasitic fungi that attack nematodes by means of adhesive or ingested spores, egg- and cyst-parasitic fungi that can colonize nematode eggs and cysts and fungi that produce nematotoxic compounds (Jansson *et al.*, 1997). During the infection process, penetration of the nematode cuticle or egg shell is believed to be the result of a combination of enzymatic action and mechanical pressure by the invading fungus.

The genus *Verticillium* is considered to be polyphyletic, based on phylogenetic data (Bidochka *et al.*, 1999). The nematophagous species originally placed in *Verticillium* appear to have evolved independently along several different routes, and they are now classified in a number of different genera. They can be grouped ecologically into species that: (i) become attached to the surface of free-living nematodes, which is followed by the penetration of the host (e.g. *Haptocillium balanoides, V. coccosporum*); and (ii) infect nematode cysts and eggs (e.g. *Simplicillium lamellicola, V. leptobactrum, Pochonia chlamydosporia*) (Gams, 1988). Another species previously placed in *Verticillium, Lecanicillium lecanii*, is an entomopathogen that has been used for the biological control of insects.

Paecilomyces species are common environmental soil-inhabiting, imperfect, filamentous fungi. The genus can be divided into two sections: section *Paecilomyces* contains members that are often thermophilic, with teleomorphs in the ascomycetous genera *Talaromyces* and *Thermoascus*, while section *Isarioidea* contains mesophiles, including several well-known nematophagous or entomopathogenic species such as *P. farinosus, P. fumosoroseus, P. amoeneroseus, P. lilacinus, P. javanicus* and *P. tenuipes* (Inglis and Tigano,

2006). Recent phylogenetic studies have shown that *Paecilomyces* can contain cryptic species and it is polyphyletic across two subclasses, the *Sordariomycetidae* and *Eurotiomycetidae* (Luangsa-Ard *et al.*, 2004; Inglis and Tigano, 2006). The previously described thermophilic species belong to the *Eurotiales*, while mesophilic species are mainly related to the *Hypocreales*.

Paecilomyces species can cause various infections in nematodes, insects and vertebrates – including humans. In the latter case they can be associated with keratitis and soft tissue infections in immunocompetent hosts, but may also become the cause of deep infections in immunocompromised patients (Groll and Walsh, 2001). *Paecilomyces lilacinus* is an effective parasite of plant and animal disease-causing nematodes. Under laboratory conditions it can infect all life stages of *Meloidogyne javanica* by direct penetration through the cuticle, including developing eggs, juveniles containing eggs, juveniles and females. *P. lilacinus* can also infect eggs and immature cysts of *Heterodera avenae* and eggs of *Radopholus similis* (Khan *et al.*, 2006). Among *Paecilomyces* species, *P. farinosus* and *P. fumosoroseus* are also found as ubiquitous entomopathogens (Mercadier *et al.*, 2002). These attributes suggest that *Paecilomyces* species may have considerable potential as biocontrol agents for nematodes and insects. Although the virulence among *P. lilacinus* species is variable, they are currently applied in the field worldwide.

The Role of Extracellular Protease Enzymes in Mycoparasitic and Nematophagous Fungi

Proteases can be subdivided into two major groups: exopeptidases that cleave the peptide bond proximal to the amino or carboxy terminal of the substrate, and endopeptidases that cleave bonds distant from the termini (Rao *et al.*, 1998). Proteases can be further classified into four groups depending on the functional group at the active site: serine proteases, aspartyl proteases, cysteine proteases and metalloproteases. Proteolytic enzymes can also be characterised as alkaline, neutral or acidic proteases depending on their pH optima.

The extracellular proteases of mycoparasitic fungi have been shown to have important roles during the biocontrol process. In addition to the well-studied involvement of β-1,3-glucanolytic and chitinolytic enzyme systems (Vazquez-Garciduenas *et al.*, 1998; Kubicek *et al.*, 2001), *Trichoderma* proteases are also involved in the penetration of the target fungi and they are important in the degradation of the protein components of the host cell wall, as well as in the utilization of protein components of the cell content. It has also been suggested that *Trichoderma* proteases have a role in the inactivation of enzymes produced by the pathogen during the plant infection process (Elad and Kapat, 1999). As a result there has been increased interest in recent years in characterizing the proteolytic system of *Trichoderma* species used in biocontrol (Kredics *et al.*, 2005).

The proteinaceous nematode eggshell and the tough, collagen-containing cuticle of parasitic nematodes act as mechanical barriers against pathogenesis by soil microorganisms. Nematophagous fungi must be able to penetrate these barriers for their successful establishment in the host. Electron microscopic studies have shown the involvement of various lytic enzymes in this penetration process, including extracellular proteases that hydrolyze the proteins of the cuticle and eggshells of plant-parasitic nematodes (Jansson *et al.*, 1997).

Mycoparasitic and Nematophagous Fungal Proteases

Several extracellular proteases have been purified and characterized from mycofungicides and myconematicides. The majority are serine proteases; molecular weights, isoelectric points and specific inhibitors of some of these enzymes are summarized in Table 16.1.

Geremia *et al.* (1993) purified PRB1, an alkaline serine protease, from the supernatant of *T. atroviride* grown in the presence of *Botrytis cinerea* mycelia as a sole carbon source (see Table 16.1). The enzyme had a preference for the chymotrypsin substrate *N*-succinyl-Ala-Ala-Pro-Phe-*p*NA, whereas substrates for trypsin and elastase were not hydrolysed. The pH range of PRBl activity was from 7.0 to 11.0, with a maximum between 8.0 and 9.5. A serine protease active towards Z-Ala-Ala-Leu-*p*Na, the substrate of subtilisin-like proteases, was found in *T. harzianum* by Dunaevsky *et al.* (2000) (see Table 16.1). The enzyme had a temperature optimum of 40°C, two distinct activity maxima at pH 7.5 and 10.0 and was stable between pH 6.0 and 11.0. A further serine endopeptidase has been purified from the culture filtrate of *T. viride* (Uchikoba *et al.*, 2001; Table 16.1). Two carboxyl sites at Arg22 and Lys29 of the oxidized insulin B-chain and peptidyl-*p*-nitroanilide substrates with Lys or Arg at the P1 position were hydrolysed by the enzyme, suggesting that its specificity was similar to that of trypsin. A 18.8 kDa alkaline protease has been purified by De Marco and Felix (2002) from a strain of *T. harzianum*, capable of controlling the witches' broom disease of cocoa plants caused by *Crinipellis perniciosa*. Maximal activity of the protease towards a casein substrate was at 37°C and pH 7.0–8.0, and the N-terminal amino acid sequence of the enzyme shared no homology with any other protease.

PRA1, a serine protease responsible for the main proteolytic activity in culture filtrates of a strain of *T. harzianum* grown on fungal cell walls or chitin, has also been purified to homogeneity (Suárez *et al.*, 2004; Table 16.1). PRA1 had a preference for the trypsin-like protease substrate *N*-acetyl-Ile-Glu-Ala-Arg-*p*NA, and showed only weak and no activity towards chymotrypsin-like and elastase-like protease substrates, respectively. The enzyme had an optimum temperature close to 35°C and an optimum pH in the range of 7.0–8.0. Suárez *et al.* (2005) subsequently analysed the extracellular proteins secreted in the presence of different fungal cell walls by this isolate, and identified P6281, a novel pepsin-like aspartic protease (see Table 16.1).

Recently, a trypsin-like serine protease with substrate preference towards *N*-*p*-tosyl-Gly-Pro-Arg-*p*NA and *N*-benzoyl-Phe-Val-Arg-*p*NA was purified from the ferment broth of a cold-tolerant *T. atroviride* strain (Kredics *et al.*, 2008; Table 16.1). The thermodependence and stability of the purified enzyme

Table 16.1. Properties of extracellular proteases purified from mycoparasitic and nematophagous fungi.

Organism	Name of protease	Type of protease	Molecular weight (kDa)	Isoelectric point	Inhibitor(s)	Reference
Mycofungicides						
Trichoderma atroviride	PRBI	Subtilisin-like serine	31	9.2	PMSF	Geremia *et al.* (1993)
T. atroviride	T. atroviride trypsin	Trypsin-like serine	57	7.3	TLCK, leupeptin, benzamidine	Kredics *et al.* (2008)
T. harzianum	T. harzianum subtilisin PRA1	Subtilisin-like serine	73	5.35	NR	Dunaevsky *et al.* (2000)
T. harzianum	P6281	Trypsin-like serine	28	4.8	PMSF	Suárez *et al.* (2004)
T. harzianum	T. viride trypsin	Pepsin-like aspartic	33	4.3	NR	Suárez *et al.* (2005)
T. viride		Trypsin-like serine	25	7.3	DIFP, TLCK, aprotinin, Anti-pain	Uchikoba *et al.* (2001)
Myconematicides						
Arthrobotrys oligospora	PII	Trypsin-like serine	32	4.6	PMSF, *p*CMB, anti-pain, chymostatin	Tunlid *et al.* (1994)
A. oligospora	Aoz1	Serine	38	4.9	PMSF, *Streptomyces* subtilisin inhibitor	Zhao *et al.* (2004)
Pochonia rubescens	P32	Subtilisin-like serine	32	NR	PMSF, *p*CMB, aprotinin, leupeptin	Lopez-Llorca (1990)
P. chlamydosporia	VCP1	Subtilisin-like serine	33	10.2	PMSF, TPCK	Segers *et al.* (1994)
Paecilomyces lilacinus	PIP	Subtilisin-like serine	33	>10.2	PMSF	Bonants *et al.* (1995)
Lecanicillium psalliotae	Ver112	Serine	32	NR	PMSF	Yang *et al.* (2005a)
Monacrosporium microscaphoides	Mlx	Serine	39	6.8	NR	Wang *et al.* (2006)
Hirsutella rhossiliensis	Hnsp	Subtilisin-like serine	32	NR	PMSF	Wang *et al.* (2007)
Clonostachys rosea	PRC	Serine	33	> 10.0	PMSF	Li *et al.* (2006)

PMSF, phenyl-methylsulfonyl fluoride; TLCK, *N*-tosyl-L-lysine chloromethylketone; NR, not reported; DIFP, diisopropyl fluorophosphate; *p*CMB, *p*-chloromercuric benzoic acid; TPCK, tosyl-L-phenylalanine chloromethyl ketone.

exhibited the characteristics described for cold-active enzymes, i.e. high activity in the low temperature range, maximum activity at 15°C and weak thermostability. Psychrophilic enzymes may play an important role in the adaptation of cold-tolerant strains belonging to the otherwise mesophilic genus *Trichoderma*. Moreover, the ability to produce cold-active hydrolases makes cold-tolerant *Trichoderma* strains promising candidates for the biocontrol of fungal plant pathogens at low temperatures.

Tunlid *et al.* (1994) purified PII, an extracellular serine protease from the nematode-trapping fungus *Arthrobotrys oligospora* (see Table 16.1). The purified enzyme had a preference for the substrates *N*-benzoyl-Phe-Val-Arg-*p*NA, casein, BSA and gelatin. Another serine protease, Ac1, with a molecular mass of 35 kDa and optimum activity at pH 7.0 and 53.2°C, has been purified from *A. conoides* (Yang *et al.*, 2007). Ac1 was able to degrade a broad range of substrates including casein, gelatin, BSA, collagen and nematode cuticles. A further serine protease P32, obtained from the egg-parasitic nematophagous fungus *Pochonia rubescens* (syn. *V. suchlasporium*), was found to degrade casein, gelatin and haemoglobin (Lopez-Llorca, 1990; Table 16.1). Another egg-parasitic fungus, *Po. chlamydosporia*, produces a chymoelastase-like serine protease, VCP1, which can degrade *N*-succinyl-Ala-Ala-Pro-Phe-*p*NA, casein, albumin, elastin and gelatin. VCP1 has been identified in several isolates of *Po. chlamydosporia* and *L. lecanii*, pathogens of nematodes and insects, respectively, but has not been reported in any related plant-pathogenic species, suggesting that VCP1 or similar enzymes may play a role in the infection of invertebrates (Segers *et al.*, 1994; Table 16.1). A comparative study of VCP1 and Pr1 from *Po. chlamydosporia* and the entomopathogenic fungus *M. anisopliae* has shown that these two proteins are functionally and serologically closely related (Segers *et al.*, 1995).

Lecanicillium psalliotae, a nematophagous fungus with commercial potential for the biocontrol of root knot and cyst nematodes, was shown to produce an alkaline serine protease, Ver112, during infection of the saprophytic nematode *Panagrellus redivivus* (Yang *et al.*, 2005a; Table 16.1). The protein gave a single 32 kDa band in SDS-PAGE, and the purified protease degraded a broad range of substrates including casein, gelatin and nematode cuticle. The N-terminal amino acid residues of Ver112 shared a high degree of similarity with other cuticle-degrading proteases from nematophagous fungi, which suggests a role in nematode infection.

Liu *et al.* (2007) found that the 3-D structural models of the three cuticle-degrading proteases, Pr1, VCP1 and Ver112, were very similar to each other and to the template structure of proteinase K. This observation suggests that these three cuticle-degrading enzymes belong to the proteinase K-like serine protease family.

Culturing *Pae. lilacinus* under inducing conditions in a special medium containing vitellin, chitin and chicken egg yolk allowed Bonants *et al.* (1995) to isolate the serine protease PIP, which had optimum proteolytic activity at pH 10.3 (see Table 16.1). Khan *et al.* (2003) also purified a similar 37 kDa serine protease from *Pae. lilacinus*, and its activity was drastically reduced at temperatures > 60°C and pH >12.0.

Mlx, a neutral serine protease, has been purified from the culture filtrate of the nematophagous fungus *Monacrosporium microscaphoides* (Wang *et al.*, 2006; Table 16.1). The enzyme had optimum activity at pH 9.0 and 65°C and a broad substrate specificity, hydrolysing casein, skimmed milk, collagen and BSA. A neutral serine protease, Hnsp, with optima at pH 7.0 and 40°C and affinity towards the substrate *N*-succinyl-Ala-Ala-Pro-Phe-*p*-nitroanilide, was purified from *Hirsutella rhossiliensis*, an endoparasite on vermiform nematodes (Wang *et al.* 2007; Table 16.1). PrC, a further extracellular serine protease, has been purified and characterized from *Clonostachys rosea* (syn. *Gliocladium roseum*), a soil saprophyte with fungicidal and nematicidal potential (Li *et al.*, 2006; Table 16.1). PrC showed optimum activity at pH 9.0–10.0 and 60°C, and could degrade casein, gelatin and nematode cuticle.

Cloning Extracellular Protease Genes from Mycoparasitic and Nematophagous Fungi

Since the beginning of the 1990s, several genes encoding extracellular protease enzymes have been cloned from fungal biocontrol agents. Currently, expressed sequence tags (ESTs: mRNA populations transcribed under specific conditions and cloned as cDNAs) are proving to be very useful for large-scale identification of active fungal proteases. Table 16.2 provides an overview of the extracellular protease-encoding genes currently known from mycofungicides and myconematicides.

The single copy gene that encodes the subtilisin-like serine protease PRB1 of *T. atroviride* (*prb1*) has been identified by Geremia *et al.* (1993). Comparison of the cDNA and genomic sequences showed that the gene contained two introns, one of 58 bp and another of 74 bp. The deduced amino acid sequence of 409 amino acids (aa) had a calculated molecular mass of 42.3 kDa and a calculated pI value of 6.44. These were different from the values determined for the purified protein, suggesting that the protease is synthesized as a pre-pro-enzyme. The peptide from Met-121 to Ala-101 had characteristics common to known eukaryotic signal peptides. The predicted pI and molecular weight of the mature protein were in agreement with those obtained from the purified PRB1. The *prb1* gene has also been cloned from *T. hamatum*, and the sequence of its promoter region has been analysed in detail (Steyaert *et al.*, 2004).

Tvsp1, the homologue of *prb1* in *T. virens*, has also been isolated (Pozo *et al.*, 2004; Table 10.2). The full *tvsp1* sequence encodes a 409 aa polypeptide with an estimated molecular weight of 42 kDa and an isoelectric point of 6.6. Two potential *N*-glycosylation and two *O*-glycosylation sites were present in the predicted polypeptide sequence, and multiple phosphorylation sites were also suggested. Analysis of the protein sequence suggests that this protein may also be directed to the endoplasmic reticulum by a signal peptide of 20 aa. The protein had typical subtilisin-type serine protease features at its active site (a catalytic triad formed by the functional residues

Table 16.2. Characteristics of extracellular protease-encoding genes from mycoparasitic and nematophagous fungi.

Organism	Type of protease	Gene	Length of ORF (bp)	Length of deduced amino acid sequence (aa)	Length of putative signal peptide (aa)	Predicted properties of the protein: Length (aa)	Calculated Mw (kDa)	Calculated pI	Reference
Mycofungicides									
Trichoderma asperellum	Aspartic	*papA*	1297	405	20	354	NR	5.45	Viterbo *et al.* (2004)
T. harzianum	Chymotrypsin-like serine	*prb1*	1227	409	20	289	31.0[a]	9.20[a]	Geremia *et al.* (1993) Suárez *et al.* (2004)
T. virens	Trypsin-like serine	*pra1*	777	258	20	229	25.0	4.91	Suárez *et al.* (2007)
	Serine	*p7480* *p5431* *p7129* *p8048* *p1026* *p5216*	765–2637	254–878	15–24	NR	25.5–92.5[b]	5.1–6.9[b]	
	Aspartic	*papA*	1212	404	20	352	36.7	4.35	Delgado-Jarana *et al.* (2002)
	Aspartic	*p1324* *p7959* *p6281*	1119–1398	372–465	16–21	NR	39.6–49.5[b]	4.1–5.5[b]	Suárez *et al.* (2007)
	Metallo	*p7455*	1044	347	16	NR	36.7[b]	6.0[b]	Suárez *et al.* (2007)
	Aminopeptidase	*p2920*	1197	398	20	NR	43.9[b]	4.8[b]	Suárez *et al.* (2007)
	Subtilisin-like serine	*tvsp1*	1368	409	20	289	29.0	8.98	Pozo *et al.* (2004)
Myconematicides									
Arthrobotrys oligospora	Serine	*PII*	1287	408	16	288	28.5	4.6[a]	Åhman *et al.* 1996)
A. oligospora	Serine	*aoz1*	1281	426	NR	303	38.4	4.9[a]	Zhao *et al.* (2004)
A. conoides	Serine	*Ac1*	1296	411	21	288	29.7	NR	Yang *et al.* (2007)
Lecanicillium psalliotae	Serine	*ver112*	1440	382	15	280	32.0[a]	NR	Yang *et al.* (2005b)
Pochonia chlamydosporia	Subtilisin-like serine	*VCP1*	1347	388	19	281	33.0[a]	10.2[a]	Morton *et al.* (2003)

aa, amino acids; NR, not reported.
[a]Values of the purified enzyme; [b]calculated from the entire length of the protein.

Asp, His and Ser at positions 161, 192 and 254, respectively). A further serine protease gene, *pra1* from *T. harzianum*, that encoded for the enzyme PRA1 was cloned by Suárez *et al.* (2004; Table 10.2). Three possible *O*-glycosylation sites were identified in the putative mature protein.

The gene for an aspartyl protease (*papA*) in *T. harzianum*, was isolated and characterized by Delgado-Jarana *et al.* (2002; Table 10.2). The deduced amino acid sequence contained no potential post-translational modification signals, in contrast to that of a homologous gene isolated from *T. asperellum* (Viterbo *et al.*, 2004; Table 10.2), where three potential sites for *N*-glycosylation were found. Suárez *et al.* (2007; Table 10.2) cloned and sequenced the gene encoding for the pepsin-like aspartic protease P6281, which has been implicated in mycoparasitism by *T. harzianum*. The amino acid sequence data for P6281 and its biochemical properties clearly differentiated this aspartic protease from the previously described PAPA. An incomplete sequence with similarity to metallo-endopeptidases has also been found to be expressed during the mycoparasitic interactions of *T. hamatum* (Carpenter *et al.*, 2005).

Recently, an expressed sequence tag (EST) approach has been applied to identify genes encoding extracellular peptidases in *T. harzianum* grown under several biocontrol-related conditions (Suárez *et al.*, 2007). The newly identified extracellular peptidases included six serine endopeptidases (four subtilisin-like, one trypsin-like and one aorsin proteases), two aspartic endopeptidases, one metallo-endopeptidase and one aminopeptidase.

Åhman *et al.* (1996) analysed the sequence of a cDNA clone containing the gene encoding the cuticle-degrading serine protease PII from the nematode-trapping fungus *A. oligospora* (see Table 10.2). Southern blot analysis of the genomic DNA of the fungus suggested the presence of several further subtilisin-like serine proteases. Analysis of the cDNA and corresponding genomic sequence of *Aoz1* from *A. oligospora* gave 97% identity with *PII*, suggesting that *Aoz1* is probably an ortholog of *PII* (Zhao *et al.*, 2004). The deduced amino acid sequence of the gene encoding the Mlx protease of *M. microscaphoides* contained the conserved catalytic triad of aspartic acid–histidine–serine and showed high similarity with both *PII* and *Aoz1* of *A. oligospora* (Wang *et al.*, 2006). Characterization of the gene encoding the Ac1 protease of *A. conoides* showed it to contain one intron of 60 bp, and the deduced polypeptide sequence also showed a high degree of similarity to PII and Mlx (Yang *et al.*, 2007; Table 10.2).

The gene encoding VCP1 of *Po. chlamydosporia* has been sequenced from six strains isolated from different nematode hosts (Morton *et al.*, 2003; Table 10.2). The gene proved to be similar to *pr1* of *M. anisopliae*. A polymorphism has been found among the isolates at amino acid positions 65 and 99, indicating a difference between isolates from cyst and root nematodes. The authors suggested that such host-related genetic variations might contribute to host preference.

Yang *et al.* (2005b) isolated and characterized *Ver112*, a cuticle-degrading serine protease gene from three isolates of the nematophagous fungus *L. psalliotae* (see Table 10.2). Comparison of translated cDNA sequences of the three isolates showed one amino acid polymorphism at position 230. The full sequence comprised an ORF, which contained three introns and four exons and encoded a 39.6 kDa polypeptide. Comparison of Ver112 with other cuticle-degrading serine proteases from nematophagous fungi showed it to be typical of fungal serine proteases, with a pre-pro-peptide structure. The deduced amino acid sequence of Ver112 showed 58.2% identity with VCP1 from *Po. chlamydosporia*.

The cDNA clone of the serine protease PIP from *Pae. lilacinus* has also been isolated and sequenced; Southern hybridization indicated that the gene was present in the genome as a single copy (Bonants *et al.*, 1995).

Protease Regulation of Mycofungicides and Myconematicides

The basic protease PRB1 of *T. atroviride* is strongly induced when the fungus is grown in either the presence of mycelia from different phytopathogens, cell walls of *Rhizoctonia solani* or chitin (Geremia *et al.*, 1993), but is not detected when *Trichoderma* is grown in the presence of a combination of an inducer and glucose. The promoter sequence of the *T. atroviride* gene *prb1* possesses potential AreA and CreA sites for nitrogen and carbon regulation, respectively, and also four putative mycoparasitic response elements (MYREs) (Cortes *et al.*, 1998). These MYREs have also been found in the promoter of the *T. hamatum* gene *prb1* (Steyaert *et al.*, 2004). Olmedo-Monfil *et al.* (2002) suggested that either one or both of MYRE boxes 1 and 2 might be required for the induction of *prb1* during mycoparasitism. The gene was shown to be active when *T. atroviride* was grown in media containing chitin or *R. solani* cell walls (Geremia *et al.*, 1993).

High-level expression was detected in dual cultures even if contact with *R. solani* was prevented by cellophane membranes, demonstrating that the induction of *prb1* is contact-independent (Cortes *et al.*, 1998). This study showed that a diffusible heat- and protease-resistant molecule produced by the host is the signal that triggers the expression of the gene. The *prb1* gene was found to be repressed by glucose (Geremia *et al.*, 1993) and subject to nitrogen catabolite repression (Olmedo-Monfil *et al.*, 2002). It has also been shown that induction of transcription of the gene by both *R. solani* cell walls and osmotic stress requires a release from the repressed condition, which is determined by nitrogen availability, and the response of *prb1* to nutrient limitation depends on the activation of conserved mitogen-activated protein kinase (MAPK) pathways (Olmedo-Monfil *et al.*, 2002).

The promoter of the *T. virens tvsp1* gene also possesses potential AreA and CreA sites, as well as a PacC site for pH regulation and four MYREs almost identical to those described in *prb1* (Pozo *et al.*, 2004). Expression of *tvsp1* could be induced by cell walls of plant-pathogenic fungi in liquid cultures. Northern blot analysis showed that no transcript was present in mycelia grown with either glucose or sucrose as a carbon source. In contrast to *prb1*, a mitogen-activated protein kinase was found to be a negative element in the expression of *tvsp1* under nitrogen limitation or simulated mycoparasitism (Mendoza-Mendoza *et al.*, 2003). Expression of *T. harzianum pra1* is also induced by fungal cell walls subjected to nitrogen and carbon regulation and affected by pH in the culture media (Suárez *et al.*, 2004).

The promoter sequence of *T. harzianum papA* contains potential AreA and PacC sites, but no potential CreA sites (Delgado-Jarana *et al.*, 2002). Expression of the *papA* gene is pH regulated, repressed by ammonium, glucose and glycerol and induced by organic nitrogen sources. Dual plate confrontation assays with *R. solani* gave a fourfold mRNA induction of *T. asperellum papA* in the presence of the pathogen before any physical contact, suggesting that aspartyl proteases may also be involved in mycoparasitism. *T. asperellum papA* was also induced in response to attachment to plant roots, which suggests a possible role in plant colonization by *Trichoderma* strains as opportunistic plant symbionts. Studies on the regulation of the pepsin-like aspartic protease gene *p6281* of *T. harzianum* showed that the expression profile obtained for this gene was comparable to that obtained for genes encoding proteins related to antagonistic activities in *Trichoderma* spp., including subtilisin-like protease PRB1 and trypsin-like protease PRA1 previously characterized in *T. atroviride* and *T. harzianum*, respectively (Suárez *et al.*, 2005).

The transcription of *p6281* was found to be only weakly triggered by carbon or nitrogen starvation, and this may suggest that P6281 does not function as part of a general response to nutrient deprivation, and a more specialized role for this protein during the early stages of *Trichoderma* mycoparasitic activity was hypothesized. Analysis of the recently identified novel extracellular peptidases of *T. harzianum* has shown that the genes within a family are differentially regulated in response to different culture conditions (Suárez *et al.*, 2007). Some genes encoding for peptidases of the same families, such as subtilisin-like, pepsin-like and trypsin-like peptidases, showed essentially different expression patterns, which might suggest that they have diverse functional roles.

The information available on the regulation of extracellular proteases produced by nematophagous fungi is limited. The regulation of the *pII* gene of *A. oligospora* has been studied by Åhman *et al.* (1996). Northern blot analysis showed that this gene is expressed under nitrogen and carbon limitation, while the expression of the gene could be enhanced by the addition of different soluble and insoluble proteins as well as nematode cuticle fragments. Conversely, ammonia, nitrate, amino acids and glucose all repressed *pII*.

Extracellular Proteases and Biocontrol

Several research groups have studied the possible roles played by extracellular proteases in biocontrol by fungi, and it has been shown that mechanisms involved in the biological control of both plant-pathogenic fungi and nematodes includes the action of extracellular enzymes. In the new millennium, increased attention has been focused on the specific role of proteases as virulence factors of biocontrol fungi.

Transformation systems have been developed to increase the copy number of *T. atroviride prb1* (Flores *et al.*, 1997; Goldman and Goldman, 1998). Transformants exhibited increased control of *R. solani*, suggesting that *prb1* is a mycoparasitism-related gene (Flores *et al.*, 1997). Overexpression of *tvsp1* in *T. virens* also resulted in increased biocontrol activity against *R. solani* (Pozo *et al.*, 2004). Results of these studies suggest that the overexpression of protease-encoding genes is a powerful tool for strain improvement. However, Flores *et al.* (1997) reported that transformants with extremely high protease levels were not the best biocontrol agents. Under conditions of extremely elevated protease production, partial or full proteolysis of other proteins important in the mycoparasitic process may occur, and so the isolation of mutants with only a moderate increase in extracellular enzyme concentrations has been suggested as a preferable tool for improving the biocontrol capabilities of mycoparasitic *Trichoderma* strains (Rey *et al.*, 2001).

Zaldivar *et al.* (2001) isolated a mutant by *N*-methyl-*N*-nitro-*N*-nitrosoguanidine treatment from *T. aureoviride*. The mutant had enhanced production of lytic enzymes, including proteases and a more complex native PAGE banding pattern of protease activities, than that of the wild-type strain. UV-mutagenesis with selection for *p*-fluorophenyl-alanine resistance or colony morphology mutants was used by Szekeres *et al.* (2004) to isolate protease-overproducing strains from *T. harzianum*. Certain mutants were better producers of extracellular trypsin- and chymotrypsin-like proteases exhibiting considerably higher levels of activities than the wild-type strain. The increase in the proteolytic activities of these mutants was low when compared with transformants overexpressing the protease gene *prb1* (Flores *et al.*, 1997), but they proved to be much better antagonists of plant pathogens than the parental strain. Certain extracellular trypsin-like protease izoenzymes from *Trichoderma* are produced in greater amounts in the presence of copper, suggesting that the antagonistic abilities of some strains could be enhanced by adding sublethal amounts of $CuSO_4$ (Kredics *et al.*, 2004). Consequently, an appropriate level of crop protection could be ensured by the application of reduced amounts of copper-containing fungicides, in combination with *Trichoderma* biocontrol strains, within an integrated pest management framework.

Trichoderma proteases may also have an indirect role in the control of plant-pathogenic fungi, as a protein similar to serine peptidases secreted by *T. virens* can elicit a plant defence response against fungal plant pathogens (Hanson and Howell, 2004).

The importance of proteases during the infection of nematodes by the nematode-trapping fungus *A. oligospora* has been demonstrated by treating the fungus with various protease inhibitors (Tunlid and Jansson, 1991). A significant decrease in the immobilization of captured nematodes was seen when trap-bearing mycelia of the fungus were incubated with the serine protease inhibitor PMSF. In a subsequent study with purified PII protease, it was suggested that the protease activity was involved in the immobilization of nematodes (Tunlid *et al.*, 1994). Bedelu *et al.* (1998) studied the relation of protease production to nematode-degrading ability in two *Arthrobotrys* spp.

The difference in the nematode-destroying capability between the two strains was suggested to be solely the result of the difference in the amount of extracellular proteases: the more proteases that were produced, the better the efficiency of the fungus in killing nematodes. They concluded that higher protease-producing strains of nematode-trapping fungi had more potential for biological control. Åhman *et al.* (2002) found that the transcript of *A. oligospora PII* cannot be detected during the early stages of infection (adhesion and penetration), but high levels were expressed concurrent with the killing and colonization of nematodes. Strains containing additional copies of the *PII* protease gene developed more infection structures and were faster at capturing and killing nematodes compared with the wild type, demonstrating the role of this protease as an important pathogenicity factor. The Ac1 protease of *A. conoides* was found to immobilize both the free-living nematode *Panagrellus redivivus* and the pine wood nematode *Bursaphelenchus xylophilus*, indicating that this protease may be involved in the infection process (Yang *et al.*, 2007).

The protease VCP1 from *Po. chlamydosporia* hydrolyses the outer vitellin layer of the eggshell of the host nematode *M. incognita* (Segers *et al.*, 1994). Extracellular proteases have also been shown to be the major enzyme activities involved in the infection process and pathogenesis of various life stages of plant parasitic nematodes by *Po. chlamydosporia* (Segers *et al.*, 1996). Lopez-Llorca and Robertson (1992) used immunocytochemistry to determine the localization of the *Po. rubescens* P32 protease during the infection of nematode eggs. Increased fluorescence was seen in fungal appressoria on eggshells of the plant-parasitic nematode *Heterodera schachtii*, indicating the presence of P32, and suggesting that it had a role in pathogenicity toward nematode eggs. Lopez-Llorca *et al.* (2002) also found a reduction in the infection of *M. javanica* eggs by *Po. rubescens*, *Po. chlamydosporia* and *L. lecanii* on addition of PMSF or other serine protease inhibitors.

The effect of *Pae. lilacinus* PIP protease on the development and hatching of *M. hapla* eggs has been investigated, and it was found that adding the purified enzyme to suspended nematode eggs resulted in prematured hatching (Bonants *et al.*, 1995). The newly hatched larvae were not visibly affected, suggesting that this protease might be involved in the fungal penetration of the nematode eggshell. The Mlx protease of *M. microscaphoides* was found to immobilize the nematode *Panagrellus redivivus in vitro* and to degrade purified cuticle, suggesting that Mlx could serve as a virulence factor during infection (Wang *et al.*, 2006). Similar results have been obtained for crude Hnsp protease from *H. rhossiliensis*, which was able to kill nematodes and degrade their cuticle *in vitro* (Wang *et al.*, 2007).

A *T. harzianum* transformant containing multiple copies of the protease-encoding gene *prb1* showed improved biocontrol activity against *M. javanica*, indicating that, in addition to its role against plant-pathogenic fungi, this protease may also be important for the control of nematodes by *Trichoderma* (Sharon *et al.*, 2001).

Conclusions and Future Perspectives

A common feature of mycofungicides and myconematicides is the involvement of extracellular hydrolases – including proteolytic enzymes – in the biocontrol process. Identification of the proteases involved in pathogenicity in different mycofungicides and myconematicides helps to provide information on host preference, substrate selection and pathogenicity-related domains at the protein level. Such information will be important for improving the efficacy of fungal biocontrol agents. Several extracellular proteases have been purified and characterized from mycofungicides and myconematicides and, for some of them, the complete sequences of the associated genes are already available. The majority of the enzymes so far described have been classified in the serine protease family; however, the expression of other enzyme types – including aspartic proteases, metalloproteases and aminopeptidases – under biocontrol-related conditions was also demonstrated. Detailed knowledge of the protease genes and enzymes involved in the biocontrol process of mycofungicides and myconematicides will assist further strain improvement by potentially allowing the overexpression of genes encoding for effective proteases.

An increasing number of fungi are being used in genome sequencing programmes, and genomic approaches such as ESTs have proved very useful for the large-scale identification of fungal genes. It can therefore be expected that more data on extracellular proteases will become available in the near future, and that this will provide more insight into the role of host protein degradation by the antagonists in various biocontrol processes.

Acknowledgement

LK is a grantee of the János Bolyai Research Scholarship (Hungarian Academy of Sciences).

References

Åhman, J., Ek, B., Rask, L. and Tunlid, A. (1996) Sequence analysis and regulation of a gene-encoding cuticle-degrading serine protease from the nematophagous fungus *Arthrobotrys oligospora*. *Microbiology* 142, 1605–1616.

Åhman, J., Johansson, T., Olsson, M., Punt, P.J., van den Hondel, C.A.M.J.J. and Tunlid A. (2002) Improving the pathogenicity of a nematode-trapping fungus by genetic engineering of a subtilisin with nematotoxic activity. *Applied and Environmental Microbiology* 68, 3408–3415.

Bedelu, T., Gessesse, A. and Abate, D. (1998) Relation of protease production to nematode degrading ability of two *Arthrobotrys* spp. *World Journal of Microbiology and Biotechnology* 14, 731–734.

Benítez, T., Rincón, A.M., Limón, M.C. and Codón, A.C. (2004) Biocontrol mechanisms of *Trichoderma* strains. *International Microbiology* 7, 249–260.

Bidochka, M.J., St Leger, R.J., Stuart, A. and Gowanlock, K. (1999) Nuclear rDNA phylogeny in the fungal genus *Verticillium* and its relationship to insect and plant virulence, extracellular proteases and carbohydrases. *Microbiology* 145, 955–963.

Bonants, P.J.M., Fitters, P.F.L., Thijs, H., den Belder, E., Waalwijk, C. and Henfling, J.W. (1995) A basic serine protease from *Paecilomyces lilacinus* with biological activity against *Meloidogyne hapla* eggs. *Microbiology* 141, 775–784.

Carpenter, M.A., Stewart, A. and Ridgway, H.J. (2005) Identification of novel *Trichoderma hamatum* genes expressed during mycoparasitism using subtractive hybridisation. *FEMS Microbiology Letters* 251, 105–112.

Chet, I. and Inbar, J. (1997) Fungi. In: Anke, T. (ed.) *Fungal Biotechnology*. Chapman & Hall, Weinheim, Germany, pp. 65–80.

Cortes, C., Gutierrez, A., Olmedo, V., Inbar, J., Chet, I. and Herrera-Estrella, A. (1998) The expression of genes involved in parasitism by *Trichoderma harzianum* is triggered by a diffusible factor. *Molecular and General Genetics* 260, 218–225.

Delgado-Jarana, J., Rincón, A.M. and Benítez, T. (2002) Aspartyl protease from *Trichoderma harzianum* CECT 2413: cloning and characterization. *Microbiology* 148, 1305–1315.

De Marco, J.L. and Felix, C.R. (2002) Characterization of a protease produced by a *Trichoderma harzianum* isolate which controls cocoa plant witches' broom disease. *BMC Biochemistry* 3 (http://www.biomedcentral.com/1471-2091/3/3).

Druzhinina, I.S., Kopchinskiy, A.G. and Kubicek, C.P. (2006) The first 100 *Trichoderma* species characterized by molecular data. *Mycoscience* 47, 55–64.

Dunaevsky, Y.E., Gruban, T.N., Beliakova, G.A. and Belozersky, M.A. (2000) Enzymes secreted by filamentous fungi: regulation of secretion and purification of an extracellular protease of *Trichoderma harzianum*. *Biochemistry (Moscow)* 65, 723–727.

Elad, Y. and Kapat, A. (1999) The role of *Trichoderma harzianum* protease in the biocontrol of *Botrytis cinerea*. *European Journal of Plant Pathology* 105, 177–189.

Flores, A., Chet, I. and Herrera-Estrella, A. (1997) Improved biocontrol activity of *Trichoderma harzianum* by over-expression of the proteinase-encoding gene *prb*1. *Current Genetics* 31, 30–37.

Gams, W. (1988) A contribution to the knowledge of nematophagous species of *Verticillium*. *Netherlands Journal of Plant Pathology* 94, 123–148.

Geremia, R., Goldman, G.H., Jacobs, D., Ardiles, W., Vila, S.B., Van-Montagu, M. and Herrera-Estrella, A. (1993) Molecular characterization of the proteinase-encoding gene *prb*1, related to mycoparasitism by *Trichoderma harzianum*. *Molecular Microbiology* 8, 603–613.

Goldman, M.H.S. and Goldman, G.H. (1998) *Trichoderma harzianum* transformant has high extracellular alkaline proteinase expression during specific mycoparasitic interactions. *Genetics and Molecular Biology* 21, 329–333.

Groll, A.H. and Walsh, T.J. (2001) Uncommon opportunistic fungi: new nosocomial threats. *Clinical Microbiology and Infectious Diseases* 7, 8–24.

Hanson, L.E. and Howell, C.R. (2004) Elicitors of plant defense responses from biocontrol strains of *Trichoderma virens*. *Phytopathology* 94, 171–176.

Inglis, P.W. and Tigano, M.S. (2006) Identification and taxonomy of some entomopathogenic *Paecilomyces spp.* (Ascomycota) isolates using rDNA–ITS Sequences. *Genetics and Molecular Biology* 29, 132–136.

Jansson, H.-B., Tunlid, A. and Nordbring-Hertz, B. (1997) Nematodes. In: Anke, T. (ed.) *Fungal Biotechnology*. Chapman & Hall, Weinheim, Germany, pp. 38–50.

Khan, A., Williams, K., Molloy, M.P. and Nevalainen, H. (2003) Purification and characterization of a serine protease and chitinases from *Paecilomyces lilacinus* and detection of chitinase activity on 2D gels. *Protein Expression and Purification* 32, 210–220.

Khan, A., Williams, K. and Nevalainen, H. (2006) Infection of plant-parasitic nematodes by *Paecilomyces lilacinus* and *Monacrosporium lysipagum*. *Biocontrol* 51, 659–678.

Komoń-Zelazowska, M. Bissett, J., Zafari, D., Hatvani, L., Manczinger, L., Woo, S., Lorito, M., Kredics, L., Kubicek, C.P. and Druzhinina, I.S. (2007) Genetically closely related but phenotypically divergent *Trichoderma* species cause world-wide green mould disease in oyster mushroom farms. *Applied and Environmental Microbiology* 73, 7415–7426.

Kredics, L., Antal, Z., Dóczi, I., Manczinger, L., Kevei, F. and Nagy, E. (2003) Clinical importance of the genus *Trichoderma. Acta Microbiologica et Immunologica Hungarica* 50, 105–117.

Kredics, L., Hatvani, L., Antal, Z., Szekeres, A., Manczinger, L. and Nagy, E. (2004) Protease overproduction in the presence of copper by a *Trichoderma harzianum* strain with biocontrol potential. *IOBC–WPRS Bulletin* 27(8), 371–374.

Kredics, L., Antal, Z., Szekeres, A., Hatvani, L., Manczinger, L., Vágvölgyi, C. and Nagy, E. (2005) Extracellular proteases of *Trichoderma* species – a review. *Acta Microbiologica et Immunologica Hungarica* 52, 169–184.

Kredics, L., Terecskei, K., Antal, Z., Szekeres, A., Hatvani, L., Manczinger, L. and Vágvölgyi, C. (2008) Purification and preliminary characterization of a cold-adapted extracellular proteinase from *Trichoderma atroviride. Acta Biologica Hungarica* 59, 259–268.

Kubicek, C.P., Eveleigh, D.E., Esterbauer, H., Steiner, W. and Kubicek-Pranz, E.M. (eds) (1990) *Trichoderma reesei Cellulases: Biodiversity, Physiology, Genetics and Applications.* Royal Society of Chemistry Press, Cambridge, UK, 114 pp.

Kubicek, C.P., Mach, R.L., Peterbauer, C.K. and Lorito, M. (2001) *Trichoderma*: from genes to biocontrol. *Journal of Plant Pathology* 83(S2), 11–23.

Li, J., Yang, J, Huang, X. and Zhang, K.-Q. (2006) Purification and characterization of an extracellular serine protease from *Clonostachys rosea* and its potential as a pathogenic factor. *Process Biochemistry* 41, 925–929.

Liu, S.Q., Meng, Z.H., Yang, J.K., Fu, Y.X. and Zhang, K.Q. (2007) Characterizing structural features of cuticle-degrading proteases from fungi by molecular modeling. *BioMed Central Structural Biology* 7, 33.

Lopez-Llorca, L.V. (1990) Purification and properites of extracellular proteases produced by the nematophagous fungus *Verticillium suchlasporium. Canadian Journal of Microbiology* 36, 530–537.

Lopez-Llorca, L.V. and Robertson, W.M. (1992) Immunocytochemical localization of a 32-kDa protease from the nematophagous fungus *Verticillium suchlasporium* in infected nematode eggs. *Experimental Mycology* 16, 261–267.

Lopez-Llorca, L.V., Olivares-Bernabeu, C., Jansson, H.B. and Kolattukudy, P.E. (2002) Pre-penetration events in fungal parasitism of nematode eggs. *Mycological Research* 106, 499–506.

Luangsa-Ard, J.J., Hywel-Jones, N.L. and Samson, R.A. (2004) The polyphyletic nature of *Paecilomyces sensu lato* based on 18S-generated rDNA phylogeny. *Mycologia* 96, 773–780.

Mendoza-Mendoza, A., Pozo, M., Grzegorski, D., Martinez, P., Garcia, J.M., Olmedo-Monfil, V., Cortes, C., Kenerley, C. and Herrera-Estrella, A. (2003) Enhanced biocontrol activity of *Trichoderma* through inactivation of a mitogen-activated protein kinase. *Proceedings of the National Academy of Sciences of the USA* 100, 15965–15970.

Mercadier, G., Saethre, M.-G. and Vega, F.E. (2002) First report of the fungal entomopathogen *Paecilomyces* in Norway. *Norwegian Journal of Entomology* 49, 71–73.

Morton, C.O., Hirsch, P.R., Peberdy, J.P. and Kerry, B.R. (2003) Cloning of and genetic variation in protease VCP1 from the nematophagous fungus *Pochonia chlamydosporia. Mycological Research* 107, 38–46.

Olmedo-Monfil, V., Mendoza-Mendoza, A., Gomez, I., Cortes, C. and Herrera-Estrella, A. (2002) Multiple environmental signals determine the transcriptional activation of the mycoparasitism-related gene *prb*1 in *Trichoderma atroviride. Molecular Genetics and Genomics* 267, 703–712.

Pozo, M.J., Baek, J.M., Garcia, J.M. and Kenerley, C.M. (2004) Functional analysis of *tvsp1*, a serine protease-encoding gene in the biocontrol agent *Trichoderma virens. Fungal Genetics and Biology* 41, 336–348.

Rao, M.B., Tanksale, A.M., Ghatge, M.S. and Deshpande, V.V. (1998) Molecular and biotechnological aspects of microbial proteases. *Microbiology and Molecular Biology Reviews* 62, 597–635.

Rey, M., Delgado-Jarana, J. and Benítez, T. (2001) Improved antifungal activity of a mutant of *Trichoderma harzianum* CECT 2413 which produces more extracellular proteins. *Applied Microbiology and Biotechnology* 55, 604–608.

Segers, R., Butt, T.M., Kerry, B.R. and Peberdy, J.F. (1994) The nematophagous fungus *Verticillium chlamydosporium* produces a chymoelastase-like protease which hydrolyses host nematode proteins *in situ*. *Microbiology* 140, 2715–2723.

Segers, R., Butt, T.M., Keen, J.N., Kerry, B.R. and Peberdy, J.F. (1995) The subtilisins of the invertebrate mycopathogens *Verticillium chlamydosporium* and *Metarhizium anisopliae* are serologically and functionally related. *FEMS Microbiology Letters* 126, 227–231.

Segers, R., Butt, T.M., Kerry, B.R., Beckett, A. and Peberdy, J.F. (1996) The role of the proteinase VCP1 produced by the nematophagous *Verticillium chlamydosporium* in the infection process of nematode eggs. *Mycological Research* 100, 421–428.

Sharon, E., Bar-Eyal, M., Chet, I., Herrera-Estrella, A., Kleifeld, O. and Spiegel, Y. (2001) Biological control of the root-knot nematode *Meloidogyne javanica* by *Trichoderma harzianum*. *Phytopathology* 91, 687–693.

Steyaert, J.M., Stewart, A., Jaspers, M.V., Carpenter, M. and Ridgway, H.J. (2004) Co-expression of two genes, a chitinase (*chit42*) and proteinase (*prb1*), implicated in mycoparasitism by *Trichoderma hamatum. Mycologia* 96, 1245–1252.

Suárez, B., Rey, M., Castillo, P., Monte, E. and Llobell, A. (2004) Isolation and characterization of PRA1, a trypsin-like protease from the biocontrol agent *Trichoderma harzianum* CECT 2413 displaying nematicidal activity. *Applied Microbiology and Biotechnology* 65, 46–55.

Suárez, M.B., Sanz, L., Chamorro, M.I., Rey, M., González, F.J., Llobell, A. and Monte, E. (2005) Proteomic analysis of secreted proteins from *Trichoderma harzianum.* Identification of a fungal cell wall-induced aspartic protease. *Fungal Genetics and Biology* 42, 924–934.

Suárez, M.B., Vizcaíno, J.A., Llobell, A. and Monte, E. (2007) Characterization of genes encoding novel peptidases in the biocontrol fungus *Trichoderma harzianum* CECT 2413 using the TrichoEST functional genomics approach. *Current Genetics* 51, 331–342.

Szekeres, A., Kredics, L., Antal, Z., Kevei, F. and Manczinger, L. (2004) Isolation and characterization of protease overproducing mutants of *Trichoderma harzianum. FEMS Microbiology Letters* 233, 215–222.

Tunlid, A. and Jansson, S. (1991) Proteases and their involvement in the infection and immobilization of nematodes by the nematophagous fungus *Arthrobotrys oligospora. Applied and Environmental Microbiology* 57, 2868–2872.

Tunlid, A., Rosén, S., Ek, B. and Rask, L. (1994) Purification and characterization of an extracellular serine protease from the nematode-trapping fungus *Arthrobotrys oligospora. Microbiology* 140, 1687–1695.

Uchikoba, T., Mase, T., Arima, K., Yonezawa, H. and Kaneda, M. (2001) Isolation and characterization of a trypsin-like protease from *Trichoderma viride. Biological Chemistry* 382, 1509–1513.

Vazquez-Garciduenas, S., Leal-Morales, C.A. and Herrera-Estrella, A. (1998) Analysis of the β-1,3-glucanolytic system of the biocontrol agent *Trichoderma harzianum*. *Applied and Environmental Microbiology* 64, 1442–1446.

Verma, M., Brar, S.K., Tyagi, R.D., Surampalli, R.Y. and Valéro, J.R. (2007) Antagonistic fungi, *Trichoderma* spp.: panoply of biological control. *Biochemical Engineering Journal* 37, 1–20.

Viterbo, A., Harel, M. and Chet, I. (2004) Isolation of two aspartyl proteases from *Trichoderma asperellum* expressed during colonization of cucumber roots. *FEMS Microbiology Letters* 238, 151–158.

Wang, B., Wu, W. and Liu, X. (2007) Purification and characterization of a neutral serine protease with nematicidal activity from *Hirsutella rhossiliensis*. *Mycopathologia* 163, 169–176.

Wang, M., Yang, J. and Zhang, K. (2006) Characterization of an extracellular protease and its cDNA from the nematode-trapping fungus *Monacrosporium microscaphoides*. *Canadian Journal of Microbiology* 52, 130–139.

Yang, J., Huang, X., Tian, B., Wang, M., Niu, Q.H. and Zhang, K.Q. (2005a) Isolation and characterization of a serine protease from the nematophagous fungus, *Lecanicillium psalliotae*, displaying nematicidal activity. *Biotechnology Letters* 27, 1123–1128.

Yang, J., Huang, X., Tian, B., Sun, H., Duan, J., Wu, W. and Zhang, K. (2005b) Characterization of an extracellular serine protease gene from the nematophagous fungus *Lecanicillium psalliotae*. *Biotechnology Letters* 27, 1329–1334.

Yang, J., Li, J., Liang, L., Tian, B., Zhang, Y., Cheng, C. and Zhang, K. (2007) Cloning and characterization of an extracellular serine protease from the nematode-trapping fungus *Arthrobotrys conoides*. *Archives of Microbiology* 188, 167–174.

Zaldivar, M., Velasquez, J.C., Contreras, I. and Perez, L.M. (2001) *Trichoderma aureoviride* 7-121, a mutant with enhanced production of lytic enzymes: its potential use in waste cellulose degradation and/or biocontrol. *Electronic Journal of Biotechnology* 4(3), 160–168.

Zhao, M.L., Mo, M.H. and Zhang, K.Q. (2004) Characterization of a neutral serine protease and its full-length cDNA from the nematode-trapping fungus *Arthrobotrys oligospora*. *Mycologia* 96, 16–22.

Index

Page numbers in **bold** refer to tables or figures.